Innovation Policy in a Knowledge-Based Economy

Patrick Llerena · Mireille Matt
Editors

Innovation Policy in a Knowledge-Based Economy

Theory and Practice

With Contributions by
Arman Avadikyan · Laurent Bach · Patrick Cohendet ·
Olivier Dupouët · Jakob Edler · Jean-Alain Héraud·
Rachel Lévy · Stéphane Lhuillery · Patrick Llerena ·
Chantale Mailhot · Mireille Matt · J. Stanley Metcalfe ·
Frieder Meyer-Krahmer · Véronique Schaeffer ·
Eric Schenk · Stefania Trenti · Sandrine Wolff

Professor Patrick Llerena
Professor Mireille Matt
BETA-ULP
61 avenue de la Forêt Noire
67085 Strasbourg
France
E-mail: pllerena@cournot.u-strasbg.fr
E-mail: matt@cournot.u-strasbg.fr

With 15 Figures and 15 Tables

Cataloging-in-Publication Data
Library of Congress Control Number: 2005925134

ISBN 3-540-25581-8 Springer Berlin Heidelberg New York

Springer is a part of Springer Science+Business Media
springeronline.com

Printed in Germany

Cover design: Erich Kirchner
Production: Helmut Petri
Printing: Strauss Offsetdruck

SPIN 11416296   Printed on acid-free paper – 42/3153 – 5 4 3 2 1 0

Table of Contents

0 Introduction

Patrick Llerena and Mireille Matt

BETA, Strasbourg, E-mail: pllerena@cournot.u-strasbg.fr
BETA, Strasbourg, E-mail: matt@cournot.u-strasbg.fr

0.1 Why Analyze Innovation Policies From a Knowledge-Based Perspective?

It is broadly accepted that we have moved (or are moving) to a knowledge-based economy, characterized at least by two main features: that knowledge is a major factor in economic growth, and innovation processes are systemic by nature. It is not surprising that this change in the economic paradigm requires new analytical foundations for innovation policies. One of the purposes of this book is to make suggestions as to what they should include.

Underpinning all the chapters in this book is a conviction of the importance of dynamic and systemic approaches to innovation policy. Nelson (1959)[1] and Arrow (1962)[2] saw innovation and the creation of new knowledge as the emergence and the diffusion of new information, characterized essentially as a public good. The more recent theoretical literature regarded the rationale for innovation policies as being to provide solutions to "market failures". Today, however, knowledge is seen as multidimensional (tacit vs. codified) and open to interpretation. Acknowledging that the creation, coordination and diffusion of knowledge are dynamic and cumulative processes, and that innovation processes result from the coordination of distributed knowledge, renders the "market failure" view of innovation policies obsolete. Innovation policies must be systemic and dynamic.

The first part of the book provides the theoretical background for the later, more empirical contributions. The three chapters in Part 1 present some analytical propositions that emphasise either the systemic dimension (and the notion of "systemic failures") or the role of the nature of knowl-

---

[1] Nelson R.R. (1959) The simple economics of basic scientific research. Journal of Political Economy, 67: 323-348.

[2] Arrow K.J. (1962) Economic welfare and the allocation of resources for invention, the rate and direction of inventive activity. Princeton University Press, Princeton, 609-625.

edge (in particular tacit vs. codified knowledge). The importance of learning as a knowledge creation process, the coordination of disseminated knowledge and hence the systemic view of the innovation process, and the incentives to produce, diffuse and acquire knowledge are recurring themes in the analysis of different policy actions.

One of the features of the "market failure" approach, based on the concept of Pareto optimality, is that it is a normative approach. The dynamic approaches proposed here are by their very nature not normative, which is one of their main advantages. They offer opportunities for different interpretations and types of analysis. This means that different contexts, i.e. each the institutional and economic systems, can be considered as specific situations, in which history and institutions matter. A dynamic approach emphasizes the importance of analyzing policies in terms of their influence on dynamic processes, and emphasizes the role of policy design. In a dynamic environment, where it is necessary, but far from sufficient, to define policy targets and objectives, policy design becomes critical. The strength of a positive approach is that it enables comparative analysis of different policy designs. Adopting a different analytical approach does not require different policy tools. The same tools can be (and are) used, but they produce different interpretations, targets and results. The contributions in this book explain, based on dynamic arguments, why some classical policy tools, such as incentives to innovate, or public procurement, were successful (or not).

The second and third parts of the book provide some interesting examples of these two types of policy. In Part 2 of the book, three chapters analyse the development or diffusion of a specific technology, developed within the framework of a procurement policy. They explain the success of mission-oriented policies (the development of digital switching systems in the telecommunication sector, the development of high-speed trains, and the diffusion of military technologies), on the basis of the learning abilities of actors, the coordination of innovative activities, and time (the analyses span several decades). The three chapters that constitute Part 3 explore the impact of incentive tools (research and development (R&D) tax credits, R&D cooperative agreements, and university–industry collaboration) on the innovation potential of firms and economic systems (regions).

Consideration of policy objectives and also policy design make the diversity of behaviour of the actors in the innovation systems very relevant. These actors are heterogeneous, particularly in terms of their strategic behaviour and their competences. Also, within a dynamic perspective, these differences are time dependent and subject to reinforcement. Policy design should take account of these actors and exploit their diversity. The originality of the contributions in this part of the book lies in showing that policy design should be based on a better understanding of the strategic positions of the economic actors.

The chapters in the last part of the book are all based around the question of how is it possible to design an innovation policy that will be applicable throughout Europe, bearing in mind the diversity of opinions in relation to innovation? One chapter analyzes the variety of cooperative agreements, that are entered into by firms, on the basis of their individual strategic positions, and underlines the specificities of government-sponsored R&D partnerships. The second chapter in Part 4 describes why it is important that policy makers design actions that encompass the growing internationalization of research and innovation, but also take account of firms' strategies. The last chapter shows that policy makers should not promote a single "university model", but should exploit the differences in the types of universities to enhance training and research in Europe.

0.2 The Rationales Behind Innovation Policies: Dynamic Approaches

The three chapters in Part 1, which provide the theoretical background to the book, are complementary in at least two respects. First, they all take the neoclassical framework as a starting point for explaining the complexity of the innovation process. However, their positions *vis-à-vis* this framework differ. Metcalfe (Chapter 2) rejects neoclassical theory because it "misreads the nature and the role of competition in modern societies through its failure to realize that capitalism and equilibrium are incompatible concepts and that innovation and enterprise preclude equilibrium". In other words, the model of perfect competition does not reflect modern capitalism, and thus cannot be used as a basis for the design of and justification for policy instruments. Cohendet and Meyer-Krahmer (Chapter 3) consider the neoclassical framework as one particular case within a more general approach based on a knowledge-oriented view of the innovation process. They assume that both these views (the neoclassical and the knowledge-based) identify the same instruments, but that the way they are interpreted, designed and implemented differs. Bach and Matt (Chapter 1) concur with this view, and consider the two frameworks to be complements: the neoclassical approach mainly dealing with the problem of allocation of resources and incentives to innovate, and the evolutionary–structuralist approach focusing on the problem of the creation of resources, and the coordination of learning processes. The allocation and creation of resources are important constituents of the innovation process.

Second, the authors of these three chapters also take the view that approaches focusing on knowledge creation and coordination, and learning processes allow a better understanding of the innovation process and hence of policy specifications. However each chapter develops a particular aspect of innovation. Chapter 1 (Bach and Matt) can be seen as an introduction to the other two chapters in Part 1, in the sense that it analyzes the evolutionary-structuralist framework which encompasses various other approaches: the evolutionary approach, the systemic approach (cf. Metcalfe, Chapter 2) and the knowledge-based approach (cf. Cohendet and Meyer-Krahmer, Chapter 3). Bach and Matt believe that these approaches have strong common features. They all recognize the cognitive ability of individuals and groups of agents. Cognition corresponds to the capability to create new knowledge through changing beliefs, routines, etc. Knowledge is intrinsically linked to the cumulative, irreversible and specific cognitive capacity of agents. Technological development, therefore, is context dependent and varies across firms, regions and countries. The innovation system follows trajectories within a paradigm, and its evolution is guided by diversity and selection processes. The virtuous circle of evolution depends on the cognitive and the coordinating abilities of the actors in the system. Public intervention is mainly justified by the existence of individual and collective "learning failures". It should be aimed at facilitating the development and orientation of the learning abilities in the system. Policy actions should be adapted to the specificities of different contexts (geographical, market-oriented). The authors show how the different instruments are implemented and how they act on the system and compare this with the neoclassical approach. They also underline possible "government failures". This very general picture acknowledges that researchers investigating innovation systems and defending knowledge-based economics emphasize specific aspects of the innovation process.

Metcalfe (Chapter 2) focuses on the systemic view of innovation. He sees competition as an evolutionary process in which innovation plays a central role in explaining the differences between firms, and their competitive advantages. Increasing complementarity between different types of knowledge, and increasing dissimilarity between these bodies of knowledge characterize the innovation process. In other words, the internal and external management of knowledge becomes crucial, and means that innovation needs to be considered in a systemic context. Knowledge must be coordinated and correlated across individuals and organizations. The system of knowledge is constructed around multiple minds, in multiple organizational contexts. The firm, which is embedded in the market process along with customers, suppliers and rivals and in interaction with another set of actors (universities), plays a unique role in this system. An innovation system is defined by its components, by the information flow and the connections between these components and their evolution. The innovation system is a device used to correlate and communicate knowledge, and to coordinate access to complementary knowledge. Public intervention, therefore, should be aimed at facilitating the emergence of an innovation system. It should provide the framework within which the system can organise itself. Policy instruments should both increase innovation opportunities and capabilities, and address areas where there are missing components or connections, or misplaced boundaries. In the absence of such a framework self-organization may fail because different agents in a diversity of organizations have different agendas, and their perceptions of the problems involved are also different. The state could design means for bridging between these different agendas such as collaborative research programmes, incubators, science parks, clusters, technology transfer offices, etc.

Cohendet and Meyer-Krahmer (Chapter 3) base their chapter on developments in the relatively new field of knowledge-based economics. A knowledge-oriented policy (KOP) should take into account the specific characteristics of knowledge. The ways knowledge is assimilated and acquired are as important as the conditions of its production. Markets and organizations can no longer be considered to be the only active players in knowledge production. Knowledge intensive communities are playing an increasingly large part in the process of generation, accumulation and distribution of knowledge. Epistemic communities are groups of people who interact with one another to create new knowledge, and the role of communities of practice is to conduct an activity in which knowledge creation is an unintended spillover (cf. patients' associations). The existence of knowledge intensive communities may help to avoid some of the market failures and learning traps that can arise. These communities are seen as the building blocks of knowledge formation. Learning by communities is the foundation for public policy in a knowledge-based economy. Cohendet and Meyer-Krahmer offer some food for thought in relation to KOPs. For instance, they show that from a KOP perspective, patents, although still a mechanism to protect innovators, can be exploited in other ways. Patents may play a strategic role in some negotiations, be considered as the first sign of a cooperation, or be used as a signalling device. Strong patents may hamper the diffusion and production of knowledge. In some cases of excessive fragmentation of the protected knowledge, no agent, or group of agents, may be able to assemble all the pieces necessary to develop the next step in an innovation (this is especially true in the case of biotechnology). The role of a KOP is to enable the construction of a cognitive web that allows different communities to communicate effectively. States should encourage the association of scientific research and lay knowledge through KOPs that promote "hybrid forums" that bring together in an innovative way the insights of both communities. This requires the creation of a cognitive architecture, the establishment of common rules and procedures, and the construction of interfaces. The value of lay knowledge must be acknowledged in public policy. From a knowledge-based perspective, policy makers should pay attention to the co-evolution of the absorptive and the emission capacities of the different communities of actors.

0.3 New Technology Procurement: Knowledge Creation, Diffusion and Coordination

The second part of the book focuses on specific technologies developed within the frame of a procurement policy. Chapter 4 (Llerena and Schenk) analyzes the development of high-speed trains in Germany. Chapter 5 (Llerena, Matt and Trenti) compares the outcome of the public programmes for digital switching systems in France and Italy, and Chapter 6 (Avadikyan, Cohendet and Dupouet) focuses on the diffusion of military technologies. All three technologies are very costly to develop; standard industrial methods cannot be used; their development involves a very small number of firms; and a single user (a public authority) purchases the final product. These three chapters underline that the success of technology development or diffusion depends on the abilities of actors to create knowledge, and on actions being coordinated.

In Chapter 4, Llerena and Schenk explore the impact of learning in the competition between the Wheel/Rail technology (ICE train) and the MagLev technology (the Transrapid) and look more generally at how learning occurs in these types of technological development. There is frequently a first phase of exploration in which the performance of a variety of technological options is investigated. As a result some options are eliminated and the selected one(s) enter the exploitation phase in which the performance of the chosen option is enhanced. Exploration and exploitation may occur in an experimental setting that reflects the representativeness of the real environment. The learning environment may be strategic and involve trade-offs (cost of experimentation vs. representativeness of results). The authors also highlight that "doing" (practice) is important when learning is taking place in an unstable environment. They demonstrate how the ICE technology had an advantage in that it had similarities with the existing system and could use the existing rail network. The innovation was incremental and was located within established technological boundaries that allowed a rapid learning curve. The degree of predictability of the outcome and the rapid commercial exploitation of the ICE allowed learning based on real experience. All these advantages kept development costs down and allowed rapid diffusion. The ICE technology benefited from first mover advantage. Finally, the lead user Deutsche Bahn (DB) was extremely committed to this development and its support was crucial in providing commercial credibility. The development of the Transrapid was experienced many difficulties and its implementation was delayed. The MagLev technology, which it used was a breakthrough technology and required a full sequence of learning, and exploration of various options, which was long and costly. There was no compatibility with the existing network. The development of the Transrapid depended on the implementation of a new high-speed network. There was no lead user involved in its development and DB's position in relation to this technology was unclear. The Transrapid was never able to demonstrate its technical feasibility or economic viability; it did not have credibility.

Chapter 5 compares Italian and French procurement policy in relation to the development of digital switching systems in the telecommunications sector. The chapter describes how the coordination of various actors impacts on the success of mission-oriented programmes. In large technological programmes, the policy maker has a clear vision of the technological goals to be achieved and the institutional proximity between the policy maker and the firms involved is substantial. The companies concerned are characterized by significant initial knowledge, coordination skills and learning abilities. In programmes such as these, the coordination mode is dependent on the technological competences of the policy maker. If these are high, then the preferred mode will be vertical coordination. The success of the digital switching system in France was in part due to CNET (the National Centre of Telecommunication Studies) a powerful research centre, working closely with the policy maker. It allowed vertical coordination of two technological options and facilitated cross-fertilization. The breakthrough technology was thus initially well supported and the learning effects were positive. In 1975, changes in the political system had significant implications: CNET lost its leading role; the breakthrough technology lost its priority; and the telecommunications industry was reorganized (emergence of a duopoly). Eventually, the superiority of the new technology became evident and the incremental innovator was pushed out of the market. These political decisions increased costs and development time. Nevertheless, France was the first country to introduce a digital switching system based on time-division technology. Italy's policy was unsuccessful because there was no institution in Italy similar to CNET, and no horizontal coordination between the firms. This lack of coordination was very destructive given that there were both complementarities between firms and also a shortage of high skilled personnel. There were three technological options and the experimentation phase was lengthy. Finally an Italian firm collaborated with an American company and there was a pooling of national resources, which resulted in the emergence of a new digital system based on a modular architecture. Although a viable technology was eventually developed, the coordination failures during the first phase had entailed tremendous costs – both financially and in terms of time.

In Chapter 6, Avadikyan, Cohendet and Dupouet look at the diffusion of military technology since the end of World War II and highlight the shift from a spin-off to a spin-in paradigm. They show that the relationships between military and civil technologies depend on the nature of the technology, the industrial organization and the nature of user networks. The authors describe the opportunities and constraints to diffusion along these dimensions. The dual nature of the enabling technologies gives spin-in a particular relevance: it facilitated linkages between the military and civil sectors and the creation of a virtuous circle of innovation. The military usually has sufficient resources to generate breakthrough technologies, whose diffusion is linked to their more or less generic nature and to their degree of maturity. Military projects to develop new products or systems that can be diffused to the civil sector usually require capabilities necessary to combine diverse technologies that exploit a variety of knowledge bases. The diffusion of military technologies depends also on the existence of organizational forms favouring knowledge circulation: often military projects are extremely complex and many firms do not have the right organization to promote diffusion. The knowledge developed within highly integrated and very specialized companies, or within a very small and hierarchical network, does not circulate outside the military sphere. The need for secrecy and the existence of a limited number of users are also not conducive to interaction. More recently, however, large military groups have begun to sub-contract to civil and military SMEs and this may enable greater knowledge diffusion. The main obstacles to diffusion can be summarized as follows. The reduction in basic research expenditure by the military reduces the possibilities for radical innovations. Defence firms increasingly have to rely on the civil sector for technological developments and scientific research (outsourcing to universities). Although military firms have to maintain a high absorptive capacity to exploit this externally developed knowledge, this does not necessarily include the ability to diffuse the knowledge. Secondly, there is a big difference in the life cycles of defence and civil products. Military products generally have long life cycles and high functionality resulting in very different dynamics in terms of competencies between the sectors. The reform of DGA (Délégation Générale de l'Armée) in 1996 has positively influenced the diffusion of technology in France from the military to the civil sector by encouraging public-private partnerships, enabling intensive cooperation in all phases of the military product life cycle, and encouraging the development of technologies that are applicable more widely.

0.4 The Impact of Incentives Tools on Systemic and Learning Failures

Part 3 focuses on how traditional incentives (tax credits, university–industry relations) may improve the learning abilities of firms and reduce coordination failures. The first chapter in this part (Chapter 7, Héraud and Lévy) analyzes the French CIFRE[3] system. This system facilitates university–industry coordination, and aims to increase knowledge transfer between the two worlds by supporting doctoral studies that are conducted partly within a company. The authors construct a typology of regions that details the different innovative actors involved and their propensity to collaborate. The next chapter (Chapter 8, Lhuillery) provides an in-depth study of the specificities of the national Research and Development Tax Incentives (RDTI). It focuses on the different targets and efficiency of these instruments. The third chapter in this part (Chapter 9, Bach and Matt) uses a method developed by BETA, to evaluate the economic benefits generated by actors participating in public R&D cooperative programmes. The authors highlight the influence of university–industry interaction and how partnerships can be designed to increase the economic performance of both firms and academic actors, and enhance the benefits to SMEs of these collaborations.

The French CIFRE system described in Chapter 7 by Héraud and Lévy is shown to be an important research training device that links the scientific and the industrial spheres. The PhD students, working in companies as part of their doctoral study, in recombining different types of knowledge and competences create new knowledge. The CIFRE system has proved to be an effective way of promoting collective learning, and the development of science-based activities in firms. It facilitates the coordination of different type of actors and reduces learning failures in the economic system. The chapter examines the regional systems of innovation within the French system. The authors use the CIFRE system, KIBS (Knowledge Intensive Business Services), and classical indicators such as scientific and technological density, to empirically define a regional system of innovation

[3] CIFRE : Convention Industrielle de Formation par la Recherche

(RSI). RSI should encompass a complete and balanced set of interconnected innovative actors. The authors demonstrate that in France, which is a centralized country, few RSI exist. Some regions are specialized in the production of academic knowledge; others are characterized by a network of efficient companies; some have neither of these features. Regions contribute to the development of a national system of innovation, but are not themselves autonomous systems with relevant competences and links. The CIFRE system is undoubtedly improving the learning abilities of firms encouraging connections with universities, but will not, on its own, ensure the formation of a RSI.

In Chapter 8, Lhuillery provides a detailed comparative study of various national RDTI systems. In summarizing these national schemes, he concludes that such incentive schemes are becoming more and more common in R&D intensive countries and are not restricted only to the OECD countries. Even though RDTI systems are being more widely used they are not sufficient to significantly increase R&D investment. There are three major fiscal mechanisms that sustain firms' R&D investment: accelerated depreciation, special allowances for R&D investments, and R&D tax credits (RDTC). Lhuillery defines four types of RDTC systems: volume mechanisms, incremental mechanisms, a combination of the two, or the firm choosing its preferred system. He underlines that definition of the tax base is imperative to protect the system from becoming subject to opportunistic behaviour from firms. He details the elements that a fiscal innovation policy should encompass, i.e. he describes how R&D activities are defined and computed in various countries. Some countries have extended the tax credit system to include the costs of innovation. Since corporate R&D expenditure is not limited to in-house R&D some countries include external R&D services, R&D cooperative agreements, R&D within a group, and international financial flows. To ensure the efficiency of innovative fiscal tools certain regulations are necessary. Imposing a ceiling on tax credits, smoothing tax credits and punishing firms for reducing their R&D spending are among such provisions. Some tax credit systems include incentives targeted at specific firms: small and new companies, and companies with few financial resources; specialist firms and high-tech firms located in specific regions (federal systems). The author analyzes RDTI in comparison to other R&D policy tools and tax incentives and to corporate taxation in general. He shows for instance that in France firms taking advantage of RDTI are less likely to receive direct R&D subsidies. He also underlines the existence in some countries of tax incentive mechanisms devoted to fostering technology diffusion, acquisition, transfer and training. RDTI, then although they are exploited in different ways in different countries, are not sufficient to promote innovation: direct R&D support is also needed. Achieving the right balance, and positive interaction between different R&D tools in innovation policy, has still to be accomplished.

Chapter 9 by Bach and Matt sets out to analyze how the coordination of different types of actors in publicly funded cooperative arrangements influences the learning activities of actors, and thus their economic performance. The chapter underlines how SMEs benefit from public R&D cooperative programmes. The chapter opens with a description of the BETA evaluation method, its relevance, its main methodological features and the different studies that have been performed based on this method. The BETA method evaluates the direct and indirect economic effects generated by the actors participating in public R&D programmes. Direct effects are those directly related to the objectives defined at the beginning of the project. The nature of the direct effect will depend on the type of public policy (procurement vs. diffusion). Indirect effects are those benefits that accrue that were not initially defined as being objectives. These benefits include transfer of knowledge within the company, and application of what has been learned through the project to other activities within the firm. Indirect effects cover such aspects as: technological and organizational learning, networking, reputation, management, increased competences, etc. The originality of the BETA evaluation method is that it enables an in-depth analysis of how participation in a public programme affects the learning processes of the actors. In this chapter the authors focus on the outcome of public R&D cooperative programmes and analyze how the design of the partnership influences the performance of participating organizations and the economic performance of SMEs in particular. They find that collaborations that combine scientific knowledge with technological competence induce higher economic performance and speed up the innovation process. The combination of users and producers or particular scientific disciplines with specific industrial sectors, or the combination of different sectors, produces distinct impacts on the innovative performances of the participating actors. SMEs face particular barriers and constraints that mean that they do not perform as well as large companies. The extent of the benefits they derive depends on their organization, for instance whether they belong to a group, whether they are independent firms or start ups, etc.).

0.5 The Relevance of R&D Strategic Management in Policy Design

The three chapters in Part 4 underline the importance in designing policy of the R&D strategies of the various actors it is aimed at. Policy makers must be aware of the behaviours and strategies within the economic system in deciding the objectives of their policy actions. Chapter 10 by Matt and Wolff underlines that the strategic importance assigned to a particular R&D activity will influence the kind of agreement that firms will enter into. Publicly financed cooperative R&D agreements are generally related to peripheral activities and the policy framework should include specific incentives, coordination modes and types of learning. Chapter 11 by Edler and Meyer-Krahmer analyzes the increasing internationalization of multinational companies' (MNC) R&D activities. It underlines the different reasons why firms internationalize their R&D, based on their corporate strategies, and raises some issues for European policies. In the final chapter, (Chapter 12), Mailhot and Schaeffer highlight the need for universities to implement strategic management of their three missions. The authors emphasize that the new challenge for policy should be to exploit the diversity of universities rather than imposing a single model.

In Chapter 10, Matt and Wolff theoretically analyze the organizational specificities of alliances sponsored by the European Commission and compare them with an ideal type of agreement entirely financed by partners. They use a tri-dimensional grid of analysis that explores the incentives to cooperate, the learning that occurs within an agreement, and how the coordination is arranged. The analysis is based on a review of the literature on strategic alliances and on empirical information gathered during interviews with participants in the Brite–Euram programme (cf. Chapter 9). The specificity of Brite–Euram projects is related to the existence of subsidies and with the requirement to reveal public information. These types of projects act as signalling strategies and allow new technological options in peripheral activities to be explored. Spontaneous agreements on the other hand, are entirely financed by the partners and their main objective is to develop strategic knowledge close to their key competences, and which is often kept secret. Publicly funded partnerships generate mainly unilateral learning and thus redeployable knowledge whereas spontaneous agreements are characterized by the creation of non-redeployable specific assets. In Brite–Euram, the presence of pre-defined rules and the existence of an arbitrator facilitate the coordination of partners and reduce opportunistic behaviour, but also impose certain rigidities in terms of learning. In spontaneous alliances the rules must be created: this generates some flexibility, but increases the risks of opportunism and the danger of premature endings. In terms of policy, knowledge complementarity should be the primary aim, with cost-sharing issues taking second place. The promotion of networks is an appropriate way to increase the coordination of complementarities, but should not become the main objective of firms. As these types of agreements differ in strategic terms, they should be seen as complements and not potential substitutes. The State should not hesitate over subsidizing projects that firms would have implemented without public support in order to develop key competences, but should emphasize that the overriding objective is to sustain the exploration of new technological options.

Chapter 11 sets out to show that despite obvious trends towards internationalization in the R&D performed by MNC, national policy makers have not devised appropriate tools. Edler and Meyer-Krahmer underline the variety of contexts that apply in different countries and draw lessons for national and European policy makers. The growing internationalization in science and technology takes place in three dimensions: the international exploitation of nationally generated knowledge, international science and technology collaborations, and the generation of knowledge. A complex mix of motives and the role played by lead markets underlie this increasing phenomenon. In Europe, the pattern of internationalization is not uniform, but in terms of hosting international R&D the role of Europe is decreasing. The strategic motives of MNC to invest in R&D abroad include knowledge exploitation (the knowledge is generated at home, but exploited abroad to meet local market requirements); knowledge augmentation (the international arena is used to create new knowledge by employing scientists participating in international networks.); and other factors such as vertical cooperation, following competitors, research costs, public RTD policy, etc. Edler and Meyer-Krahmer highlight that policy makers have been slow to respond to the growing internationalisation of innovation. Five major initiatives can be identified: attraction of foreign scientists, attraction and integration of foreign industrial R&D, improvement of access to foreign knowledge and lead markets, targeted learning from practice abroad, and support for international networking. As a consequence of this in-depth study, Europe and its individual nation states should orient their policies to take account of the strategic motives of firms to locate their R&D abroad. European policies should identify possible lead markets (pharmaceuticals, communications, fuel cell technology, etc.) to attract foreign companies. Direct policy measures should not be discriminatory and should facilitate cooperation between foreign companies keen to exploit a lead market and partners or lead users. One of the greatest challenges will be to render countries scientifically or technologically attractive. This could be achieved by encouraging scientific excellence for instance, or by maintaining a wide scope of scientific and technological competences. Attracting the prime players to a market may attract followers; facilitating vertical cooperation may also be an attractor. In sum, European policies should aim at establishing the market and knowledge generation conditions that will attract foreign companies and take account of their different strategic behaviours.

Chapter 12 (Mailhot and Schaeffer) highlights the convergence of scientific policies and the emergence of a unique university model: the entrepreneurial university. The authors show that this global model is disconnected from the real world in which different types of universities co-exist within one country, and in which the academic system differs between countries. They provide management and policy recommendations and argue that policies should exploit the existing diversity among universities. The first part of the chapter describes how the missions of universities have evolved in line with various socio-economic constraints on their research orientation. Since the 1990s, academic research has been determined by social and economic needs and is evaluated in terms of its contribution to national objectives. An entrepreneurial university model has emerged as a result of the pressure imposed by science and technology policies. Pressures to make money from in-house research are forcing universities to increase their interdisciplinarity, to establish new links with industry and to adopt active intellectual property rights (IPR) policies. To cope with these new challenges universities need to implement strategic management, i.e. to manage the conflicts induced by the contradictions that emerge among their different missions. Universities must develop strategies taking account of their own particular constraints, opportunities, competences, value system and declared objectives. It is thus impractical to impose the same set of objectives on all universities: realistic objectives will take account of a university's specific assets. In other words applying a unique model increases the gap between those universities that fit within the entrepreneurial model, and those that do not. Current policies are not aimed at exploiting the existing diversity of universities: the challenge for policy is to take advantage of this variety. If the aim is to foster knowledge diffusion, then different kinds of universities will fulfil different roles. The entrepreneurial university should be in the best position to develop innovations with companies, while training-oriented universities could play a more societal role. The presence of a university in a region may have an impact on the population in terms of financial inputs and taxes, and may foster urban and network (of students, academics and industry) developments. In designing policy, policy makers, should have a greater appreciation of how the presence of a university affects the economic environment, and also take into consideration the different strategies of universities.

Part I The Rationales Behind Innovation Policies: Dynamic Approaches

1 From Economic Foundations to S&T Policy Tools: a Comparative Analysis of the Dominant Paradigms

Laurent Bach and Mireille Matt

BETA, Strasbourg, E-mail: bach@cournot.u-strasbg.fr
BETA, Strasbourg, E-mail: matt@cournot.u-strasbg.fr

1.1 Introduction

The objective of this chapter is to analyze the economic rationale behind science and technology (S&T) policies. Reference will be made to concepts and ideas stemming from works in other fields, such as management sciences, sociology, etc. For the purpose of this analysis, we first identify the main theoretical frameworks on the basis of which innovation related phenomena are currently analysed in economic terms. Next, we identify for each framework, the justifications for State intervention as well as the main forms that this intervention might take.

In order to simplify the presentation and the very subtle, complex, and sometimes controversial scholarly debate (Lundvall and Borras 1997)[1], two main frameworks are distinguished: the neo-classical (NC) and the evolutionary-structuralist (ES), which adopt different approaches that highlight specific aspects. The following questions are addressed in relation to each framework:

- main features, especially regarding innovation;
- the "circumstances" in which the innovation processes do not work well or fulfil the role they are designed for, and the consequences of these so-called "failures";
- the principles of State intervention designed to remedy these failures, illustrated by the most representative types of S&T policy action that can be adopted. (Frameworks are heuristic tools rather than the basis for di-

[1] See also works by Metcalfe (1995, 1998), Metcalfe and Georghiou (1998) and Lipsey and Fraser (1998).

rectly operational policy advice; a detailed description of these is beyond the scope of this present work);

– the main problems raised by these principles when they are implemented into real actions; government failures are included here.

Broadly, it will be assumed that each framework provides rationales for science policy, technology policy, and also, more generally, for innovation policy (and even for other types of policy, e.g. competition policy, trade policy, education policy, and so on). In other words, the hypothesis is that within each framework, the rationales behind every policy are the same. For this reason (from an analytical point of view) we have not separated science policy from technology policy. However, the combinations of these different policies and the frontiers between them vary from one framework to another.

In Sections 1 and 2, we present the two frameworks and their implications for S&T policy. As most of the elements of this analysis are well documented in the literature, we focus only on the main aspects, or those aspects that are not always highlighted. Tables 1.1 to 1.2 summarize the analysis. Table 1.1 presents the main features of each framework; Figure 1.1 presents the types of "failures" connected with each. framework as well as the basic principles of S&T policy, central to each framework, that are designed to remedy such failures; Table 1.2 shows how the main types of S&T policy actions can be seen as specific applications of these principles. (The way this is presented allows comparison of the underlying principles on the basis of which real, although archetypal, policy actions are formulated.) Section 3 is devoted to one key point underlying the question of State intervention: the additionality problem. We conclude by offering some comments about the complementarity of the two frameworks and a comparison of the policy principles and policy actions resulting from each.

1.2 The NC Framework

1.2.1 Allocation of Resources, Technology as Information, and Market Failures

Of the two frameworks proposed, the NC is probably the one that exhibits the highest level of internal coherence, because of its inherent linear logic. This exists within its foundations and also in some measurement tools. For this reason, and probably also for historical reasons, it is the dominant framework, although it has been strongly challenged by the competing paradigm from the late 1980s. Without going into too much detail, the key point in relation to innovation and technological progress is how they have been endogenized by the neo-classical approach (roughly since Arrow (1962)), whereas before they were treated as exogenous to the economic rationale (more precisely they were considered as "given", and their origin was not questioned; they were included in the choice parameters of agents or seen as equally influencing all of these agents). In line with the "input/output" neo-classic way of reasoning, innovative activity is performed by an individual agent (the innovator) using inputs to produce a particular good, i.e. the technology, which is regarded as information (see Table 1.1). The line of argument put forward on theoretical grounds by Arrow (1962) and Nelson (1959) is roughly as follows: the peculiar activity of innovation and the peculiar good resulting from it do not show the "adequate" properties that the theory requires to optimize the decision of the agent. Namely, there are some indivisibilities in both the inputs and the outputs; the output is uncertain and may take a long time to realize and, being a non-rival and non-excludable good, it is non-appropriable.

The consequence is the well-known "lack of incentive" to innovate on the part of the innovator. The activity is costly, mostly because of its indivisibility, and is risky, because of uncertainty, on the one hand as regards its final outcome, and on the other as regards the level of demand resulting from the problem of price determination (according to the so-called "paradox of information", the buyer does not know the value of the information unless he buys it). Moreover, the economic gains are difficult to appropriate since they may benefit: i) consumers or clients, who have access to better products without necessarily being charged a corresponding increased price; this is the basis of the consumer surplus and "market" externalities; ii) competitors and the rest of the economy could use the technology produced by the innovator without paying anything, giving rise to "knowledge" and "network" externalities (see, for instance, Griliches (1979) on market and knowledge externalities, and Jaffe (1996) for a clear exposition of the links between the three types of externalities). In other words, the private rate of return to the potential innovator is too low for him to make further investments, although the social rate of return for the rest of the economy may be high. In this situation, the resource allocation mechanisms that are at the heart of the neo-classical approach do not work so as to generate the socially optimal situation: the investment in innovative activity is inferior to its socially optimal level because of these "market failures".

1.2.2 S&T Policy Principles and Actions

To remedy this situation, the State can ground its S&T policy actions on some basic principles. For instance, the State can:

- try to provide (or help to circulate) better information to reduce uncertainty and to give the demand (supply) side better information on supply (demand);
- substitute wholly or partially for the market either on the supply side (by itself carrying out innovative activity, or contributing to the firm's investment in research and development (R&D) by means of subsidy, tax credits, grant, etc.), or on the demand side (by ordering innovative outputs – products, processes, techniques or whatever – to firms, or helping agents to buy such outputs), in order to reduce, or more evenly distribute, the uncertainty and the risk and to reduce the cost for innovative firms. By "substitute", we mean that public action takes over from the private action that would have been required in order to reach the social optimum. The basic assumption here is that the cost savings for firms will compensate for their losses from externalities; the amount invested by the State, therefore, should not be larger than the sum of the externalities;
- promote mechanisms or regulations to remove or diminish externalities or facilitate their internalization in the agent's optimizing calculations:
    - to provide a property right to the innovator on his technology as a compensation for generating knowledge externality (this rewarding role of patents being closely linked to their protection role, which is probably more important);
    - to promote cooperation between users and producers of technology (vertical cooperation) to share market externalities, and costs and diminish the uncertainty;
    - to promote cooperation between producers of technology (horizontal cooperation) to share knowledge externalities and share the costs and the risks associated with the production of technology.

Table 1.2 and Figure 1.1 set out these basic principles and include a list of the corresponding S&T policy actions.

According to the NC approach, these corrections to market failures will lead the optimizing rationality of agents to allocate resources through market mechanisms in such a way that a "second-best" equilibrium can be reached. The "first-best" equilibrium is not possible due to these market failures; but State intervention allows the system to achieve a "second-best" equilibrium (inferior to the "first-best", but better than the situation would be without any public policy). If State intervention leads to such a "second-best" equilibrium, there is "additionality" compared to a situation in which there is no government action.

Obviously, this is an oversimplified picture, and since Arrow's seminal paper many new arguments have developed, which have made it more complex, although, from our perspective, they are all within the same general framework.

Perhaps one of the most important developments at the boundary of the NC paradigm is the so-called "new economics of science and technology" developed by, among others, Dasgupta, Stoneman, and David (Dasgupta and David 1994; Foray 1991). On the one hand, it helps to explain the implications of the public-good properties of technology considered as information, as well as the inherent and specific properties of information (for instance, in relation to its value, its cumulative and combinatory nature, the high costs of its production compared to the low costs of its duplication, the network externalities associated with it, etc.). These features contribute in a sense to reinforcing the NC conclusion mentioned earlier. On the other hand, the new economics of S&T proposes that a new line be drawn between science-related and technology-related activities, and their respective outputs. A tentative summary of the proposed distinction would be that it relies not so much on the nature of the outputs (both fundamentally subject to non-rivalry and at least partial non-excludability) than on:

- the practices of diffusion associated with incentive schemes (openness or free access to scientific results, with priority to the inventor associated with social rewards – this is largely inspired by Robert Merton; closedness or property rights on technology outputs associated with economic reward, i.e. appropriation of the economic gains from innovation through market mechanisms);
- the possibility to choose the optimal level of codification taking into account the reward system;
- the greater uncertainty in the production and use of scientific results;
- the results of basic research being considered mainly as an information input for applied research.

One fundamental outcome of these new developments is that they allow for a better justification of the distinction between S&T policies in the NC framework. To a certain extent, the justification for State intervention is stronger for science, especially if one includes the indivisibility of the production process (with its comparatively higher costs for science than for technology) and the more generic usefulness of scientific outputs, thus allowing for larger externalities. In the field of science, State intervention is generally direct, through funding, while for technology it is indirect (accomplished through co-funding and the property rights system). In this respect, science policy is almost always acknowledged to be indispensable, even though some argue (probably too simplistically given the analysis developed above) that the type of results produced by publicly supported basic research are not useful for industry, because if they had been, the industry would have done the research itself (Kealy 1996). But both science and technology policies still rely on the same analytical grounds. More generally, following the linear model of innovation, education policy is aiming at providing the system with "good inputs" (researchers), and diffusion policy is aimed at helping agents to adopt innovation and circulating information about existing or potential needs and resources.

From another standpoint, the endogenization of technology in economic analysis has also been the ambition of macroeconomic analysis, in the field of endogenous growth theory or New Growth Theory (NGT). However, here also, despite various technical refinements, only some types of development, accumulation and diffusion of information have been introduced, guided by rather simple forms of appropriability regimes and incentives, and most frequently associated with other classical hypotheses related to behaviour and the search for equilibrium (see, for instance, Firth and Mellor (2000))[2]. It is questionable, therefore, whether these approaches have provided much more than an illustration at macro-level and a formalization of the innovation-related phenomena already identified, which, however, is obviously a very useful achievement. To move further along this line, it must also be acknowledged that learning phenomena are not completely absent from the neo-classical perspective, especially in NGT. Certain forms of "learning by doing" might, for instance, be compatible with this perspective.

However, NGT and the new economics of S&T are at the frontier of both frameworks, especially the latter approach, which, in many ways, avoids this fundamental feature of the NC paradigm, i.e. that, "… the closest we get to something called *learning* is *information acquisition*" (Lundvall and Borras 1997). More precisely, one of the main (or the most important) differences with the second framework resides in this point, as outlined below.

---

[2] This is not to say that specialists of NGT, such as Romer, ignored the tacit dimension of knowledge, but they did not fully incorporate this dimension into their models.

1.2.3 About Empirical Problems of Applications and "Government Failures"

Without entering into major debate about the relevance of the NC view and its compliance with reality, the main problems raised by the application of the principles of State intervention should be outlined.

The first problem is related to the difficulty involved in identifying the situations in which "market failures" occur, and even more so the difficulty of determining exactly how to correct these failures in practice: which firm to support, which project to finance, which information to diffuse, how far to extend the property rights, and so on. The NC framework, as such, does not provide the tools to directly express real phenomena using the limited number of variables that are required to operate the optimizing calculation.

The second problem is inherent in the method adopted to correct the market failures. A good example here is provided by the proposed solution to the lack of appropriability, namely the property rights system. While on the one hand, the information should be the innovator's property in order to urge the innovator to innovate, this, on the other hand, introduces asymmetries of information between actors when the information does not diffuse to all actors. The "pure neo-classical" axioms stipulate that all agents have the same information, and this specific feature leads the system to the social optimum. Any departure from this hypothesis entails at best that the achievable equilibrium is only "second-best". In other words, there is a trade-off between guaranteeing the property rights to the innovator and diffusing the technology throughout society, and the right balance is not easy to find. The patent system is one solution to this problem, but it generates a distortion in the price mechanisms to the detriment of consumers because of the monopoly position the patent secures to the innovator: this, in turn, leads to a non-optimal social surplus. Moreover, the patent system may induce a duplication of innovative efforts by firms competing to be the first and to receive protection. Over-investment could then result, or, conversely, under-investment could re-emerge if one firm dominates the "patent race". Neither result would be socially optimal.

The same argument, more or less, as for the subsidy principles can be put forward: subsidy causes asymmetry for the one that benefits, which runs fundamentally against the "pure neo-classical" axioms. The fact that, in theory, the amount of subsidy should not exceed the amount of externalities generated by the innovation only ensures that a "second-best" optimum can be reached. The same argument is relevant for the promotion of cooperation, which restricts the diffusion of information and economic gains to a limited number of agents, and thus creates failures in the distribution of information.

The argument underlying these trade-offs is that solving one market failure always makes another emerge. Thus, it is the balance between the positive effects resulting from the correction of the first and the negative effects of the second that is crucial. There is sometimes a risk that corrective public actions create other market failures that are worse than the original ones, i.e. that the increase in social surplus obtained as a result of the corrective action is inferior to the decrease in social surplus induced by the created market failure[3]. When this occurs it is termed "government failure" or "policy failure".

Obviously there are certain remedies to overcome this problem, for instance, in the case of the patent system, limitation in time and scope of the patent protection, the obligation to concede licences, a buy-out of the patent by the State, or even the replacement of the patent system by an ex-post reward system. But, these solutions raise other market failures (see, for instance, a discussion of the reward/buy-out vs. patent systems in the light of the role of patents in firms' strategy (Penin 2003)). In any case, the balance between the positive and negative effects of State intervention is hard to assess, and thus often as much a matter for policy decision as an economic one[4].

Another possibility is that the benefit effects are less important than the cost of intervention, the latter including the cost of researching information about the presence of market failure and the evaluation of the actual and "corrected" situation. These are both typical cases of "government failures" associated with the NC framework.

It is not surprising that these two fundamental difficulties are related to the set of information owned by the State, and its capacity to acquire and use new information. For a "pure" NC theory, by definition there can be no such thing as government failure since the perfect information hypothesis holds. If one rejects this hypothesis, the consequences (in terms of strategies, incentives and modes of coordination of agents) of uncertainty, im-

---

[3] On the basis of this argument, other types of public policies may be required to limit the damage caused by technology progress on health or the environment.

[4] We should mention here the basic exceptions to competition policy accepted by public authorities:

- phenomena as exceptions to monopoly regulation: patents; financial support to the supply side in the case of high cost and natural monopoly caused by indivisibility and related economies of scale;
- agreement as an exception to cartel and agreement regulation: R&D cooperation (with the idea of "pre-competitive" cooperation fully coherent with the distinction between science and technology mentioned above).

perfect information and related asymmetries between agents should be investigated in depth (Laffont and Tirole 1993). This leads to a complexification of the different situations of market failure and allows for fine tuning of the applications of State policy actions listed above to specific contexts in terms of the information at the disposal of actors. If this is properly done, again, at least in theory, there is no possibility of further government failure.

Similarly, at the basis of another extension of the NC analysis, and particularly important for explaining the role of cooperation, is the question of the transaction costs associated with market relations (but not included in prices) and the influence of these costs on the determination of the optimal mode of coordination. But, again, most of the theoretic renewal in this field falls within the scope of the same framework, because it does not essentially preclude considering technology as information and favouring optimizing rationality. Therefore, we would be tempted to state that these approaches have not profoundly modified the NC framework.

However, it is arguable that the State has more information than the market, and, moreover, that it is more able than the market to adapt its information structure. To follow this line of argument, it should also be recalled that other developments in economic theory favour a more dynamic idea of competition, which would be seen as based on the continuous creation and exploitation of asymmetries, especially regarding information. As long as technology is treated as a set of information, and the choice of agents is seen as an optimization taking into account existing or anticipatable alternatives, these approaches still remain within the boundaries of the NC paradigm. Despite the progress of analytical tools and the continuous refinements to and complexification of contractual schemes in this field, there is still room for more dynamic approaches to take account of these asymmetries of information.

Another instance of "government failure" emphasized by some researchers in the field of public administration theory, is related to the process of decision-making in public bodies, reflecting private interests, lobbying, etc. and raising the unsolvable – at least using the NC analytical apparatus – problem of the aggregation of individual preferences.

1.3 The Evolutionary-Structuralist Framework

1.3.1 Creation of Resources, Knowledge and "Learning Failures"

Under this heading falls a constellation of approaches that have some strong common features. Three main approaches can be distinguished – very artificially and for pedagogical purposes: the evolutionary approach, the systemic approach, and the knowledge-based approach. Often inspired by other disciplines than economics (sociology, history, psychology, management, epistemology, biology, etc.), they claim to be part of the Schumpeterian heritage, and have for a long time been built against the NC framework. Over the last ten to twenty years, they have been developing and been being refined at a great rate, and have probably been gaining in coherence with the emergence of the knowledge-based approaches (although the full implications of these for S&T policy are still to be explored).

Again, it is beyond the scope of this study to specify the main features of this framework. It remains only to underline that most of them can be seen as being opposite to the features of the NC framework (see Table 1.1). What is probably more relevant here is that this framework fully acknowledges the learning capacity, or, more generally, the cognitive capacity of agents and groups of agents (individual as well as collective capacity). This does not only embrace a capacity to learn something that would exist somewhere, but also the capacity to create new knowledge, especially by changing ways of thinking, beliefs, visions, routines, etc. In terms of innovation, cognitive capacity concerns not only scientific and technical knowledge, but also the complementary knowledge required along the innovation process (organizational, management, etc). The key points are that on the one hand knowledge cannot be reduced to pieces of information, but is a mix of tacit and codified knowledge, while on the other, knowledge is intrinsically linked to the cognitive dimension[5]. This has some decisive consequences on appropriation and diffusion phenomena: for example, learning is obviously a cumulative and collective, rather than a purely individual process. This logically leads to an acknowledgement that it is a context-dependent process, which varies from one agent, group

[5] The economics of science and knowledge also recognizes this tacit dimension of knowledge, but without emphasizing the cognitive dimension associated with knowledge (see an analysis of problems raised by this point of view and of the related debate about the codification of knowledge in Ancori et al. (2000) and Cowan et al., (2000)).

of agents, firm, industry, clusters of industries, regions, institutions, etc. to another.

The three approaches that we consider to be at the basis of ES thinking (respectively the evolutionary, the systemic and the knowledge-based approaches), more or less explicitly possess all the features mentioned above. But each focuses on some specific aspects and, therefore, has brought some decisive concepts and analytical tools to the global framework. It is remarkable to note that these concepts, most often with quite minor adaptations, have been adopted by the other approaches, which use them as indispensable to their own conceptual construction.

In following the logic of our description of the NC framework, we are tempted to argue that the evolutionary, the systemic (including national or local systems of innovation, clusters, and the like) and the knowledge-based approaches have provided respectively the "general logic" of the evolution, the "how it works", and the "basic engine" of evolution. But the main problem lies in defining what would correspond to the NC optimal situation. Surely there is neither a static nor a dynamic equilibrium-like situation, which would have very little meaning, if any, in this learning-oriented framework. Correspondingly, there is nothing like optimality, even though it is sometimes possible to assess *ex post* if one situation is preferable to another. Referring to Schumpeter's cycle analysis reinterpreted and enriched, the only apparent consensus on this would be to assume that:

- the system follows some trajectory induced by a paradigm, and therefore it must be able to exploit a "good trajectory" as well as to ensure a "good transition" from one paradigm to the other;
- in order to be able to do so, in the whole system and at all levels of the system, there should be sufficient diversity to allow the selection processes to perform satisfactorily;
- this should be accomplished without too much loss of cohesion, which would unbalance the system and/or prevent it from evolving well;
- the basic engine that allows for all of this would be the maintenance of, or increase in, the cognitive capacity of all agents or groups of agents at all levels of the system.

With an approach in which all phenomena and processes are so complex and intrinsically related, it is rather difficult to identify and especially isolate "failures", as we did for the NC paradigm. Indeed, the term failure is adopted for the sake of simplicity, but may be misleading. In the "market-oriented" framework, as already stated, there is always an implicit or explicit reference to an "optimal situation" that would be reached if all theoretical conditions were fulfilled, that is, if markets and behaviours were perfectly similar to the prediction. It is through reference to this mythical "optimum" that something is seen as going wrong, as "failing". Within the alternative framework, this reference to an optimal situation does not exist and it is thus not exactly appropriate to consider "failures". One should rather consider "traps", "dysfunctions", "gaps", or "holes", leading to "dilemma and trade-offs" between existing forces driving the system, rather than "dilemma and trade-offs" between two possible states of the system (as choosing between two "second-best" situations). Various authors have already pointed to some of them (Malerba 1996; Smith 1996; Lundvall and Borras 1997; Metcalfe 1998; Teubal 1998), but there is no unified and unanimously accepted list of failures in the evolutionary constructivist framework.

We will assume here that what matters are, generally speaking, "learning failures", i.e. problems that limit (or constrain the use of) the cognitive capacity of agents and groups of agents. A series of failures falls within this type of failure; they are expressions of learning, and cognitive problems in different contexts and at different levels of analysis. We propose then to distinguish between:

- exploration/exploitation failures: misallocation of efforts and of cognitive attention to one activity to the detriment of another;
- selection failures: technology, practice, firms (infant firms, for instance) or other sorts of "species", among which selection is at work, are eliminated too rapidly (or maintained too long), or maintained based on inappropriate criteria;
- system failures: lack of coordination and complementarity between the cognitive activities of agents and groups of agents; rigidity of cooperative structures; lack of appropriate institutions allowing a collective creation and diffusion of knowledge; bad adjustment and desynchronization between the evolution of institutions and technological evolution;
- knowledge processing failures: codification problems (lack of standards and platforms, rigidity linked to excess of standardization, appropriability of codes, etc.); lack of/limitation of/absence of/control over absorptive and emitting capacity; lack of capability to articulate knowledge coming from different sources (for instance, external and internal to a firm); structure of knowledge badly adapted to appropriate sharing and distribution.

As suggested earlier, the negative consequences of these different types of failures (and thus basically resulting from a deficit of cognitive capacity) are principally lock-in to "bad trajectories", lack of diversity in the system, difficulty in creating new paradigms and in warranting a transition from the old to a new paradigm, and the existence of "gaps" in terms of knowledge, networks, institutions, economic and social conditions, and so on, all of which unbalance the evolution of the system.

1.3.2 S&T Policy Principles and Action

Based on the above, the basic principle of State policy should be to help, by all possible means, the development and the orientation of the cognitive capacity of actors and provide conditions conducive to the use of this capacity. The different approaches developed within the ES framework will put the emphasis on different aspects of this question. For instance, researchers investigating innovation systems will defend the role of institutions, infrastructures and collective interactions. The supporters of knowledge based economics will debate the necessity to help with the codification process or to support the development of a knowledge infrastructure allowing for a better use of the increasing amount of codified knowledge within the whole society. Therefore, these basic principles of State policy can be activated in very different ways, such as the promotion of norms, platforms, or other knowledge-related infrastructures, support to communities and agents of knowledge, reinforcement and adaptation of the education system, renewal of the property rights system to take account not only of the cumulative nature of knowledge creation, but also the nature of the knowledge and of its other modes of appropriation, support for infant firms at their different stages of development, etc. Above all, action must be adapted to contexts defined according to geographic, industrial, sectoral, market-related, and institutional dimensions. Table 1.2 and Figure 1.1 set out these basic principles and include a list of corresponding S&T policy actions. Table 1.2 shows only the main S&T policy actions, which can be combined, adapted, refined, etc., and thus promote a wide range of practical initiatives.

To close the loop, the adequate use of an appropriate cognitive capacity is then supposed to provide the conditions for appropriate selection processes, with sufficient diversity within which the selection mechanisms operate, and without too many "gaps" in the system. In turn, this would guarantee the evolution of the system along satisfactory trajectories and through relevant paradigms. This rather naïve picture could obviously be made more sophisticated by orienting it towards some specific aspects involved in learning. But, fundamentally, it would be within this argument, even though this argument may appear somewhat simplistic.

In this framework, the frontiers between science policy, technology policy, and other innovation-related policy become blurred for various reasons. These include the fact that the actors are interacting at different levels within the system, the practices of diffusion, the incentives, and the fact that the activities of these actors are not strictly differentiated between science and technology, and the nature of the knowledge produced is not fundamentally different. Moreover, knowledge is created everywhere in the system, and thus there is not an absolutely clear functional separation between activities and between actors as regards creation of knowledge, transformation of knowledge into innovations, and diffusion. For instance, the clear-cut distinction between science and technology proposed by Paul David and others (see above) is called into question when analyzing the recent trends in the way research is performed in modern economies, as described in Gibbons et al. (1994). Therefore, coordination, coherence, and complementarity of policy actions are crucial in order to make the overall system better able to learn. For instance, recent work by Sherer (2001) shows that this is probably a combination of different policy tools that favours the re-dynamization of the US economy, especially as regards its innovative activity.

1.3.3 About Empirical Problems of Applications and "Government Failures"

Nevertheless, we still face a situation in which it is as easy to identify, in the real world, traces, examples, partial assessments, and finally pieces of evidence about the fundamental relevance of this approach (see the vast number of case studies, monographs and other empirical studies produced since the 1980s) as it is difficult to define metrics and tools to operationalize and measure all the concepts that have emerged over 20 to 30 years. The problem is complex and located at different levels: it concerns the "measurement" of some aspects, even in qualitative terms (for instance, diversity, learning capacity), the definition of what would be a good or satisfying level of the corresponding variable, and the desirability to reach some degree of homogeneity in the metrics, scales or analytical tools. With the complexity of the approach, due to its systemic and constructivist nature, this is probably the main problem encountered within this framework, mainly because it prevents analysts and policy makers from really envisaging the policy options that would favour the right balance between all the trade-offs.

A very important difference between both frameworks lies in the very "place" of the State. As underlined in Part 1.1.3, in the NC framework the State is normally considered as "outside" the market system. Therefore, if markets and behaviours are consistent with the theory then the State is needless. Only when market failures occur, does the State "appear". Then questions of public/private rate of return, crowding effects, substitution, and the like, arise and finally questions traditionally related to additionality. In the second framework, the perspective is quite different. The systemic, path-dependent, and cumulative approach fully accepts State-related institutions as being "part of the game" and recognizes their influence on institutional, technological, social, and economic changes. State-related institutions learn roughly according to the same basic rules as other actors. They do not necessarily have more knowledge, or greater cognitive capacity, or a broader vision, etc. than other parts of the system. Therefore, most of the "learning failures" listed above and which affect society, affect public bodies as well, and logically entail "government failures" (Malerba 1996). In this respect, the main source of "government failure" probably lies in the desynchronization of the speed of adaptation of public institutions and the speed of technological and scientific change in the system. This default in the speed of adjustment may have some negative countercyclical effects. Because of their specific role, public institutions are urged to develop integrative and coherent policy visions, tools and instruments, and to adapt constantly to the new requirements and trends in the economy. In this perspective, the "adaptive policy maker" should also continuously try to implement experimental policies, to use different policy instruments, to change the mix of instruments, and to make use of benchmarking approaches as policy learning mechanisms.

Policy learning should also encompass "diagnostic learning": it is also the ability to identify the changes in the environment and the changes in the relative position of actors (firms, countries, etc.) in this environment that is at stake. The recent debate about the "European innovation paradox" clearly shows the importance of this learning capacity upstream from the innovation policy itself (Muldur 2001).

1.4 The Issue of Additionality

1.4.1 General Remarks

The question of additionality is obviously at the heart of the justification for State intervention in the field of S&T, and, thus, is intrinsically linked with the rationale for S&T policy. But it is also linked with evaluation and assessment problems. In this section, we try to focus on the first point, but inevitably there will be some links with evaluation that will be addressed.

Following Buisseret et al. (1998), the additionality problem could be expressed as: what difference does State intervention make? And this question should immediately be linked to a second one: does this difference justify State intervention?

Additionality then is directly linked to the consequences of policy action: but policy action aims at objectives that are defined according to the rationale behind each framework. This raises a double problem:

- difference could be assessed in the light of the targeted objectives, but other differences may occur that are beyond the scope of the targeted objectives, and may be unexpected;
- these unexpected differences can be coherent or not with the framework that gave birth to the objectives; if they are not, it will be necessary to adopt the theoretical view of the other framework to identify them and, if possible, to evaluate them.

An essential dimension of the additionality problem is related to the situation that must be seen as an alternative to State intervention. The temporal dimension is essential here. Two possibilities can be envisaged. The "null hypothesis" stipulates that everything would continue as before the public policy. The "counterfactual scenario" is a fictive construction about what would happen if there were no such State intervention (or, ex-post, what would have happened had there not been any policy implementation). Then *ex ante*, additionality is between the targeted situation (e.g. the objectives) and the forecast alternative scenario. *Ex post* additionality is between the actual situation and the alternative scenario, but this actual situation may be better or worse than the one the objective.

Directly linked to this question is at what level the alternative scenario is under consideration: project level, firm level, programme level, policy level, etc. As this alternative and hypothetical situation must be comparable with an evaluation of the actual or anticipated situation "with" State policy, both must logically be analyzed at the same level.

Another question relates to the temporal dimension: the time horizon over which the additonality is examined. "One-off differences" or short-term differences are one thing; probably what is more crucial is the persistence of these differences (decreasing or increasing) over time.

1.4.2 Different Concepts of Additionality

For a long time, debate ranged about how to analyze policy action as a means of providing inputs to the innovative process resulting (or not) in outputs from the process. This follows NC thinking about analysis of the activity of the agents, which is particularly appropriate when the policy action takes the form of providing input to the innovation process Within this perspective generally two types of additionality most often quoted are input additionality and output additionality. It is also necessary to concentrate on the process itself, introducing the concept of behavioural additionality. Finally, the recent development of knowledge-based economics (and its influence on the ES framework mentioned earlier) suggests that a fourth type of additionality might be envisaged, that is, cognitive capacity additionality. Each of these four types of additionality brings something to the global problem of additionality, but in isolation cannot address this global problem. Nor is the sum of the four types equal to global additionality. The four types of additionality are briefly defined and discussed in the light of the two frameworks proposed earlier.

1.4.2.1 Output Additionality

At first sight output additionality is the most intuitive: would we have obtained the same outputs without the policy action? Clearly this question is related to the problem of evaluation (i.e. the definition and measurement of the impact of S&T policy actions), which is beyond the scope of this chapter. However, some brief comments can be made.

First, again, the notion of output is strongly connected to the NC framework. Products, processes, and other physical devices, patents, articles, blue prints, and other forms of S&T products, can be more or less compatible with an output perspective (see the list of outputs adapted from the COMEVAL study, for instance, in (Bach and Georghiou 1998)). Knowledge, standards and norms could be considered as outputs, but with specific properties sometimes far removed from those of information-like outputs. But, cognitive capacity and all the various types of capacities that are rooted in it, do not fit into this frame.

Second, it is not possible to identify the “changes” brought about by the output; it is necessary to assess impact in terms of use (in production activities, through market relations, etc) of these outputs, to derive a global assessment of additionality from the existence of output additionality only. Thirdly, it is not necessary to do detailed analysis of the alternative scenario since the process involved is not so important. It is only necessary to find a common definition and measure to compare the output obtained with what might have been obtained "if".

Finally, as suggested above, there might be cases where there is no additionality in terms of the outputs directly related to the objectives of a given policy action. However, there may be additionality derived from other types of outputs from this same policy action. This highlights the importance of choosing the relevant "array" of outputs that are considered in assessing additionality.

1.4.2.2 Input Additionality

The question here is: does public action add to the inputs dedicated by the agents to the innovative process or does it partially or completely displace these inputs? There are many arguments to explain why there could be displacement. The problem has been studied in depth in David et al. (2000), and extensively discussed from an ex-ante perspective. For instance, some academics have provided empirical hints (Jaffe 1996) and theoretical rules (Usher 1994), who employ an "incrementality test", to guide the choice of the appropriate public action (for instance, funding the right project). The overriding argument is that if the State financially supports actions that would have been conducted anyway by the agents, then these agents will be tempted to use their resources for other activities. Therefore, there is only additionality when the State supports actions that would not have been carried out by the agents (i.e. to which they would not have dedicated inputs), provided that the actions are socially desirable. In terms of the NC framework, this means that these actions result in a social rate of return higher than the private rate of return of the agent (this difference is often called the "spillover gap"), whereas the private rate of return from the same action without State support is inferior to the minimum required by the agent.

Since innovation is a risky activity, this ex-ante problem obviously is further complicated by the fact that the investment choice criteria become more like a trade-off between rate of return, and risk. Public funds could motivate an agent to carry out activities with high risk, but a high rate of return, leading to complementarity rather than displacement. Apart from the problem of uncertainty inherent in this type of investment, there is also the possibility that in some instances the State favours projects that are profitable from a private point of view, in order to demonstrate some kind of success from its policy.

But there could be other ways through which, at least, partial displacement may occur. The main example is the impact of State support for innovative activities on the supply of inputs to innovative activities; if it is inelastic, then prices could increase and make the cost of these activities so high that the agents reduce their efforts in this field. The support given by the State to one agent could convince its competitors to reduce their own efforts because of the fear that they will be disadvantaged in the competition. State support may be seen as a revenue from, rather than an input to, innovation. Or the funds provided by the State may persuade other funders to reduce their requirements in terms of profitability; this is the sharing of risk argument.

It is often argued that the additionality problem is envisaged from this input perspective because there is a lack of available materials to elucidate the output additionality problem: the variables (basically the respective public and private resources invested in innovative activities) are more easily defined than those required to assess output additionality. We would rather say that it is only when we accept the relevance of the input/output type of analysis that this "proxy" analysis can apply.

It is quite clear that the application of input additionality as a parameter of public choice between the actions to be taken is closely linked to the neo-classical approach described earlier. In particular, it more or less explicitly involves the following assumptions:

- there is a clear link between input and output of the innovation activities;
- divisibility and constant return to scale of the innovative activity exist;
- the nature of the output generated by public funds and private funds is the same.

Naturally, advocates of the ES framework contest these views. On the one hand, they maintain that analysis of public financial support should always take into account how it is provided and the context in which it is used (beyond the structure of information asymmetries). On the other hand, they defend the complementarity of inputs rather than their substitutability related to crowding effects. Apart from the question of increasing returns or threshold effects in the production of innovative output, other arguments more deeply rooted in the ES framework are, for instance, that public funds could help to develop the knowledge base of the agents and their absorptive capacity, which could even allow them to reduce their own investment while increasing their profitability. Another line of argument is that thanks to the cumulative nature of knowledge creation, public support could increase the efficiency of future innovative activity by increasing the cognitive capacity of the firm. Therefore, input additionality has no general application in the alternative framework: we can say that some public money displaces, complements, or adds to private money, but it is only when looking at the specific context in which public money is used that we can reach any conclusions about its additionality.

Note, that input is sometimes difficult to define in the case of policy actions that do not correspond to some sort of funding of innovative firms; obviously all policy actions are costly, so they always involve financial outlay on the part of the State, but it is not clear that this would be the best variable to take into account. Perhaps displacement/additivity of resources would be a more meaningful concept. For instance, in the case of property rights, comparing State investment in the legal system to the resources devoted by the agent to establishing some type of protection; in the case of cooperation, comparing State investment in the funding of networks with the resources invested by firms, etc.

1.4.2.3 Behaviour Additionality

The concept of behaviour(al) additionality relies on the possibility that State policy has an influence on the behaviour of agents: in the absence of public action, the agents might have acted differently during the time corresponding to the period of the policy action (for instance, during a State-supported project); the project could have been less ambitious, involved different partners, taken longer, etc.

One problem is that investigating the behaviour of agents does not directly give information about whether the behaviour "with" State action is better or worse than the behaviour "without" State action, i.e. if there is additionality or not. Also, to a certain extent, examining behaviour additionality seems redundant in the attempt to build the alternative scenario mentioned above, even though it might result in a much more detailed scenario than in the case of output additionality. From an analytical point of view, examining behaviour additionality only provides an explanation for the existence or the absence of input and output additionality. For instance, if one assumes that the objective of actions supported by the State would have been achieved (although perhaps later) and in a different way, even without State intervention, the difference looked for in the additionality analysis lies in the behaviour of the agents. In other words, by this means one could enrich the evaluation with an analysis of the innovation process itself, which could eventually help the policy maker to refine the way he implements his programme. To this extent, it does not add much to the solution of the additionality problem from either a neo-classical, or an evolutionary-structuralist perspective. However, it may allow us to compare different ways of reaching the same result, in particular different learning processes, if we adopt the second framework.

1.4.2.4 Towards a Cognitive Capacity Additionality?

As we have seen, the first two types of additionality (and to a lesser extent the third) are strongly connected with the NC framework and the related input/output method of analyzing the innovative activities. Based on the arguments developed in Section 2 of this chapter another focus might be the changes affecting the agents themselves, or, more generally, the changes affecting the system that runs the innovative activity. Following the different approaches to understanding organisations and systems, this could lead to the definition of organizational additionality, structural additionality, institutional additionality, behavioural additionality (in a quite different sense from that described above, dealing with the change in the agents' behaviour after State intervention[6]), etc. From an ES point of view, we would argue that the fundamental issue would be that of cognitive additionality. Does the policy action change the different dimensions of the cognitive capacity of the agent? It is obvious that cognitive capacity additionality is linked to those types detailed above. It depends on certain types of physical devices, on certain types of codebooks and codified knowledge (patents, publications, norms, etc.) and on certain types of explicit procedures (project management methods, quality control, etc.), that all to some extent could be considered to be outputs of the activity supported by the State. But whereas such outputs are treated as independent objects in output additionality, they are here context-dependent and combined with other dimensions of the cognitive capacity of the system affected by the policy action. Also, it can be argued that the changes in cognitive capacity will determine the future capacity to produce new outputs. But they cannot be reduced to some sort of discounted value of future outputs, since the cognitive capacity encompasses supplementary dimensions related to creativity and adaptation that allow determination of (adaptation to) future situations that cannot be envisaged.

It must be acknowledged that the difficulty of putting this concept into practice relates to the difficulty in defining all the dimensions of cognitive capacity and of the changes to it. Only some pieces of the puzzle can be identified. Some important dimensions could concern the absorptive capacity of the agent, its ability to master the codes used to articulate the existing and emerging knowledge, its capacity to interact with its environ-

[6] The agent will use different routines (project management, research activities, etc.) or will use existing ones differently; he will interact differently with his environment, etc.

ment, etc. Obviously, this is also related to the concepts developed by the competence-based approaches.

Finally, it is obvious that the problem of cognitive additionality in the alternative scenario is more complex than output additionality, and probably much too complex to cope with. The null hypothesis, although already very difficult to define, would be the only solution.

1.5 Conclusion: Beyond an Oversimplified Antagonism Between the Rationales for S&T Policy

The description in this chapter has painted a picture of a radical opposition between the NC and the ES frameworks. However, these frameworks should be envisaged in terms of their complementarity, each focusing on one particular aspect, that is, the problem of the allocation of resources (for the NC framework) and the problem of the creation of resources (for the ES framework). The fact that, according to the frameworks, one aspect is often privileged to the detriment of the other should not be seen as a problem. For instance, the NC framework can help us to understand how to solve incentive problems by enhancing resource allocation, but does not explain how this will result in scientific and technological progress nor how it will affect economic development. And, whereas the ES framework proffers some ideas about networking and the evolution of research and innovation systems, it does not help in determining the level of resources that should be allocated to the system and its components. This complementarity between the two frameworks is also clear when we look at the role of markets and the role that the State can play in terms of market creation. Although obviously central to the NC framework, the role of the market is not completely denied by the alternative framework, being acknowledged, for instance, to be a decisive element in the selection process. However, in the NC framework, the creation of markets requires the creation of demand or supply, the definition of property rights on the good that is exchanged, and the general conditions of market operation. In the ES framework the main focus is on the creation of an infrastructure and knowledge capacity, which are required to make the market exist. In other words, this framework tends to adopt a sociological view according to which the market is a social construct involving cognitive capacity, and not just a natural way of organizing the economy.

In terms of policy analysis and policy design, this complementarity first entails that market failures can coexist with learning failures, and probably these two aspects are more intrinsically connected than has been analyzed so far (possibly one reinforcing the other). But a second, and equally important conclusion, is that the S&T policy actions derived from both analytical frameworks are neither necessarily different nor antagonistic, and may be also complementary. What is crucial is that a given policy action (for instance, subsidizing firms' R&D activities), seemingly common to both frameworks (although a much more detailed analysis would obviously reveal some differences in the practical applications of this given policy principle), is differently justified by each framework, and is effected for different purposes. For instance, property rights tools will be constructed with a view to optimizing incentives for potential innovators (in NC oriented policy), while they will probably be designed in order to help knowledge sharing and combination (in ES oriented policy). Similarly, cooperative R&D will be publicly supported to reduce risk and achieve cost sharing following a NC orientation, but will be supported to create or reinforce networking and the creation of collective knowledge in an ES perspective.

Much conceptual and empirical work remains to be done to thoroughly articulate the perspectives provided by both frameworks, and to benefit from a "dual use" of policy actions. Related to this question of duality is the question of complementarity and coherence between different policy actions. Facing different sorts of failures, the policy-maker is never able to choose one single action, but rather has to define a policy mix. This issue certainly needs deeper investigation.

Another consequence of this dual dimension of most, if not all, policy actions is that logically the impact of any one policy action should not be evaluated in the same way in each framework. In other words, as *S&T policy principles* are based on different rationales related to different theoretical frameworks (leading to different objectives), *evaluation techniques and tools* are based on different evaluation perspectives related to different theoretical frameworks (leading to different "objects" of evaluation). One must take account of this necessary coherence when launching any evaluation exercise. In this respect, the different, and surely complementary, dimensions of the additionality concept are particularly interesting, since they highlight how various rationales lead to various understandings of the possible "differences" generated by public intervention.

Since the mid-1980s, policy options have also been analyzed as favouring either the horizontal or, conversely, vertical dimensions. Again, in each framework, the alternatives could be justified, and one could not claim that a given option stems analytically from a particular framework. Neoclassical externalities could be higher between than within sectors, or market failures could be more prominent and damaging in certain sectors (Martin and Scott 2000), and the collective cognitive capacity could be more or less relevant to different sectors. Another classification of S&T policy orientations is to distinguish between mission and diffusion-oriented policies (Ergas 1987). At first sight, mission-oriented policy appears to be more connected to the NC framework, while diffusion-oriented policy is more rooted in the evolutionary constructivist one. However, there may be circumstances in which the concentration of support for a small number of technologies, and for larger firms, is the best way to develop cognitive capacities and enlarge the knowledge base. Thus, the so-called large programmes that are frequently associated with mission-oriented policies can find justification within both frameworks. Conversely, diffusion of information and cooperation are not absent from the neo-classical perspective (Cantner and Pyka 2001). The two distinctions – horizontal vs. vertical policy and mission vs. diffusion policy – are a rather empirical orientation of policy, which always borrows implicitly or explicitly from both frameworks.

In both frameworks, there is also a need to combine S&T policy with other policies (anti-trust, commercial, education, etc.). In the NC framework, the distinction between the different policies is straightforward, since each of them can act on a limited and *a priori* defined set of variables of the NC model (such as incentives, price, market structure, etc.). In the ES framework, the differences are less evident and almost all policies can impact on the whole system because of all the interactions occurring at all stages. Recent works (Koelliker 2001) demonstrate the combined impact of S&T and anti-trust policies on innovation.

More broadly, it must be stressed that the present overview only deals with the theoretical basis of S&T policy. Obviously, actual policy-making is not a simple application of these theoretical recommendations, and is largely influenced by other rationales, for instance, related to politics, administration, lobbying, etc. A better knowledge of the coherence between those two sets of principles should certainly be developed in order to make the decision-making process more efficient and more beneficial for the whole of society.

1.6 References

Ancori A, Bureth A, Cohendet P (2000) The economics of knowledge: the debate about codification and tacit knowledge. Industrial and Corporate Change 9: 255-287.

Arrow KJ (1962) Economic welfare and the allocation of resources for invention, the rate and direction of inventive activity. Princeton University Press, Princeton, 609-625.

Bach L, Georghiou L (1998) The nature and scope of RTD impact measurement A discussion paper for the International Workshop on Measurement of RTD Results/Impact, http://www.cordis.lu/fp5/monitoring/studies.htm. Brussels, 28 29 May.

Cantner U, Pyka A (2001) Classifying technology policy from and evolutionary perspective. Research Policy 30: 759-775.

Cowan R, David PA, Foray D (2000) The explicit economics of knowledge codification and tacitness. Industrial and Corporate Change 9: 211-253.

Dasgupta P, David P (1994) Towards a new economics of science. Research Policy 23: 487-521.

David PA, Hall BH, Toole AA (2000) Is public R&D a complement or substitute for private R&D ? A review of the econometric evidence. Research Policy 29: 497-529.

Ergas H (1987) Does technology policy matter in BR Guile & H Brooks (ed.) Technology and global industry companies and nations in the World Economy. National Academy Press, Washington DC, 192.

Firth L & Mellor D (2000) Learning and the new growth theories: policy dilemma. Research Policy 29: 1157-1163.

Foray D (1991) Economie et politique de la science : les développements théoriques récents, Revue Française d'Economie, VI(4): 53-87.

Gibbons M, Limoges C, Nowotny H, Schwartzman S, Scott P, Trow M (1994) The New Production of Knowledge. Sage Publications, London.

Guellec D, van Pottelsbergue de la Potterie B (2001) The effectiveness of public policies in R&D. Revue d'Economie Industrielle (numéro spécial), 94: 49-68.

Griliches Z (1979) Issues in Assessing the Contribution of R&D to Productivity growth. Bell Journal of Economics 10: 92-116.

Jaffe AB (1996) Economic analysis of research spillovers - Implications for the advanced technology program. http://www.atp.nist.gov/

Kealy T (1997) The economic laws of scientific research. St Martin's Press, New York.

Laffont JJ, Tirole J (1993) A Theory of Incentives in Procurement and Regulation. MIT Press, Cambridge.

Koelliker A (2001) Public aid to R&D in business enterprises : the case of the US from an EU perspective. Revue d'Economie Industrielle (numéro spécial), 94: 21-48.

Lipsey RG, Fraser S (1998) Technology policies in neo-classical and structuralist-evolutionary models. OECD STI Review 22: 31-73.

Lundvall B-Å (1992) National System of Innovation, Towards a Theory of Innovation and Interactive Learning. London, Pinter & Publisher.

Lundvall B-Å, Borrás S (1997) The globalising learning economy: Implications for innovation policy, EUR 18307 EN, TSER/Science, Research and Development/EC, Luxembourg.

Malerba F (1996) Public Policy and Industrial Dynamics: an Evolutionary Perspective In: research project final report on Innovation Systems and European Integration (ISE), funded by the TSER/4th FP, DG XII/EC (contract SOE1-CT95-1004, DG XII SOLS).

Martin S, Scott JT (2000) The nature of innovation market failure and the design of public support for private innovation. Research Policy 29: 437-447.

Metcalfe JS, Georghiou L (1998) Equilibrium and evolutionary foundations of technology policy. OECD STI Review 22:75-100.

Metcalfe JS (1998) Innovation as a policy problem: new perspectives and old on the dof labour in the innovation process, SME and innovation policy: Networks, collaboration and institutional design. 13th November, Robinson College, Cambridge.

Metcalfe JS (1995) The economic foundations of economic policy: equilibrium and evolutionary perspectives. In: Stoneman P. (ed) Handbook of the economics of innovation and technological change. Blackwell Handbooks in Economics, Oxford UK and Cambridge USA.

Muldur U (2001) Is capital optimally allocated in the overall process of European innovation. Revue d'Economie Industrielle (numéro spécial), 94: 115-153.

Nelson RR (1959) The simple economics of basic scientific research. Journal of Political Economy, 67: 323-348.

Penin J (2003) Patents verus ex-post rewards: a new look. Working paper BETA, L. Pasteur University and CNRS, Strasbourg.

Sherer FM (2001) US government programs to advanced technology. Revue d'Economie Industrielle (numéro spécial), 94: 69-88.

Smith K (1996) Systems approaches to innovation: some policy issues In: Research project final report on Innovation Systems and European Integration (ISE), funded by the TSER/4th FP, DG XII/EC (contract SOE1-CT95-1004, DG XII SOLS).

Teubal M (1998) Policies for promoting enterprise restructuring in NSI: triggering cumulative learning and generating system effects. OECD STI Review 22: 137-170.

Usher D (1994) The collected papers of Dan Usher. Edward Elgar Publishing, Aldershot, UK.

Appendix A

Table 1.1. The economic foundations of the two dominant paradigms

Standard/Neo-classical framework: main features	Evolutionist structuralist framework: main features
Market: unique mode of coordination and selection	State is part of the game
Equilibrium	No equilibrium
Static analysis	Dynamic analysis/Path dependency
Optimizing rationality	Other forms of rationality
Input-output perspective/linear model of innovation	Inter-active model of innovation
Central focus: optimal allocation of resources	Central focus: creation of resources, key resource is knowledge + knowledge is different from information
Normative reference: welfare/Pareto analysis	Unclear normative reference: "adequate" system/process/cognitive capacities? Environment ensuring
Research (S, T, I) as input-output system producing information + information as an input for downstream activities	Knowledge coming from anywhere in the system (not only research)

Appendix B

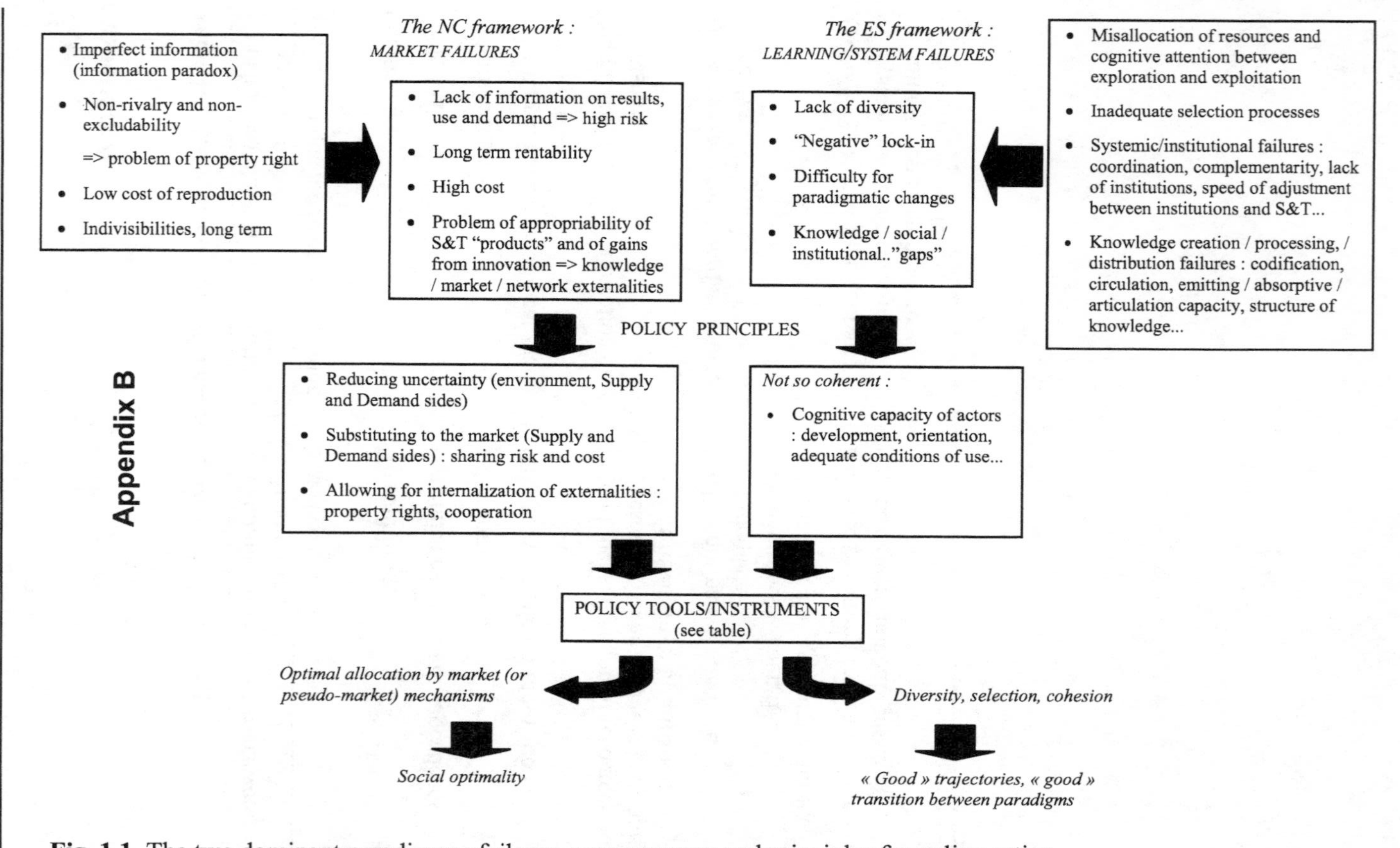

Fig. 1.1. The two dominant paradigms : failures, consequences and principles for policy action

Appendix C

Table 1.2. Policy tools and instruments in the two dominant paradigms

Interpretation in the NC framework	Basic tools and instruments of S&T policy	Interpretation in the ES framework
Reduce uncertainty and a-symetries	Diffusion of Information Knowledge	Change the available knowledge-base; involves codification; change distribution of knowledge
idem	Public intermediaries of Information Knowledge	Idem; Reinforce coordination
Substitute to private investment for production of scientific output considered as public good	Public labs in Science	Increase and change the available knowledge-base by reinforcing exploration; involves codification; change emitting/absorptive capacity of labs
Partially substitute to private investment for production of technology considered as non-rival and partly excludable good	Subsidy to R&D activities of firms	Increase and change the available knowledge-base by reinforcing exploration; involves codification; change emitting/absorptive capacity of firms
Substitute to private demand (limited in time)	Public procurement	Orient selection process by reinforcing exploitation
Full guarantee of appropriability of technology considered as non-rival and partly excludable good	Property rights	Partial change of emitting/absorptive capacity
Internalize externalities : monetary (vertical cooperation), knowledge (horizontal cooperation); diffusion of information; risk/cost sharing	Cooperation Firms, All types firms and public labs	Change distribution and sharing of knowledge; reinforce coordination and complementarity; change emitting/absorptive capacity
Substitute to private investment for production of human capital	Education	Increase cognitive capacity
	Emergence of standards and plateforms	Orient selection process; involves codification
	Norms, regulations	Orient selection process; involves codification
	Other related policies	Orient selection

2 Systems Failure and the Case for Innovation Policy

James Stanley Metcalfe

ESRC, Manchester, E-mail: Stan.Metcalfe@man.ac.uk

2.1 Introduction

The topic of this chapter is the rationale for innovation policy in advanced market economies. Since innovation and its associates, invention and the diffusion of innovation, play such a central role in the performance of modern economies, indeed they constitute a defining element in the claim that they are knowledge-based, it is hardly surprising that this rationale should be an indispensable part of economic policy more generally. The Barcelona accord on research and development (R&D) spending[1] suggests how important this issue is for European governments, and raises the question of whether policy frameworks and instruments exist to reach the objectives of this accord. In particular, will it be possible to protect any sustained increase in innovation expenditures from the effects of diminishing marginal returns in the short run and in the long run[2]? We shall argue that new perspectives are needed on innovation policy if innovation is to be stimulated in Europe while avoiding the spectre of diminishing returns. We also suggest that the traditional rationale for innovation policy, market failure, is flawed in its understanding of the innovation process and, more fundamentally, flawed in is understanding of the wider process of competition in the modern world. The reasoning behind this claim is that processes of innovation depend on the emergence of innovation systems connecting the many actors engaged in the innovation process, and that these systems are essentially self organizing. Innovation systems do not exist naturally, but have to be constructed, instituted for a purpose, usually but not uniquely to facilitate the pursuit of competitive advantages by firms. To

---

[1] To raise European R&D to 3% of GDP by 2010 with at least two-thirds of this contributed by industry.

[2] Diminishing returns to the economic payoff, not diminishing returns to the growth of scientific and technical knowledge.

anticipate the conclusion, innovation policy should be about facilitating the self-organization of innovation systems across the entire economy, not only in 'new' sectors. In sustaining this claim, we shall argue that innovation is one element, perhaps the most important, of the general class of investment activities in an economy, that it is complementary with other classes of investment undertaken by firms and other organizations, and that it requires much more than expenditure on science and technology for its realization. A functioning science and technology (S&T) policy is in the first instance a stimulus to invention, in the process it facilitates innovation, but the connection between the two is essentially a matter of investment, of present commitment in anticipation of future return, and it is equally important that policy promote the general process of investment if innovation is to flourish. Thus R&D spend may be a necessary underpinning for innovation but it is certainly not sufficient, other complementary investments in skills, productive capacity and markets are also required. As an innovation policy lever on its own, S&T policy leaves much to be desired. Moreover, all investment is uncertain in its consequences, but investment in innovation is particularly prone to the unexpected and the unintended consequences of action, precisely because innovation is a major source of business uncertainty. In exploring the limits of the market failure doctrine we also draw attention to the general limitations of an equilibrium approach to the analysis of innovation and competition and suggest that an adaptive evolutionary process view is a far sounder framework for understanding and policy guidance. Innovation involves the growth of multiple kinds of knowledge including knowledge of how to organize and knowledge of the market opportunities, and these different kinds of knowledge, complementary to scientific and technological knowledge, are gained inseparably from the competitive market process. Innovation is a route to competitive advantage, but the converse is true also, that competition shapes the innovation process; the two phenomena are inseparable. In developing the argument, we will amplify the idea of innovation systems but not from a national perspective. Rather we emphasize the local character of innovation systems and the need for policy to deal with the issues surrounding their birth, growth, stabilization and, if necessary, decline. National arrangements influence the ecology of organizations and the institutional rules of the game that enable innovation systems to be formed but innovation systems are not intrinsically national. Indeed a central implication of the unification of the European market is that 'local innovation systems' will cross national boundaries, with the prospect that national policies develop inconsistencies that are inimical to innovation performance. Thus, we argue that a systems failure perspective allied with notions of evolutionary competition will enable governments to form and implement effective policies. Since innovation requires the development of new knowledge, typically within firms but more broadly in contributing innovation systems, the stimulus of innovation cannot ignore the conditions that facilitate the growth of knowledge and the communication of information. This epistemic dimension turns out to be of quite crucial importance in the innovation systems failure perspective. Finally, we say little directly about existing policies and instruments and direct the reader to a recent paper by Luke Georghiou for a detailed elaboration and evaluation of European level policies[3].

2.2 Attributes of the Innovation Process

We begin with a brief statement of the relevant attributes of the innovation process. Innovation is, first and foremost, a matter of business experimentation, the economic trial of ideas that are intended to increase the profit of or improve the market strength of a firm. This occurs in two broad sets of conditions, defined by innovation in existing enterprises and innovation by new enterprises, and the two are quite different contexts for innovation experiments. Innovation, in this regard, is the principal way that a firm can acquire a competitive advantage relative to its business rivals. As a process of experimentation, a discovery process, the outcomes are necessarily uncertain; no firm can foresee if rivals will produce better innovations nor can it know in advance, even when all technical problems are solved, that consumers will pay a price and purchase a quantity that justifies the outlay of resources to generate a new or improved product or manufacturing process. This is not a matter of calculable risk, for probabilities cannot be formed in respect of unique events, events that change the conditions under which future events occur. There is an inevitable penumbra of doubt that makes all innovations blind variations in practice, and the more the innovation deviates from established practice the greater the fog of irresolution. Perhaps the fundamental point is that innovations are surprises, novelties, truly unexpected consequences of a particular kind of knowledge-based capitalism. This does not mean that innovation is irrational behaviour, firms are presumed to innovate in ways to make the most of the opportunities and resources at their disposal; however, neither the opportunities nor the resources available can be specified with precision in advance. Innovation is a question of dealing with the bounds on human decision making, it is to a substantial degree a matter of judgment, imagination and

[3] See Raising EU R&D Intensity, European Commission, 2003, the report of an expert group under the chairmanship of Professor Georghiou.

guesswork, and the optimistic conjecturing of future possible economic worlds. Consequently, policy instruments must be subject to the same penumbra of doubt in terms of their effects on the innovation process; there will be unanticipated consequences of innovation policy and great difficulty in tracing cause effect relationships in the evaluation of policy.

The second attribute of the innovation process is the necessity for new beliefs and knowledge to emerge before innovation is possible. Moreover, innovation requires the drawing together of many different kinds of information, on the properties of a device or method, on the way to organize production and the perceived needs of the market. It is the combination of these elements that matters and the only locus of combination in capitalism is the firm[4]. Thus while many agencies may provide information valuable to the innovation in question, only the innovating firm can combine them into a "plan" for innovation. Neither universities, nor government laboratories nor knowledge consultancies, which play an increasingly important role, have this final combinatorial responsibility, in this, the for-profit firm is unique. The corollary of this is that multiple kinds of knowledge are typically required to innovate and many of the sources of this knowledge will lie outside the firm, which has to extract the necessary information and integrate it into its own knowledge (Gibbons et al. 1994). Consequently, the external organization of the firm and the management of its internal processes are essential elements in the innovation process and this insight is the foundation of the innovation systems perspective.

The third attribute of innovative activity is its embeddedness or more accurately its instituted position in the competitive market process. Not only do firms innovate to generate market advantages relative to their perceived rivals, so that the functioning of markets shapes the return to innovation, but market processes and their wider instituted context of law, custom and regulation greatly influence the outcomes of innovation and the ability to innovate. Innovation is not a matter of market processes acting in isolation but of the interdependence between market and non market, and public and private spheres of action. Moreover, the instituted context is broadly based, for example, the way users respond to an innovation and the ability of a firm to raise capital and acquire skilled labour and components necessary to an innovation are essential market process determinants of innovation activity. Yet, the fundamental test for successful innovation is not that it works but that it is profitable *ex post*, and this is a matter of market process. If markets are inefficient and distorted, this can only harm

---

[4] Broadly defined to include not for profit organizations that produce goods and services, such as hospitals, as well as the traditional for profit business organization.

the innovation process and when incumbents and conservative users unduly control the relevant markets, the effect will be similar. It follows that competition policy and an efficient markets policy, more generally, are necessary elements in innovation policy. Conversely, a pro-innovation policy is perhaps the most effective contribution to a strong competition policy.

Since innovation entails the acquisition of new knowledge, we need to be clear what is meant by knowledge, and the processes by which it is generated and diffused. Knowledge has a unique property, it always and only ever exists in the minds of individuals and it is only in individual minds that new innovative concepts and thoughts can emerge. This is fundamental, it is why we recognize the entrepreneur and the prize-winning scientist – they are different as individuals – and from it follows the fact that knowledge is always tacit it is never codified as knowledge. What is articulated and codified is information but information is only ever a public representation of individual knowledge, sometimes virtually a perfect representation, but in many significant cases not. As Polanyi (Michael not Karl) expressed it, we know more than we can say and can say more than we can write. Since economic activity in firms and beyond depends on the ability of teams of individuals to coordinate their actions, it follows that processes must exist for correlating the knowledge of the individual members so that they understand and act in common. In regard to innovation, the internal organization and business plan of the firm are the primary means of coordinating information flow and turning individual knowledge into the necessary hierarchy of understanding and actions. It may be helpful to conceive of the organization of a firm is an operator, a local network of interaction through which what the individual members of the firm know is combined to collective effect. The spread of understanding in correlated minds is essentially a social process of human interaction, however, a chief consequence of information technology is that information can be communicated at a distance and this makes possible the inclusion of a firm in wider, less personal networks, including the scientific and technological networks that communicate almost exclusively in written form. To call these knowledge networks may be understandable, but it is a mistake. The relevant networks are information networks, perhaps better expressed as networks of understanding, and could not be otherwise, and their significance is in shaping what individuals in firms, and other organizations transmit and receive as information. It is not that information is transmitted with error, it may be, rather, what matters is that information may legitimately be 'read' by recipient and transmitter in different ways. The interpretation of the message is not in the message but in the different minds of the parties concerned (Arthur 2000). Indeed the growth of knowledge depends on this possibility of divergent interpretation of the information flux. All innovations are based on disagreement, on a different reading of information much of which is currently available in the public domain. Thus, the prior knowledge state influences what is 'read' and what is 'expressed' and, as Rosenberg (1990) made clear, firms have to invest in their own understanding if they are to participate effectively in innovation information networks and this is why it is necessary for them to conduct their own R&D[5]. Thus while information is a public good, in the sense of being useable indefinitely, it is not a free good, scarce mental capacity must always be engaged to convert it to and from private knowledge (Cohendet and Meyer-Kramer 2001). Here we find one of the principal sources of variation in the innovation process, innovations are conceived in individual minds and these minds differ. It only needs a moment's reflection to recognize that if all individuals held the same beliefs there could be no growth of knowledge and no innovation and thus the beliefs in question could not have emerged in the first place. Idiosyncracy, individuality, imagination are the indispensable elements in the innovation process and the way innovation policy is framed must recognize this fact, indeed, without them entrepreneurship would not be recognizable. The obvious corollary to the policy process is that innovation cannot be planned from on high, it emerges from below.

Scholars interested in innovation have for many years drawn upon the useful Polanyian (1958) distinction between tacit and codified knowledge, the former embodied in human skill and practice, the latter in material form. Tacitness is presented as a reason why information does not flow freely, while codification, is a process to make information public. Thus, Callon (1994) is quite right to point out that the limits to excludability depend upon the way in which information is embodied in different communication media, and that access to any particular knowledge depends upon complementary assets being accumulated to give the capability to maintain and use knowledge based statements. However, it is important to recognize the point that the division of knowledge into mutually exclusive categories, codified and tacit, does not uniquely reflect properties of the knowledge itself. Rather, it is in part an economic decision dependent on the scale on which the information is to be used and the costs of codification. It is thus

[5] It is said that the British system of Industrial, Cooperative Research Associations, set up primarily in fragmented industries, failed to raise innovation performance, precisely because their target firms did not invest in acquiring their own capacity to understand the research and development carried out on their behalf.

inextricably linked with the division of labour in the economy more widely, as I shall explain below.

The distinction drawn here between information, knowledge and the understanding necessary for teamwork and cooperative endeavour, also bears on the question of property rights in knowledge and the assessment of the system of protection of intellectual property. Quite obviously, in the light of the above, knowledge is always proprietary it never leaves the minds of individuals, it is only its expression as information in a public domain that raises questions of intellectual property. Thus the oft-expressed view that secrecy is a very effective, perhaps the most effective form of protection. Where this is unavoidable, and in respect of product innovations it is nearly always so, patents protect the economic exploitation of the idea, but not its exploitation in a wider sense. Namely, the quid pro quo for economic protection is the placing of an accurate description of the invention in the public domain thus opening the possibility that others are guided to the same effect, and by a different route. This is the technological price the inventor pays for the economic right of exploitation, and rightly so. It should also be remembered that patents protect the exploitation of the knowledge in the invention and that this is only part of the knowledge required to innovate. The firm with the excellent patent record is not necessarily the firm best able to turn those inventions into profitable innovations.

A fourth implication for the innovation process is that the systemic, emergent nature of group understanding leads directly to the basis of innovation systems. There is an increasingly elaborate division of labour in the generation of knowledge, to use an old economic concept, the division of knowledge labour is becoming increasingly 'roundabout' in nature. Since Adam Smith, scholars have recognized that the knowledge contained in any economy or organization is based on a division of mental specialism. It is not simply that the division of labour raises the productivity of the pin maker, it also raises the productivity of the 'philosopher and man of speculation' and greatly augments the ability to generate knowledge in the process. When this division of labour is not contained within the firm we have the conditions for an innovation system to emerge and the necessity for the coordination of the divers minds within that system. Innovation systems are the necessary consequence of this division of knowledge; and these systems do not arise naturally, they have to be organized and are not to be taken for granted. This self-organization process is a central concern of innovation policy from a systems failure perspective. Innovation systems are, in Hayekian terms, a form of spontaneous order, that is to say they are self-organizing. Perhaps the most obvious characteristic of modern economies is the distributed nature of knowledge generation and the consequent distributedness of the resultant innovation processes across multiple organizations, multiple minds and multiple kinds of knowledge (Coombs et al. 2004). As a system, what matters are the natures of the component parts, the patterns of interconnection and the drawing of the relevant boundaries and each of these aspects forms a dimension of innovation policy, as we explore below.

Fifthly, and finally, it is helpful to group the factors that influence the ability to innovate into four broad categories, perceived opportunities, available resources, incentives and the capabilities to manage the process. In principle, we could imagine policy levers for each of these elements, but what matters is that all four need to be addressed if policy is to be effective. Thus, increasing the resources devoted to innovation is likely to run into rapidly diminishing returns if new opportunities are not perceived or if the management of innovation is weak and poorly connected with other activities in the firm (Carter and Williams 1957)

With these aspects of the innovation process in mind we turn now to an assessment of the traditional market failure rationale for policy and then contrast it with the systems failure perspective before drawing general conclusions for policy.

2.3 The Limits of Market Failure

The development of an economics of information and knowledge in the 1960s led scholars to the realization that knowledge and information are not normal economic commodities but possess attributes that do not make them natural candidates for market exchange (Nelson 1959; Arrow 1962). The market failure doctrine and the rationale it provides for innovation policy have followed from these insights. Central is the idea that markets in relation to knowledge and information have an inherent tendency to produce socially inefficient outcomes, inefficiencies that provide the justification for failure correcting public policies. The private hand is not guided to produce and use the socially optimal amount of knowledge, and the optimizing policy maker is justified in corrective intervention through the joint provision of resources and incentives at the margin. This has proved to be a powerful set of ideas for shaping policy debate, particularly concerning the public support of university based science and technology that are far from market application. I shall argue that it has been a far less useful means for designing specific innovation policies in relation to private firms. The reason is clear, the idea of a perfectly competitive allocation of resources (the doctrine of Pareto optimality) on which the idea of market failure is premised is a distorting mirror in which to reflect the operation of a restless capitalism. This doctrine seriously misreads the nature and role of competition in modern societies through its failure to realize that capitalism and equilibrium are incompatible concepts and that innovation and enterprise preclude equilibrium.

Why does the market failure doctrine fail in respect of innovation? The reasons are hidden within the properties of a perfectly competitive economy. For its realization not only must all agents be denied the power to influence prices of products and productive factors, there must also be a complete set of markets that values all consequences of all economic action in the present and in the indefinite future. In general, the set of possible markets is incomplete and serious problems follow. The consequences of action that are not priced in the market are called externalities and, from an innovation perspective, the most significant externalities relate to imperfect property rights in the exploitation of knowledge. If the works of the inventor can be copied without cost, others may turn invention into innovation, and erode the incentives to invest in invention. This has long been recognized as a justification for patent and copyright systems and rightly so. Nonetheless, the practical implications of intellectual property protection are less straightforward.

The problems are two-fold. It is not information spillovers *per se* that damage the incentive to invest in knowledge production but a presumption of instantaneous and complete spillover, an unlikely state of affairs for reasons which become clear below[6]. Absent this and the existence of many practical ways that firms have developed for protecting knowledge acquired privately, and it becomes clear that inventors and innovators may still gain an adequate return from their investments without patent protection. Secrecy and a short product lifecycle, are familiar examples and help explain why patent protection is only considered significant in a small number of industries, those with high invention costs and long lead times to market. Secondly, this doctrine is far too negative, not all information spillovers are between direct competitors or diminish innovation opportunities. The difficulty arises from thinking that all firms are the same, losing sight of the fact that they read the information flow with different 'minds'. Spillovers can, and generally will, have positive benefits in stimulating the differential creation of new knowledge, which should not be underestimated, indeed, this is why patents are designed to put inventive ideas in the

[6] I note in passing that what is spilt is information (messages) not knowledge. The knowledge content of any information flow is, of course, notoriously unpredictable as any university examiner knows only too well. That this is so is essential to the emergence of novelty.

public domain. There is no reason why an alert firm should not gain more than it loses from the unplanned flow of information and so enrich its innovative capacity. In this regard, information spillovers are to be encouraged and one might expect firms to try to manage this process through links with other knowledge generating institutions, which is precisely what we observe in practice[7]. What is interesting about the idea of property rights in commercially valuable knowledge is that they sit side by side with very imperfect property rights in economic activities more generally[8]. Copy my invention and I can pursue you in the courts. Make a better, but unrelated, equivalent and there is nothing we can do except compete. Indeed if it were otherwise, it is difficult to see how capitalism could have been the source of so much economic change and development. This means of course that competition is a painful process. Investors, whether their assets are in paper titles or human skills, are ever open to the erosion of their worth by innovations made by others, this is why innovation driven capitalism is, from a welfare point of view, an uncomfortable restless system. The fact that, on average, innovation enhances the standard of living should not blind us to this fact and to the inherently uncertain, potentially painful nature of innovation related economic processes. From a policy viewpoint, one immediate implication is that the scope of patents should not be drawn too broadly, for this simply limits the ability of others to creatively explore the design space which any patented invention has placed in the public domain. A world with no spillovers simply restricts, perhaps makes impossible, the wider and deeper growth of knowledge. Thus, broad patents have the potential to damage the creativity of the capitalist model (Merges and Nelson 1990).

Externalities do not exhaust the idea of missing markets. Perhaps more important is the absence in general of futures markets to guide investment decisions. All innovations are investments, activities that require current outlay in advance of the economic return. Yet the markets to trade these future outputs, by establishing the price today for an activity to be sold say a year hence, exist only for a narrow range of standardized commodities that are broadly speaking unaffected by the prospect of innovation. In the absence of known prices, the only recourse is to substitute the judgment of entrepreneurs. This uncertainty is intrinsic to the market process, for the

---

[7] Hence the increasing volume of work which points to the role of knowledge spillovers in productivity growth. Cf. Griliches (1998) for an authoritative treatment.

[8] It is worth noting that competition authorities in the UK have taken a dim view of firms that refuse to grant licences to exploit their patents and of attempts to use licences to distort the competitive process.

most significant sources of uncertainty relate not to whether it will rain a year from today, but whether others will have developed superior innovations by that date. It is not the game against nature that matters but the capitalist game of innovation, of rival against rival. In modern capitalism, genuine uncertainty is 'built in', as it were, and its consequences for the willingness to invest in innovation are far more difficult to cope with, because innovations, like all discoveries, are unique events for which the probability calculus is an inappropriate method of analysis. Much decision making about knowledge creation is at root an act of faith, it is a matter of the conjecture of imagined future worlds with necessarily unpredictable time delays between knowledge creation, application and market testing (Loasby 1999). Keynes's much ignored notion of animal spirits is certainly appropriate as a route to understanding innovation in capitalism. Moreover, it is not at all obvious that the process of accumulation of scientific or technological knowledge is any less hazardous than the accumulation of market knowledge (Callon 1994). A central implication of this theme is that investment becomes impossibly difficult in perfectly competitive markets as pointed out by Richardson (1961). Current market prices do not convey the information required to invest since they do not convey information about the investment plans of rivals. Consequently, firms seek other ways to tacitly or explicitly coordinate their activities whether complementary or competitive and these necessitate deviations from the atomistic competitive ideal. Although Richardson directed his analysis at investment in productive capacity, it applies equally well to investment in innovation, and one would predict a need for market imperfections if such investment is to be stimulated. One consequence of all this is that innovation processes are mediated by a range of non-market methods, primarily involving information networks and other forms of arrangement between organizations and individuals, procedures that build confidence and trust and work to limit the damaging consequences of uncertain, asymmetric information. These arrangements are precisely contrary to the idea of competition between isolated, atomistic, independent firms. Without market power, innovation becomes an unlikely occurrence, and collaborative R&D arrangements, for example, are one way of dealing with the implied coordination failures.

However, it is not at all obvious from a wider view that the missing markets constitute market failures in the narrow sense that we have used so far. Uncertainty and asymmetries in knowledge are direct consequences of a market process in which innovation is the driving force for competition. Without innovation, it is possible that a richer ecology of futures markets would come into existence, but this is not the market capitalism we know, it is rather a picture of a stationary state. It is surely perverse to label as market failures phenomena that are integral to the competitive market process and which give modern capitalism its unique dynamic properties. Nor is there any obvious way that policy could 'correct' for lack of futures markets, they are simply intrinsic to the process of innovation and economic change. The fundamental fact is that profits follow from the deployment of ideas that others do not have, with the consequence that the whole system dynamic depends upon the generation of unquantifiable uncertainty and asymmetries in information. One cannot sensibly argue that the economy would perform better if innovation related uncertainties were reduced, for the only way to reduce these uncertainties is to reduce the incidence of innovation and thus to undermine the mainspring of economic progress. This does not deny that radical uncertainty can be a justification for policy intervention. Indeed the rationale for the public support of fundamental research in science and technology lies in the fact that the links between these general categories of knowledge and market exploitation of specific innovations is often so tenuous that private firms would, quite legitimately, find no justification for investing in these kinds of knowledge. Even here, the matter is not clear-cut. For by no means all university research in fundamental science and technology is funded by government, and of that which is, a proportion is directed at meeting the mission objectives of government agencies in such areas as defence or health. Conversely, non-academic organizations carry out a substantial portion of work on fundamental science and technology; indeed large private firms, usually multinational firms, can often boast far more advanced research facilities than can universities[9].

Having dealt with the problem of missing markets, consider next the idea that perfect competition requires an absence of market power, in particular that each firm be small relative to the scale of market to make this possible. It is recognized that a major reason this condition will not be achieved is the presence of some form of increasing returns in the use of resources. Yet, fundamental to the economics of knowledge production and dissemination is the fact that the exploitation of all knowledge is subject to increasing returns: the fixed cost of producing an item of knowledge can be spread over a greater volume of output, as it is used more widely and more intensively in the production process. Since one cannot innovate on the basis of a fraction of a technology or a quarter of a scientific fact, there is necessarily an indivisible cost of creating the complete set of knowledge behind an innovation. Consequently, the costs of exploiting an

[9] Narin et al. (1997) find that of the US scientific papers cited by US industrial patents only 50% came from academic sources while 32% came from scientists working in industry.

innovation fall with the scale of exploitation, precisely the condition that removes the possibility of perfect competition. Furthermore, every investment in innovation now requires its own expected minimum scale of exploitation if an adequate return is to follow. The result of these considerations is the complete inability of the perfectly competitive model to provide guiding innovation policy principles in a world where firms are required to innovate in order to compete (Stiglitz 1994). The overhead innovation costs that firms must incur unavoidably mean that their behaviour will at best be imperfectly competitive and that there will be systematic and uneven deviations between prices and marginal production costs across the economy. The only way the fixed costs of knowledge production could be recovered, independently of prices and outputs, would be for public laboratories to develop that knowledge, or for all private research and development expenses to be fully subsidized from the public purse. This is not a model for innovation likely to commend itself outside of very special cases such as metrology and public technical standards (Tassey 1992).

Nor do missing markets and market power exhaust the difficulties in using perfect competition to reflect modern capitalism. There is also the, so-called, public good problem. All knowledge and information has the intriguing property that it is used, but not consumed in its using and that, once discovered, it is in principle useable by any individual on any number of occasions to any degree. In the terminology of economics, there is non-rivalry and non-excludability of knowledge. This terminology is not well chosen in relation to knowledge and information. We argued in the previous section, that knowledge is only ever private and is certainly excludable by choice of the knowing individual. It is a representation of that knowledge, information that is placed in the public domain, but this is only accessible to everyone in principle. In practice, and as a direct consequence of the division of knowledge labour, to gain knowledge from information requires prior background knowledge to read that information and this knowledge has not been acquired without opportunity cost. There is much more to the transfer of knowledge than the costs of communication in the narrow sense. In many cases the interchange of knowledge requires communication between correlated "like minds" only open to those who have acquired comparable abilities to understand the significance of new scientific and technological information. Self-exclusion follows from an inability to make the necessary background investments; information may be "free" but the ability to extract knowledge from it is not and it is the knowledge that matters for innovation (Mowery and Rosenberg 1989; Rosenberg 1990; Hicks 1995; Veugelers 1997).

To a degree, these different dimensions of market failure are interrelated. The public good aspect of information links directly to information spillovers and the externality problem. The fact that knowledge can be used repeatedly connects to the increasing returns dimension of the exploitation of information, whether in producing goods or, more significantly, in the further production of knowledge. In each case, we are led to deviations from the perfectly competitive market ideal, but it is not at all obvious that affairs can otherwise be arranged. All economies are knowledge based and the problems of the economics of knowledge are not an optional extra, they are intrinsic to the nature of a capitalist, market economy. When we turn to innovation policy, it is apparent that we are in difficulty in basing the rationale on a model of perfect competition. Leaving aside the well recognized imperfections that governments can be subject to when they intervene, backing the wrong horse too quickly or maintaining programmes long after the evidence against continuation is conclusive (Walker 2000), it is clear that market failure as a policy framework leaves much to be desired (Metcalfe 1995a,b). Market failure is a general rubric not a recipe for stimulating individual innovations. The logical underpinning it provides tells us nothing about the precise design of policy instruments, or their appropriate method of implementation or the firms that are most appropriately in need of support in their attempts to innovate. Is the focus to be on new knowledge, new skills or new artefacts? Is it to be concerned with design, with construction or with operation? Is it to focus on the creation of innovation or upon the diffusion of innovation? The answers to these questions could generate very different policy initiatives that bear no particular relation to specific innovation activities. The information to provide the answers is simply not available to the policy maker, nor for that matter to anyone else. The market failure framework, despite its formal elegance, is an empty box. In the presence of the apparent market distortions in relation to knowledge and information, there is no warrant for the idea that piecemeal policy can improve economic welfare, the world is simply too complicated to avoid these problems of the second best. Perhaps the problem is deeper, in that the issues of uncertainty, "spillovers", increasing returns and "publicness" are not failures at all but vital elements in the evolutionary process that is capitalism. This thought takes us to the nature of competition and the idea of innovation systems and their failure as the basis for policy.

2.4 Innovation Systems and the Competitive Process

The foundation of an alternative approach is that of competition as an evolutionary process not as a state of equilibrium. In this perspective innovation plays the central role as the source of the differences in firm behaviours that give rise to competitive advantages. Rivalry depends on differential behaviour and these differences are resolved into differences in profitability and the consequential differences in the relative growth of rival producers. If markets are working well from an evolutionary perspective, firms with superior competitive knowledge and thus practice are able to grow at the expense of less competitive rivals. This is the central dynamic of evolutionary competition as a dynamic discovery process. All that competition requires is rivalry, and two firms can be as competitive as many and have a greater impact on the long run capacity to use resources efficiently. In such a view, the role of markets is to coordinate and evaluate rival business conjectures and so guide the economic change we (partially) measure in rising standards of living. This involves adaptation to new opportunities, new needs and new resources and market institutions are to be judged not by the canon of Pareto optimality, but by their openness in stimulating innovation and adapting to change (Metcalfe and Georghiou 1998).

Thus, the central weakness of the market failure approach is not its lack of precision, but its attempt to establish a policy perspective within the confines of the static equilibrium theory of markets and industry. The market failure arguments identify significant features of the production and use of knowledge, but these features have their full impact only in relation to the dynamic nature of the competitive process. Economic progress depends on the ongoing creation of private, asymmetric knowledge, knowledge that is sufficiently reliable and defendable to justify the original investment, yet has prospective returns that are not only uncertain to the investor but create uncertainty in complementary and competitive fashion to other investors. The imperfections identified in the market failure approach are to be viewed now in a different perspective, as integral and necessary aspects of the production and dissemination of knowledge in a market economy. From this perspective, it is surely perverse to call them imperfections or market failures. This is, of course, not a new point: for those who have studied Schumpeter they are the natural features of an economic process driven by creative destruction. Another way of putting this is to say that without asymmetries of knowledge and the correlated uncertainties and indivisibilities the competitive process has nothing with which to work. The quasi-public good nature of knowledge, indivisibility and increasing returns, the inherent uncertainties of creative, trial and error processes and the imperfect nature of property rights in knowledge are essential if market capitalism is to function. They are not imperfections to be corrected by policy.

Several important themes now fit into place in a way that is impossible with the market failure doctrine. First and foremost among them is entrepreneurship a phenomenon, which has no meaning in economic equilibrium of any kind. Entrepreneurs introduce novelty into the economy, they disrupt established patterns of market activity, they create uncertainty and they provide the fuel that fires the process of economic evolution. The fact that the framework of perfect competition cannot incorporate the entrepreneur is a telling statement of its inapplicability to an innovation driven economy. Secondly, the reward for entrepreneurship is the differential economic reward that comes from introducing economic improvements relative to existing practice. Such abnormal rewards are not the consequence of market imperfections, they do not necessarily reflect the undesirable use of market power; they are instead the rewards for superior performance and are to be judged as such. It is a view that abnormal profits are the socially undesirable consequences of market concentration that is the real Achilles heel of the market failure approach and which denies it anything useful to say in the appraisal of knowledge-based, innovative economies.

Thirdly, this perspective of competition and innovation as a coupled dynamic process provides us with a framework to formulate innovation policy. Innovations create the differences in behaviour, which we identify as competitive advantages, and the possibility of competition provides the route and the incentives to challenge established market positions. Moreover, to the extent that market institutions function properly, firms with superior innovations will command an increasing share of the available scarce productive resources, the process that is the link between innovations in particular and economic growth in general[10].

This suggests that innovation policy and competition policy are complementary, indeed that a pro-innovation policy may be the surest form of competition policy, and that its broad purpose is to ensure that conditions remain in place for the continued creation and exploitation of asymmetries of knowledge. In truly competitive markets, all established positions are

[10] As an aside here, we note that competition is not to be judged by market structure. Two rivals may compete far more intensively than many. The way to judge the efficacy of competitive arrangements is to consider the degree to which rivals can gain market share at the expense of each other and the degree to which they are innovating in the pursuit of competitive advantage.

open to challenge and it is this link between innovation and competition, which has proved to be the reservoir of economic growth. Thus, capitalism is necessarily restless, occasionally kaleidoscopic, and competition is at root a process for diffusing diverse discoveries, the utility of which cannot readily be predicted in advance. The market mechanism is a framework within which to conduct innovative experiments and a framework for facilitating economic adaptation to those experiments[11]. The key issue, therefore, is how this competitive process interacts with the conditions that promote innovation.

2.5 Increasing Returns, 'Roundabout' Knowledge Production and Innovation Systems

We have referred already to the inevitable presence of increasing returns in a knowledge-based economy, the fact that the returns to investments in innovation increase with the scale of their exploitation. That this rules out a perfectly competitive allocation of resources is well understood but there is much more to the phenomena than is suggested by this partial and static perspective. The point is a more general one. As Adam Smith understood so clearly, increasing returns applies to the generation of knowledge as well as to its exploitation precisely because of the increasing specialization of bodies of knowledge and knowledge generating institutions. What we are observing in modern innovation systems is the increasing roundaboutness of production, not of material artefacts but of knowledge in general (Young 1928).

It can be argued that two features shape the modern innovation process; namely, increasing complementarities of different kinds of knowledge together with increasing dissimilarity of these bodies of knowledge, a reflection of an increasingly fine division of labour in knowledge production (Richardson 1972). Innovating firms need to draw on and integrate multiple bodies of knowledge, whether scientific, technological or market based, produced in an increasing range of increasingly specialized contexts[12]. At the same time to understand the significance of and contribute to

[11] This theme of the experimental economy has been particularly important in the work of Eliasson (1998). It has an inevitable Austrian hue, that markets are devices to make the best of our limited knowledge (Rosenberg, 1990).

[12] Cf Grandstrand et al. (1997) for evidence that large corporations are increaseingly diversified in the technological fields which they employ, and more diversified relative to their product fields. See also Kodama's (1995) work on technology fusion.

advances in these various kinds of knowledge is increasingly beyond the internal capabilities of the individual firm. Consequently, firms must increasingly complement their own R&D efforts by gaining access to externally generated knowledge and learn how to manage a wide spectrum of collaborative arrangements for knowledge generation (Coombs and Metcalfe 2000) The consequence is that innovations take place increasingly in a systemic context with respect to the use of new technologies and their generation. How they occur is a question in the coordination of the division of labour in innovation systems. This is a central difference from the market failure approach, in which innovation is treated as a problem internal to the firm. Instead, we have to enquire how groups of organizations are coordinated to give innovation processes a systemic dimension.

The essential point is the distinction made above between private knowledge and public understanding. All new knowledge arises only in the minds of individuals and if it is to have a wider effect it must not only be communicated to other minds but these minds must absorb it and reach similar understanding of the phenomena in focus. In short, knowledge must be correlated across individual minds. This is essential for any joint action and it is essential to the further growth of knowledge, as enquiring minds respond to the information that constitutes the testimony of others. The consequence of this is that what is understood is systemic, covering multiple individuals, it is combinatorial and it is emergent. Not only is understanding complicated, in the sense of the multiplicity of minds involved, it is also dynamically complex, in the sense that it its development generates novelty in unpredictable and unintended ways; this is one foundation for the uncertainty that underlies innovation-led capitalism. Capitalism is a restless evolving system precisely because its knowledge foundations are restless and adaptive too. The process of correlation of knowledge is complicated further by the fact that individuals typically express and communicate their knowledge in the context of the organizations of which they are members, and the rules and routines of these organizations shape the interplay of information both within and without that organization. Thus, all knowledge systems are constructed around multiple minds in multiple organizational contexts and here we should distinguish invention systems from innovation systems proper. The science and technology systems composed of universities and public and private research laboratories are primarily invention systems[13] and, as Schumpeter insisted, invention is conceptually distinct from innovation. Innovation systems depend on additional sets of actors in relation to the availability of productive

[13] See, in particular, Carlsson (1995,1997) for a detailed exposition of the related concept of technological systems.

inputs, the design of organization and the engagement with customers and they depend on the unique role of the firm to combine the knowledge of these elements to achieve innovation. The knowledge and ability to organize a productive activity, to identify markets and to mobilize resources, are essential elements in the innovation process; for innovations are not only about the generation of knowledge, but also the economic application of knowledge. Thus, innovation systems are embedded in the market process, with customers, suppliers and even rivals on occasion, acting as important system components (Lundvall 1986, 1992). Markets are the context in which resource problems in relation to innovation are solved and in which innovation opportunities are identified.

However, systems are not defined only in terms of their components, in this case knowledgeable individuals in organizations; the nature of the system also depends on how these individuals are connected by flows of information and the purpose that lies below the flow of information. That correlation of knowledge requires communication of information indicates the importance of the connections in the innovation system, and the need for these connections to change as the innovation problems change. In many important cases, communication requires personal interaction and its correlates of trust and empathy between the individuals. In other cases, particularly in regard to science and technology, communication can rely on communication technology so that much of the information considered reliable comes from minds that are distant and anonymous. Indeed, it is these non-social forms of communication, information technology broadly defined, that have transformed knowledge generation. By permitting connection between a far greater number of minds than is possible through personal interaction alone, information technology has been of vital importance not only to correlate knowledge more widely but also to stimulate the further growth of private knowledge within innovation systems.

Yet science and technology systems are not innovation systems, the latter are far more focused in scope and directed to business objectives, that is to say they are focused around very specific, local problem sequences reflecting the proprietorial concerns of the innovating firm. The most appropriate way to conceive of these systems is that they self-organize and that private firms take the lead in stimulating the self-organization of the knowledgeable minds in the system. This means that innovation systems are locally dynamic entities, they are born, grow, stabilize and ultimately decline and fail and that the basis for the dynamic of self-organization is the evolution of the particular innovation problem sequence. Part of the dynamic of system change is that the growth of knowledge depends on disagreement across individuals and the fact that the solution to one problem typically opens up new problems that may require different kinds of knowledge in their solution. As Cohendet and Meyer-Kramer (2001) are right to point out, innovation systems operate as recursive trial and error processes for stimulating the growth of knowledge in relation to specific problems. The consequence of this is that as the problem sequence evolves so too do the components and connections defining the particular innovation system. Thus, there seems great merit in seeing innovation systems as a form of self transforming, spontaneous order that interacts with the process of market competition outlined above. Perhaps the key point to note is that innovation systems are the constructed bridges between invention systems and market systems (Carlsson et al. 2002).

We can summarize the focus of this perspective in terms of the development of the innovation infrastructure in the economy; an information infrastructure that facilitates the intercommunication of existing knowledge and mutually shapes the future agendas of different organizations around innovation problem sequences. In short innovation systems are devices to correlate knowledge and in the process advance knowledge in regard to specific innovations. It is an infrastructure to correlate knowledge through communication and to coordinate access to complementary kinds of knowledge required to innovate and it is more than the infrastructure for science and technology (Edquist 1997; Carlsson 1997; Nelson 1993). Many organizations are involved, private firms operating in market contexts, universities and other education bodies, professional societies and government laboratories, private consultancies and industrial research associations, but only the first of these is in the unique position to combine the multiple kinds of knowledge to innovative effect. Between them there is a strong division of labour and, because of the economic peculiarities of information noted above, a predominance of coordination by networks, public committee structures and other non-market mediated methods (Tassey 1992; Teubal 1996). The division of labour is of considerable significance for the degree to which the different elements of the system are connected. Different organizations typically have different cultures, use different "languages", explore different missions, operate to different timescales and espouse different ultimate objectives. Consequently, information is "sticky", it is partially unintelligible, it does not flow easily between different institutions or disciplines and thus it is difficult to correlate knowledge to the desired degree. Therefore, there is a major problem to be addressed in seeking to achieve greater connectivity of information flow processes[14].

[14] Cf. Andersen et al. (1998) and Green et al. (1998) for further elaboration of the systems perspective. Also see Edquist (1997) for a quite excellent overview of the current state of the art.

One influential strand of thinking in this area has been to emphasize the national domain of the science and technology infrastructure, and rightly so (Freeman 1987, 1994; Lundvall 1986; Nelson 1993)[15]. Policy formulation and implementation is essentially a national process, reflected in language, law and the nature of national institutions and conventions. However, there are good reasons to elaborate the national perspective both downwards and outwards. It is important to recognize that different activities have different supporting knowledge infrastructures so that a sectoral innovation system perspective becomes essential[16]. This is simply one way of recognizing the specificity of the broad innovation opportunities facing firms (Carlsson 1995; Malerba et al. 2004). On the other hand, it is clear that the sectoral infrastructures frequently transcend national boundaries; a firm may draw on several national knowledge ecologies in its pursuit of innovation depending on where the knowledgeable individuals are located. Gibbons and colleagues (1994) draw attention to the emerging characteristics of new models of knowledge production, a view that fits exactly with the view that innovation requires many kinds of knowledge for its successful prosecution. What they term "mode-2" knowledge is produced in the context of application, seeks solutions to problems on a transdisciplinary basis, is tested by its workability not its truthfulness and involves a multiplicity of organizational actors, locations and skills. Together this entails a distributed system for innovation with no one-to-one correspondence with traditional national or sector boundaries.

To summarize the argument thus far, while nations and sectors contain the ecologies of knowledgeable individuals usually within organizations these ecologies do not constitute innovation systems. Systems require connections as well as components, and it is the formation of the connections that is the necessary step in the creation of any innovation system. Innovation systems do not occur naturally, they self-organize to bring together new knowledge and the resources to exploit that knowledge, and the template they self-organize around is the problem sequence that defines the innovation opportunity. Hence, innovation systems are emergent phenomena, created for a purpose, they will change in content and pattern of connection as the problem sequence evolves, and they are constructed at a mi-

[15] Carlsson (2004) has found 750 studies of innovation systems published in the past 15 years, half of which relate to national innovation systems.

[16] There is a growing literature on regional innovation linkages in which an attempt is made to correlate innovation clusters with the processes of university based scientific activity. See Varga (1998) for a review and empirical study of linkages in the USA. The paper by Malerba et al. 2003 is a comprehensive summary of these sectoral perspectives.

cro scale. Within these networks, firms, the unique organizations that combine the multiple kinds of knowledge to innovative effect, play the key role in the self-organization process. Science and technology systems, networks and communities of practice, are necessary parts of the innovation networks but they are not sufficient.

2.6 Policy for Systems Failure

Reflection on the above leads to a new rationale for innovation policy one that subsumes science and technology policy within its remit; this is the rationale based on system failure. It takes for granted the significance of an economic climate, with low real interest rates and stable macroeconomic and monetary conditions that encourages investment in all forms. Here the primary role of the state is to facilitate the emergence of innovation systems. In so doing it takes responsibility for the ecology of public organizations and institutions that facilitates business experimentation but recognizes that without the necessary interconnections the ecology is not a system. Since competition depends on innovation and innovation depends on the emergence of distributed innovation systems, it is clear that this provides an interesting alternative to the market failure perspective on innovation policy[17]. We call this the system failure perspective. The state is not promoting individual innovation events in this view; rather it is setting the framework conditions in which innovation systems can better self-organize across the range of activities in an economy. Moreover, whereas the market failure approach leads to instruments that allocate resources to firms in the form of R&D grants or tax incentives, the systems failure approach leads to instruments that enhance innovation opportunities and capabilities. Because systems are defined by components interacting within boundaries, it follows that a system failure policy seeks to address missing components, missing connections and misplaced boundaries. Each of these is a problem associated with the division of knowledge labour and the increasingly roundabout knowledge production processes, and the location of relevant knowledge in specialized organizations.

The availability of components is none other than the availability of knowledgeable individuals that can be allocated to an innovation process either in a firm or some other knowledge organization. The supply of knowledgeable minds to which innovating firms have access is perhaps the most crucial aspect of the innovation systems approach and of innovation

[17] Cohendet and Meyer-Kramer (2001) use the phrase knowledge oriented policies to capture much of what is meant here.

policy for it is individuals within organizations who are the elemental components of innovation systems. There availability is, in part, a general question about the wider education and capability formation process but, more specifically, it concerns the quality of the science and technology system in a country. "Are there sufficient knowledgeable individuals in relation to multiple branches of knowledge, in place or in training, on which firms can draw to solve innovation problems?" This is the question that governments need to answer. Capabilities may be weak in some areas and non-existent in others and government has a role to ensure that a sufficiently rich knowledge ecology is available from which innovation systems can be assembled.

The availability of knowledgeable individuals is a necessary but not sufficient condition for the emergence of innovation system. In relation to some kinds of knowledge, the required information may be in the public domain in published form, papers, reports, patents, in which case the transfer process is effectively anonymous and impersonal. In many other cases, the information has to be elicited in some form of implicit or explicit contractual arrangement through a direct process of personal interaction. This is the social network basis for innovation systems. In all cases, the knowledge of the existing members of the firm is crucial to the ability to identify and absorb external information (Cohen and Levinthal 1990) If the individuals are not employed by innovating firms, then only an external transfer arrangement can communicate what they know to the firm and here there is a wide spectrum of possibilities, not only in relation to the external organization of the firm, but also in relation to the external organization of other knowledge holding organizations. All the organizations in a systemic context must be consciously outward oriented if system failure is not to occur. Self-organization can fail because the different individuals are within organizations whose agendas and practices are misaligned in respect of a particular innovation problem sequence. The rules that shape each knowledge organization are often effective barriers to communication with other organizations, a natural consequence of the different purposes of each organization and the primary need to focus on internal procedures. Thus, firms and universities are remarkably different kinds of knowledge organizations, they reflect a natural division of specialization and each is to be presumed appropriate to task; consequently, it would be as inappropriate to make universities operate like firms as it would be to attempt the converse. These differences are a potent source of innovation system failure, and the systems failure policy response to this problem is the design of effective bridging arrangements, notionally between different organizations, but ultimately between individuals.

In the past two decades, policy makers in the USA and Europe have followed such an approach without perhaps realizing its systemic foundations. The current emphasis on collaborative research programmes including firms, customers, suppliers and universities, the incentives to set up science parks or university incubators, the emphasis on cluster development programmes, the establishment of technology transfer offices in universities, the funding of major industrial R&D programmes within university laboratories and the intensive national efforts at Foresight activity are important examples of bridging mechanisms[18]. Each of these is a device, whether conscious or not, to deal with a systemic failure in the innovation process, a failure in the self-organization of connection and interaction. Bridging processes are not designed to generate passive flows of information but to engage all the parties in an alignment of knowledge generating and information sharing processes, that is, to create a distributed innovation system (Coombs, Harvey and Tether 2004). Distributed innovation processes are partnerships with reciprocal obligations as well as collaborations in pursuit of shared objectives. Since firms are likely to be the lead partners in defining the innovation problem sequences it is vital that they have the internal capabilities to interact with other knowledge agencies. There is consequently little point in governments supporting S&T in universities and public laboratories in the hope that this will lead to greater wealth creation unless private firms throughout the economy have the R&D capacity to ask the right questions of external individuals. This is one reason why tax credits for R&D, for example, may be a useful complement to an innovation systems policy.

However, the fact that problem sequences evolve implies that the related innovation systems need to evolve also. Policy can only facilitate, it cannot design because design is always emergent. The members of a system and their connections will change over time and eventually any system becomes redundant as its underlying innovation opportunities are exhausted. It is important, therefore, that innovation systems are seen as transient, that they have useful lives, and that they need to be dissolved when their purpose is fulfilled. In innovation policy as elsewhere, there is an ever present danger of preserving arrangements designed and instituted for yesterday's problems not the problems of the future (Walker 2000).

Within this systems failure framework, there is a predominant emphasis on supply side measures directed primarily at the invention system with little attention given to the wider market context of the innovation process. From a policy viewpoint this misses an opportunity for a complementary

[18] See the 'Georghiou Report' (footnote 2 above) for a comprehensive summary of such policy initiatives in Europe.

demand side approach that focuses on the procurement of innovation through public expenditure programmes. To recognize this is to recognize the significant role that users and consumers play in the innovation process, they are not obviously the passive elements described in the Schumpeterian approach. In respect of health, education and transport as well as defence, public agencies account for substantial proportions of national expenditures and even a small proportion of this funding could be used to contract for innovation and provide a degree of market stability. As in the case of the USA, with the SBIR programmes, public procurement can also be used to stimulate innovation in SMEs, since they face greater difficulties in participating in innovation system arrangements by virtue of their limited managerial resources.

2.7 Conclusion

In this chapter I have reviewed recent developments in innovation policy thinking and attempted to view them through the lens of new developments in our understanding about innovation systems and the processes that form them. Here the fundamental insight is the experimental, evolutionary nature of a market and network economy. As Schumpeter aptly observed capitalism works by means of creative destruction, and we have suggested that innovation systems are created and destroyed as part of that process, a process that is played out on a global scale nowadays. Patterns of international competition are ever changing and an advanced country must be ever aware of new opportunities and threats if its standard of living is to be sustained. Central to this must be the rate of innovative experimentation and I have suggested that a consistent thread to policy has emerged in the past twenty years based around a distributed innovation systems perspective and innovation-led competition. In this new approach, it is the transient, institutionalized basis of innovation that is the focus of attention, rather than expenditure on research and development. I have called this the system failure perspective. From a political point of view this raises an interesting problem. Experimental economies experience many failures as well as successes, blind variation means that a great deal of effort is wasted, but this is a necessary part of the process of knowledge accumulation. As a general rule concerns for public accountability within the political process do not easily accommodate the notion of misdirected effort, which often appears with the benefit of sufficient hindsight. Governments must learn to be experimental and adaptive too, just like the firms and other organizations whose innovative efforts they seek to jointly stimulate. In this way they can expect to facilitate the self-organization of innovation systems that underpin the future self-transformations of the economy on which standards of living will continue to depend.

2.8 References

Andersen B, Metcalfe JS, Tether B (1998) Distributed innovation systems as instituted economic processes. In: Metcalfe JS and Miles I (eds.) Innovation systems in the service economy: measurement and case Study analysis. Kluwer Academic Publishers, Amsterdam, pp. 15-42.

Arrow K (1962) Economic welfare and the allocation of resources to invention. In: Nelson R (ed) The rate and direction of inventive activity: economic and social factors. NBER, New York, pp. 609-652.

Arthur WB (2000) Cognition: the black box of economics. In: Colander D (ed) The complexity vision and the teaching of economics. Edward Elgar, Cheltenham, pp. 47-60.

Callon M (1994) Is science a public good? Science, Technology and Human Values 19: 395-424.

Carlsson B (ed.) (1995) Technological systems and economic performance: The case of factory automation. Kluwer-Academic, Dordrecht.

Carlsson B (ed.) (1997) Technological systems and industrial dynamics. Kluwer, Dordrecht.

Carlsson B (2004) Innovation systems: A survey of the literature from a Schumpeterian perspective. Mimeo, Case Western Reserve University: Weatherhead School of Management.

Carlsson B, Jacobsson S, Holmen M, Rickne A (2002) Innovation systems: Analytical and methodological issues. Research Policy 31: 233-245.

Carter C, Williams BR (1957) Industry and Technical Progress. Oxford University Press, Oxford.

Cohen WM, Levinthal D (1990) Absorptive capacity: A new perspective on learning and innovation. Administrative Science Quarterly 35: 128-152.

Cohendet P, Meyer-Kramer F (2001) The theoretical and policy implications of knowledge codification. Research Policy 30: 1563-1591.

Coombs R, Metcalfe JS (2000) Organizing for innovation: co-ordinating distributed innovation capabilities. In: Foss N, Mahnke V (eds) Competence, governance and entrepreneurship: advances in economic strategy research. Oxford University Press, Oxford.

Coombs R, Harvey M, Tether B (2004) Analysing distributed processes of provision and innovation. Industrial and Corporate Change 12(6): 1125-1156.

Cowen RP, David P, Foray D (2000) The explicit economics of knowledge codification and tacitness. Industrial and Corporate Change 9: 211-253.

Edquist C (ed.) (1997) Systems of innovation: technologies, institutions and organizations. Pinter, London.

Eliasson G (1998) On the micro foundations of economic growth. In: Lesourne J, Orléan, A (eds), Advances in self-organization and evolutionary economics. Economica, London, pp. 296-307.

Freeman C (1987) Technology policy and economic performance. Pinter, London.

Freeman C (1994) The economics of technical change. Cambridge Journal of Economics 18: 463-514.

Gibbons M, Limoges C, Nowotny H, Schwartzman S, Scott P, Trow M (1994) The new production of knowledge. Sage, London.

Granstrand D, Patel P, Pavitt K (1997) Multi-technology corporations: why they have "distributed" rather than "core" competences. California Management Review 39: 8-25.

Green K, Hull R, Walsh V, McMeekin A (1998) The construction of the techno-economic: networks vs paradigms. CRIC Discussion Paper, No. 17, University of Manchester.

Griliches, Z, (1998) R&D and Productivity: The Econometric Evidence, Chicago University Press.

Hicks D (1995) Published papers, tacit competencies and corporate management of the public/private character of knowledge. Industrial and Corporate Change 4: 401-424.

Kodama F (1995) Emerging patterns of innovation. Harvard Business School Press, Boston MA.

Loasby BJ (1999) Knowledge, institutions and evolution in economics. Routledge, London.

Lundvall, B-Å (1986) Production innovation and user-producer interaction. Aalborg University Press, Aalborg.

Lundval B-Å (ed.) (1992) National systems of innovation: towards a theory of innovation and interactive learning. Pinter, London. Not in refs

Malerba F, Edquist C, Steinmueller WE (eds) (2004) Sectoral systems of innovation. Cambridge University Press.

Merges RP, Nelson R (1990) On the Complex Economics of Patent Scope, The Columbia Law Review, Vol.90, pp. 839-916.

Metcalfe JS (1995a) The economic foundations of technological policy: equilibrium and evolutionary perspectives. In: Stoneman P (ed.) Handbook of the economics of innovation and technological change. Blackwell, Oxford.

Metcalfe JS (1995b) Technology systems and technology policy in an evolutionary framework. Cambridge Journal of Economics 19: 25-46.

Metcalfe JS, Georghiou L (1998) Equilibrium and evolutionary foundations of technology policy. OECD Science, Technology Industry Review 22: 75-100.

Mowery DC, Rosenberg N (1989) Technology and the pursuit of economic growth. Cambridge University Press. Cambridge.

Narin F, Hamilton KS, Olivastro D (1997) The increasing linkage between US technology and public science. Research Policy 26: 317-330.

Nelson RR (1959) The simple economics of basic scientific research. Journal of Political Economy 67(3): 297-306

Nelson R (1993) National innovation systems. Oxford University Press, New York.

Polanyi M (1958) Personal knowledge. Routledge, London.
Richardson GB (1961) Information and investment. Oxford University Press, Oxford.
Richardson, GB (1972) The organization of industry. Economic Journal 82: 883-893.
Rosenberg N (1990) Why firms do basic research (with their own money). Research Policy 19: 165-174.
Smith A (1776) An enquiry into the nature and causes of the wealth of nations (Cannan edition 1904). The Modern Library, New York.
Stiglitz JE (1994) Whither socialism Oxford University Press, Oxford.
Tassey G (1992) Technology infrastructure and competitive position. Kluwer, Dordrecht.
Teubal M (1996) R&D and technology policy in NICs as learning processes. World Development 24: 449-460.
Varga A (1998) University research and regional innovation. Kluwer Academic, London.
Veugelers R (1997) Internal R&D expenditures and external technology sourcing. Research Policy 26: 303-315.
Walker W (2000) Entrapment in large technology systems: institutional commitment and power relations. Research Policy 29: 833-846.
Young A (1928) Increasing returns and economic progress. Economic Journal 38: 527-542.

3 Technology Policy in the Knowledge-Based Economy

Patrick Cohendet and Frieder Meyer-Krahmer

BETA, Strasbourg and HEC, Montréal, E-mail : cohendet@cournot.u-strasbg.fr
BETA, Strasbourg and FhG-ISI, Karlsruhe, E-mail : Frieder.Meyer-Krahmer@isi.fhg.de

3.1 Introduction

Traditional research, technology and development (RTD) policies are inspired by the vision of (partly) linear models of innovation, where the production of new knowledge can be reduced to mere information. The rationale for designing traditional policy instruments is that it is necessary to compensate for *market failures* that arise from the externalities of *new knowledge production.* When knowledge is regarded as information its easy reproduction fosters externalities (others may readily access and use it) and market failures (the ability of others to use the information reduces the incentives to invest in its creation). Several types of intervention have been proposed, and even tested, in a bid to come closer to the optimal level of research. David (1993) refers to "*the three P's*": public *Patronage* (prizes, research grants, subsidies, etc.), state *Procurement* (or Production) and the legal exclusive ownership of intellectual *Property* that shape government interventions designed to counter the failures of the market. The focus of traditional RTD policies is essentially, therefore, on the conditions of production of new knowledge, and not on the ways that the knowledge is assimilated and diffused through society: according to the traditional vision, agents are supposed to be able to assimilate new knowledge without significant costs.

We believe that these results need to be revisited and radically revised in the context of the knowledge based economy (KBE). As the KBE emerges and grows, there is an increasing need to think in terms of knowledge-oriented policies (KOP), which will take account of the specific characteristics of knowledge. As long as RTD policies based on a restricted vision of knowledge as mere information continue to be applied, the need to build new policy tools that take account of the key properties of knowledge such as tacitness, cumulativeness, path dependency, and contextualization becomes more urgent. In contrast to traditional RTD policies, KOP refer to the non-linear interactive models of innovation, in which knowledge and competencies, rather than physical resources, are the factors that are crucial for production, innovation, and competitiveness. In highlighting the interactions that promote the generation and diffusion of knowledge, this approach differs from the linear models (science-push, market-pull) in which knowledge constitutes only one input among others intervening in a sequential process. A particular, and key, difference is that in terms of KOP, the way the knowledge is assimilated and acquired by agents (the ways competencies are built), are as important as the conditions of production of new knowledge (the ways innovations are produced).

Moreover, we argue that one of the main limitations of RTD policies is that they are conceived in a universe that considers only markets and organizations as being active units of knowledge. The difficulty here is that the usual categories of markets and organizations are not fully adapted to capturing these essential characteristics of the production and circulation of knowledge in a decentralized learning society. From this perspective, we consider that in the classical frame, a key intermediate role is missing: *the role of communities*. The main hypothesis underlying the contribution in this chapter is, thus, that in a KBE context, an important and growing part of the process of generation, accumulation, and distribution of knowledge is achieved through the functioning of "knowledge-intensive communities", which complement the functioning of markets and organizations. As an example of a knowledge intensive community, which exemplifies a "stylized fact" in the literature, one can take the case of Linux, which demonstrates how knowledge can be produced, accumulated, and diffused in a competitive way through the function of non-formal interactions between members of a specific community (of "hackers" in this case). The rationale for KOP is that neither purely private forces, nor government intervention can guarantee satisfactory conditions for the production, mediation, and use of knowledge in the economy. Not only is there a risk of market failure when dealing with knowledge generation and diffusion in society, there could also be potential *learning failures* in the ways the knowledge is acquired, assimilated, and used. In this chapter, we describe how knowledge intensive communities can compensate for some of the learning failures of markets and organizations.

To be more precise, we would maintain that knowledge intensive communities play an increasing role in the KBE, because they can be responsible, through the passion and commitment of the members of the community to a common goal or practice, for *some significant parts of the "sunk costs" of the process of generation or accumulation of specialized parcels of knowledge*. These communities, thus, can be considered to be key building blocks of knowledge formation in the KBE. We propose, therefore, to focus on some of the most radical implications of public policy in placing the notion of learning by community at the very centre of the scene of production and exchange of knowledge in society. This will allow us to reconsider the foundations of public policies in the KBE through a careful examination of the role of knowledge intensive communities, as a complement to the classical functioning of markets and organizations.

One major implication of this novel approach is that the production of knowledge in society is not dichotomous: it is not marked out, on the one hand, by high-tech or elite-based production of knowledge in the scientific domain, or, on the other hand, by the day-to-day or marginal production of lay knowledge at lower levels of society. On the contrary, our argument is that in future years, there will be an increasing need as well as an opportunity to innovate through the interactions in society between scientific knowledge and lay knowledge. As Nowotny et al. (2001, pp. 246-247) argue, "as expertise becomes socially distributed" in an economy marked by the proliferation of knowledge across the social and institutional spectrum, one consequence is that synthesis and authority depend on the ability "to bring together knowledge which is itself distributed, contextualized and heterogeneous", rather than through expertise located at one specific site, or through the "views of one scientific discipline or group of highly respected researchers". In this regard, "science and society have both become transgressive; that is, each has invaded the other's domain, and the lines demarcating the one from the other have all but disappeared".

In order to define the main properties and characteristics of KOP, and to design appropriate policy recommendations in a KBE context, we propose to develop the following points. First, we define the notion of knowledge intensive communities (Section 3.2); then we propose in this new perspective to revisit the conditions of the production of new knowledge (Section 3.3), and the conditions of diffusion of this new knowledge (Section 3.4), before turning to policy issues in the KBE (Section 3.5).

3.2 Knowledge-Intensive Communities

We start by delineating precisely the type of community we are considering. The clarity of this at the level of the firm is not matched by a similar clarity at a more general level, where the idea of community has become fashionable in many disciplines. This interest can be related to a significant degree to the concept of social capital, which has taken a firm hold across the social sciences and within the public policy realm, as Bowles and Gintis (2000, p. 3) explain:

"...the social capital boom reflected a heightened awareness in policy and academic circles of real people's values, which are not the empirically implausible utility functions of *Homo economicus*, of how people interact in their daily lives, in families, neighborhoods, and work groups, not just as buyers, sellers, and citizens, and of the bankruptcy of the ideologically charged planning-versus-markets debate. [...] Perhaps social capital, like Voltaire's God [needed] to be invented if it did not exist. It may even be a good idea. A good *term* it is not. Capital refers to a thing that can be owned, even a social isolate like Robinson Crusoe had an axe and a fishing net. By contrast, the attributes said to make up social capital describe relationships among people. 'Social capital' has attracted so many disparate uses that we think it better to drop the term in favor of something more accurate. 'Community' better captures the aspects of governance that explain the popularity of 'social capital', as it focuses attention on what groups *do* rather than what people *own*. By community we mean a group of people who interact directly, frequently and in multi-faceted ways. People who work together are usually communities in this sense, as are some neighborhoods, groups of friends, professional and business networks, gangs, and sports leagues. The list suggests that connection, not affection, is the defining characteristic of a community."

Seen in these terms, community can be acknowledged as a concept that pre-dates the modern values of markets and planning, but for all that, is condemned to history as "the anachronistic remnant[s] of a less enlightened epoch that lacked the property rights, markets and states adequate to the task of governance" (*ibid*, p. 15). In particular, the parochialism of community has been considered antithetical to modern institutions, an old fashioned idea in the context of market and state institutions. However, communities have survived the emergence of modern social institutions, not least because of their important contribution to governance when market contracts (in the provision of local public goods, for example) and government fiats, have failed. Associations, neighbourhood groups, and other forms of grouping offer efficient arrangements that are not plagued by the usual problems of moral hazard and adverse selection, or by the illusion that governments have both the information and the inclination to always offset market failures.

However, an important part of the process of generation, accumulation, and distribution of economic knowledge is achieved through communities acting as "a nucleus of competence through the daily practices of the community" (Cowan and Jonard 2001, p. 19). The types of knowledge problems that communities solve, and which escape government and market solutions, are those that arise when individuals interact in forms of knowledge exchange that cannot be regulated by complete contracts or by external fiat. However, the generic value of communities lies in their ability to absorb a significant proportion of the unavoidable sunk costs associated with building and exchanging knowledge.

Epistemic communities and communities of practice are the most common types of knowledge intensive communities, since they are the place where knowledge creation occurs on a regular basis, independent of any hierarchical decisions. Epistemic communities are truly oriented toward new knowledge creation, whereas communities of practice are oriented toward the achievement of an activity. In this latter case, knowledge creation is an unintended spillover. It must be emphasized that these communities are not at all exclusive of an elitist scientific universe. The communities of scientists may be the paradigm of epistemic communities, but these communities can be found in diverse economic activities such as painting, music, cooking, etc. Similarly, communities of practice can be found in all domains of social life.

An interesting example of a community of practice concerns the role of the *community of patients associations* in advancing medical knowledge. Callon (1999b) showed that in areas of medicine, such as the treatment of rare genetic diseases which are poorly understood by medical and pharmaceutical institutions, lay associations have forced the pace and direction of research and remedy by engaging in a primitive accumulation of knowledge: researching and identifying diseases; organizing and effectively participating in the collection of DNA; producing films or compiling photo albums designed to be effective observation tools for monitoring and comparing clinical developments of the disease and establishing the effects of certain treatments; recording testimonies, which transmit life experiences; and carrying out surveys among patients, which sometimes lead to the publication of articles in academic journals (Callon 1999b, p. 90). In the area of muscular dystrophy, such lay knowledge in France, enshrined in the power of the French Muscular Dystrophy Association, has produced pioneering advances in understanding and treatment, through research commissioned by the Association, the availability of a historical records of evidence, patient experience and lay knowledge, the constant interactions between patients, doctors, and biologists, a division of labour and power play between the lay organizations and public bodies, and other aspects of what has become a model of hybrid and collective learning (Rabeharisoa and Callon 2002).

This role of communities in knowledge formation forces the public policy discussion far beyond the traditional "market versus public intervention" dualism. In what follows, we outline the general principles of a new policy approach through a reconsideration of the domain of science and technology policy in community terms. Then we consider the growing significance of lay knowledge communities and their interaction with communities of expert knowledge, a phenomenon that raises some of the most critical issues for public policy towards economic knowledge today.

3.3 The Production of Knowledge: a Renewed Vision of the Classical Frame, Which Takes Account of the Role of Communities

3.3.1 The Traditional Vision

The traditional vision of knowledge production dates back to Arrow's (1962)[1] seminal contribution dealing with knowledge creation in the firm. For Arrow, the process of invention can be interpreted as the production of new knowledge, which, in its turn, is assimilable as information. Arrow stressed that in such a context, the production of new knowledge faces the key problem of appropriability. He emphasized that it is difficult or even impossible to create a market for knowledge once it is produced, so it is difficult for producers of knowledge to appropriate the benefits that flow from it. Arrow's proposals rely on a body constituted of several hypotheses, some very explicit, some rather implicit, but important to underline for our purpose. These are detailed below:

1. Knowledge treated as "information" possesses the generic properties of a pure "public good". It exhibits the two conditions for being considered as a public good (non-rivalry and non-exclusion):

- First, codified knowledge is a *non-rival good*, that is, a good that is infinitely expansible without being diminished in quality, so that it can be possessed and used jointly by as many as care to do so.
- The second property concerns the characteristics of information as a *non-exclusive* good. A good is exclusive if it is relatively easy to exclude individuals from benefiting from the good once it is produced. A good is non-exclusive if it is impossible or very costly to exclude individuals from benefiting from the good.

[1] One could also mention Nelson's (1959) seminal contribution on the production of knowledge in basic science.

2. The only incentive that matters for the producer of knowledge is experiencing full ownership of the new piece of knowledge produced. There is no trade-off between the incentive to be the sole owner of the innovation and other forms of incentive that could influence the behaviour of the producer of new knowledge.
3. The producer of new knowledge is *solitary*. In Arrow's perspective, the producer of knowledge acts in isolation. Nothing is said about the complementary forms of knowledge that have been necessary for him to invent[2]. Nothing is said about the community of agents who supported him in the process that led to the invention. Nothing is said about the interest to him of the new piece of knowledge that has been produced. (Is it an incremental invention aiming at improving a current process? Is it a radically new invention opening the perspective of new fields of research?) Such a solitary perspective has the important consequence that the producer of knowledge is in a position to claim the totality of the invention.
4. The producer of new knowledge is *facing* the market. More precisely, the agents who may capture for free the new piece of knowledge, are anonymous. The mechanisms of externalities generated by the producer of knowledge, on which the diffusion process relies, are "isotropic". As in a market mechanism, one can refer to a "representative agent", who benefits from the knowledge spillovers emitted by the producer of knowledge.
5. The producer of knowledge is not supposed to have emitting capacities. In other words, he has no the ability to "tune" the disclosure/secrecy dimension. He is merely supposed to try to avoid loss of the integrity of the piece of new knowledge produced.
6. All the agents in the economy have the absolute capability to absorb the innovative idea emitted by the producer of knowledge. Any buyer of the knowledge can effectively destroy the market, since he can reproduce the knowledge at very low cost.
7. The epistemic content of knowledge does not matter. The content of knowledge exhibits a "cognitive" equivalence, which means that in such a context it is impossible, for instance, to distinguish between the generic and specific forms of knowledge.

[2] We could assume either that he had all the capacities to invent, or that the complementary forms of knowledge he needed to achieve his invention have been bought on a market.

These hypotheses of Arrow's show that the characteristic of knowledge-reduced-to-information as a public good implies the existence of some major positive externalities that prevent the effective functioning of a market for knowledge. Any buyer of the knowledge can effectively destroy the market, since he can reproduce the knowledge at very low cost. If the producers of knowledge cannot appropriate the benefits of that knowledge, then they have no incentive to produce it. Without external intervention, the incentives for doing research are not sufficiently high and the level of the research in society will be sub-optimal.

The consequences of this broad traditional vision, which reduces knowledge to mere information, were considerable. Since the knowledge generated by the different research activities possesses the generic properties of a public good, it cannot be optimally produced or distributed through the workings of competitive markets. Here is the justification for government subsidization of scientific, technological, and engineering research, and for innovative activity more generally (the "3Ps"). This vision shaped the conception of public intervention in R&D for decades. It justified the role and creation of public laboratories, of centres of research, of public R&D programmes, of public institutions (patent offices, for instance), of a public infrastructure for the transfer of technology. It explained why public R&D efforts were generally disconnected from applications, and why arguments about the existence of spillovers from public research programmes were so important in justifying the public money being spent on R&D. It suggested that scientific production was in fact considered as exogenous to the economic sphere, and governed by rules and behavioural norms (reputation effects, peer reviews, etc.) that are radically different from the norms and behaviours of industry (seeking profit and technical efficiency). In particular, from this perspective, academics' choice of research themes should remain independent of the objectives of industry.

3.3.2 Questioning the Traditional Vision

The idea that research produces only codified information is increasingly being questioned. Dosi (1988) and Pavitt (1984) amongst others, stated that research produces not information, but knowledge, some of which is coded and the rest is tacit. Cohen and Levinthal (1990) argued that the degree of spillovers and imitation depends on both the nature of the knowledge and the absorptive capacity of firms. All things being equal, the more codified the knowledge, the easier will be its absorption. But, even in the case of codified knowledge, the user or imitator needs a certain amount of know-how and technical ability to benefit from the knowledge. To appropriate the results of academic research, even if they are codified, one has to know the code[3].

Public policy has moved on from the top-down model of science and technology-based innovation and learning that dominated policy discourse during the 1980s. Much of this shift has been supported by the OECD, which has come to accept insights from evolutionary and institutional economics that see innovation as an embedded, path-dependent, bottom-up, and tacit process. There is now an explicit recognition of innovation and adaptability based on craft, apprenticeship, learning by doing, work routines, informal networks, employee competences, basic and applied learning, experiential knowledge, and other dimensions of learning as an ongoing and grounded process. Thus, for example, the recognition of informal local conventions and tacit knowledge in both craft industrial districts, as well as high technology regions, such as Silicon Valley, has forced a reevaluation of purely science-based models of innovation and learning.

In practical terms, this has led to policy suggestions aimed at strengthening technical and craft colleges, continual learning, employee participation, the areas between research institutes/academic organizations and the world of entrepreneurship and work, and policy learning based on reflexivity and ongoing monitoring of goals and routines. No longer is policy practice confined to support for the production of new codified knowledge, the transfer of technology, or reforms to the formal education and training systems.

For all the above reasons, the different hypotheses of the traditional model of knowledge production should be carefully reconsidered, within the theoretical framework described in the introduction, and based on the following assumptions concerning the production of knowledge in a KOP context:

1. Knowledge is not a pure public good. There is a range of situations varying from the completely appropriable to the completely public. Thus, to sum-up this discussion and try to categorize the different kinds of economic forms of knowledge, following Romer (1993), we can consider that:

[3] From this point of view, to quote Joly and Mangematin (1996) research activity has two complementary facets: It naturally contributes to the creation of information and knowledge, but it is also a learning process, which helps to increase absorptive capacity. Not only, are externalities not evenly distributed, but they increase when the knowledge bases of firms are similar. In such contexts, external research cannot be substituted for by internal research: the two are complementary.

- knowledge expressed in codified statements ("strings of bits") constitutes the prototype of a non-exclusive and non-rival good, which exhibits a completely public character. If one renders such goods appropriable, which is always possible, but which implies the costs of reconfiguring them or giving them legal protection, one would create a sub-optimal situation;
- knowledge expressed as information codes or encrypted messages is intrinsically non-rival, but exclusive;
- knowledge expressed tacitly is a rival good, with a wide range of modalities from pure personal tacit knowledge to shared tacit forms of knowledge.

2. Appropriation is not the only incentive for knowledge production. Firms do have other incentives than the direct exploitation of the monopoly rent, the sale of licences, or the advantage in negotiations, offered by patents. The willingness to keep the firm at the technological frontier[4], the search for reputation, the objective of signalling[5], the need to build absorptive capacity, and, more generally, the endeavours of agents in building competencies, are amongst the other main incentives for firms to invest in R&D.

3. The production of knowledge is not a solitary venture. Knowledge is generally produced within a community. The community could deliberately aim at producing new knowledge, as is the case in the epistemic community (Cowan et al. 2000). However, the building of knowledge could also be made within other types of communities, such as communities of practice. Networking between academic institutions and private enterprises is a growing phenomenon that takes different forms. Net-

[4] As Schumpeter argued, competition is about new products, new innovations. It is a dynamic process: "*In capitalist reality as distinguished from its textbook picture, it's not price competition which counts but the competition from the new commodity, the new technology, the new source of supply, the new type of organisation... This kind of competition is a much more effective than the others as a bombardment is in comparison as forcing a door*" (Schumpeter 1942, pp. 84-85). Therefore, it would be suicidal for a firm not to invest in knowledge production.

[5] The growing number of publications by firms can be interpreted (Hicks 1995; Meyer-Krahmer 1997) as an attempt to find new access to external knowledge and to signal the existence of tacit knowledge and other unpublishable resources. By becoming a "member of the club" of academic activities, by paying an implicit fee to access the epistemic communities of researchers, the firm clearly expects a right of access to the tacit academic knowledge in a particular field.

works can offer a way to share knowledge complementarities. They can also enable the building of collective forms of knowledge, and generate a sufficient level of trust between partners to facilitate the collective creation of knowledge.

Table 3.1. Public and private forms of knowledge

Rival \ Exclusive	YES	NO
YES	Personal tacit knowledge	Shared tacit forms of knowledge
NO	Information codes encrypted messages	Codified statements

4. The producer of knowledge is not facing the market, but a specific structure of interaction of economic agents. As Nonaka and Takeuchi (1995, p. 59) state, "organisational knowledge creation should be understood as a process that organisationally amplifies the knowledge created by individuals and crystallises it as a part of the knowledge network of the organisation. This process takes place within an 'expanding community of interaction', which crosses intra and inter-organisational levels and boundaries".

5. The producer of knowledge has emitting capacities. An agent producing new knowledge will generally operate a process of selection between communities: on the one side, he will take into account which communities the new knowledge is addressing, and, on the other side, he will give consideration to those communities that he chooses to exclude. This raises, in particular, the *disclosure/secrecy* dimension (David and Foray 1995).

6. Other agents do not possess the full absorptive capabilities to absorb the innovative ideas emitted by the producer of knowledge. First, firms cannot assimilate – absorb – knowledge without effort. In order to absorb new external information, the firm needs to develop what Cohen and Levinthal (1989, 1990) call an absorptive capacity. The principle is that we cannot understand something if we know nothing. Firms need to build a knowledge background, based on knowledge previously acquired, to be able to absorb the external knowledge ("*the two faces of R&D*"). This not only creates new knowledge, it also helps firms to as-

similate external knowledge. Thus, Cohen and Levinthal show that spillovers do not necessarily have a negative impact on R&D.

7. The epistemic content of knowledge matters. As noted by Callon (1999a), we should distinguish the case of knowledge with a high degree of generality (knowledge that can be potentially used in various contexts by a large variety of agents) from the very specific forms of knowledge that can be absorbed and used by only a few other agents.

This realigned set of hypotheses provokes the following remarks in terms of KOP. We can now begin to outline a radically different public policy model, which recognizes and supports the material practices and material cultures of learning in networks of communities. What the above suggests is that an essential part of the process of production of knowledge can be interpreted as resulting from the dynamics of the interactions between communities. These interactions can be approached through the principle of "translation/enrolment", elaborated in particular by Callon and Latour. According to these authors, the innovative diffusion of ideas (for example, from the laboratory to the market) can be interpreted as a process of progressive contagion of communities, where each community makes efforts to "command the attention" of other communities to convince them of the relevance of the knowledge it has elaborated. Callon and Latour suggest that the producer of knowledge faces not an anonymous competition, but a specific structure of interaction among economic agents. The group of agents that succeeds in expressing and formalizing an innovative idea faces a major problem: not the risk of being copied (at no cost), but the risk of being misunderstood by others (including agents belonging to the same institution). There is a risk that their procedures and experience will not be reproduced by others. Inventors will thus make considerable efforts to alert other communities in order to convince them of the utility and potential of their discoveries.

3.4 The Main Determinants of KOP in a Knowledge-Based Perspective

It must be emphasized that the hypotheses outlined above, of the conditions of production of innovative ideas in a KOP context do not invalidate the traditional hypotheses. In fact, the traditional context appears as a particular case of the more general one. This implies that the traditional policy instruments (3Ps) elaborated within the traditional context will still be valid as instruments in a KOP context. However, the ways of interpreting them, of designing and implementing them, and the ways of using them will be different. This very difference lies in the fact that the KOP hypotheses take into account both the tacit and the collective dimensions of knowledge. Thus, any "classical" policy instrument needs to be reconsidered in the KOP context.

For example, in considering the key question of "appropriability", this is resolved within the traditional technology policy perspective, through patents, which are viewed as strong property rights instruments. In a KOP context, patents naturally keep their (necessary) appropriability dimension to protect the innovator, but they reveal other dimensions. Patents, for instance, are assuming a more and more important strategic role in negotiation. Very often patents are the first sign of cooperation or knowledge exchange. Thus, they determine the strength of the rapport between the members of a network. Patents may be used as signalling devices within complex negotiations related to the building of networks. On the other side, the collective dimension of the building of knowledge introduces new ways to look at appropriability. When the incentives for efforts to build knowledge within a given community are strong enough (for instance, to gain membership to specific communities of practice), the necessity to decide, at least temporarily, on the question of appropriation could be marginal. An extreme example is the recent case of "free software" development, such as Linux. The members strongly believed that property rights were a threat to the user's freedom and the dynamic of innovation in the industry. In order to allow free use of all their software, they adopted the "copyleft" system, in contrast to copyright. In short, these developments suggest that in order to understand these recent issues requires in-depth analysis and a redefinition of the *incentives* applying in a new mode of production of knowledge.

3.4.1 Reconsidering Incentives in a KOP Context

The introduction, as well as the appropriability dimension, of other characteristics of the production of knowledge (signalling, voluntary disclosure, increase of absorptive capabilities) leads to key consequences in terms of the incentives for producing new knowledge. It underlines the key role of institutional settings in shaping the incentives to innovate. The institutional settings, as expressed by the norms, rules, and standards to be adhered to, govern, to a large extent, the incentives to produce and distribute knowledge among members of different social organizations. The institutional settings also contribute to shaping the nature of the codification processes that take place. In particular, the modes of organizing research activities strongly influence the costs of transferring the knowledge that has been produced. As an example, with regard to the differences between fundamental and applied research, Callon (1999a) recently emphasized what he considers to be two extreme visions among economists of the main ways to produce knowledge.

- On the one hand, there is the vision proposed by Romer (1993), for whom the main difference between basic research and applied research resides in the difference in the content of the knowledge produced (basic science consisting essentially in codified statements having a large degree of generality, and applied research consisting essentially in manipulating private tacit knowledge in forms of know-how incorporated in workers or equipment).
- On the other hand, there is the vision proposed by Dasgupta and David (1994) that the difference is not to be found in the content of knowledge (which, a priori, would exhibit a "cognitive equivalence" that leads to a strong substitutability between the two forms), but instead lies in the institutional settings. The incentive schemes and norms laid down by the institutional settings are the main reason why codified forms are preferred by some agents (researchers who have incentives to publish articles, theorems, treaties, etc.) while tacit forms are preferred by others (engineers working in private firms)[6].

The above considerations emphasize the role of incentives as shaping the nature of the production of knowledge. For example, when considering the status of incentives for research in the US, Stephan and Levin (1997, p. 54) asked:

"Why have researchers in the US focused so extensively on individuals as opposed to groups and why has this focus persisted despite widespread evidence that science is becoming increasingly a collaborative effort? It is virtually impossible for a scientist to survive and have a career at a university without becoming a "principal investigator" (PI) and directing a lab. The research the PI directs is collaborative, but the majority of the collaborators are graduate students and post-docs statuses which by their very definition are temporary. This individualistic vision of incentives is in ac-

[6] As expressed by David and Foray (1995), "The true nature of new knowledge does not stem from any intrinsic differences between knowledge that is scientific rather than technological, nor between basic and applied scientific knowledge. The critical factor governing the distribution and the utilization of new findings are those regarding the rules structures and behavioral norms about information disclosure that dominate in the particular social organizations within which the new knowledge is found or improved" p. 24.

cordance with the Mertonian model of scientific activity, where the individual trajectory of the researcher and his/her capacity to accumulate a stock of credibility is the main driver of the academic domain. This leads for instance to the well known "Matthew effect": as public fundings of scientific research is related to previous accomplishments, the system may give disproportionate recognition to scientists who attained early discoveries"

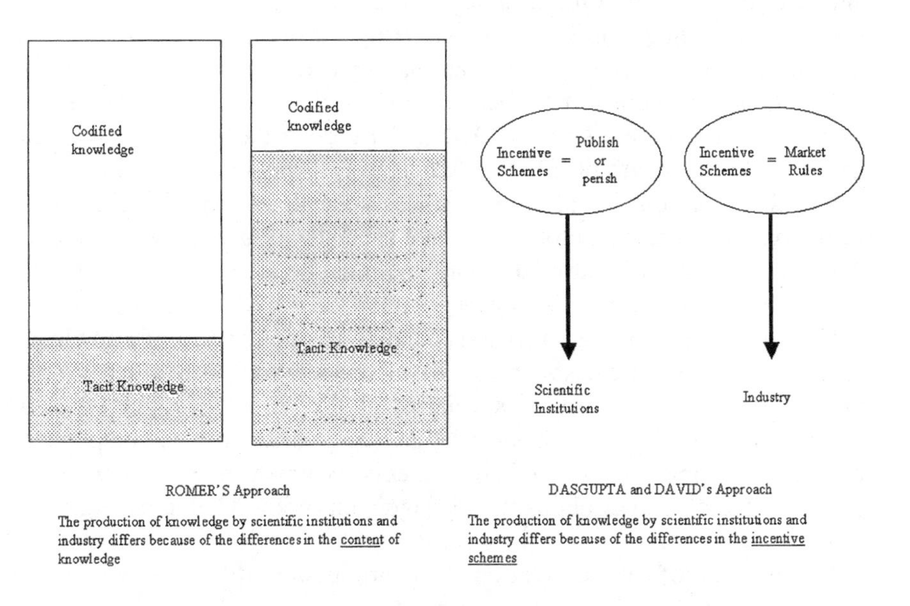

Fig. 3.1. The role of incentive mechanisms: Differences between Romer's and Dasgupta and David's approaches concerning the production of knowledge (Callon 1999a)

Callon and Foray (1997) pointed to some other types of issues related to the setting up of incentive mechanisms. In particular, they refer to a series of recent works (Hanson 1997) that raise the question of the differences between a system of awards (ex-post recognition) and a system of grants (ex-ante stimulation), and their implications for the efficiency of research.

3.4.2 The Role of Trust

Besides the institutional setting, we must consider another variable that influences the behaviour of the innovator in a KOP context: the degree of trust between the different agents involved in the production of knowledge. Taking account of the degree of trust raises an important issue, which contributes to influencing the choice between specialization and cooperation in the production of knowledge. As stated by Zuscovitch (1998, p. 256):

"Trust is a tacit agreement in which rather than systematically seeking out the best opportunity at every instant, each agent takes a longer perspective to the transactions, as long as his traditional partner does not go beyond some mutually accepted norm. Sharing the risks of specialization is an aspect of co-operation that manifests an important trust mechanism in network functioning. Specialization is risky business. One may sacrifice the 'horizontal' ability to satisfy various demands in order to gain 'vertical' efficiency in an effort to increase profitability. Any specializing firm accepts this risk, network or not. A risk-sharing mechanism is essential because, while aggregate profits for participating firms may indeed be superior to the situation where firms are less specialized, the distribution of profits may be very hazardous. To make specialization worthwhile, the dichotomous (win-lose) individual outcome must be smoothed somehow by a cooperative principle of risk sharing."

The choice made by agent A to specialize in one domain of knowledge (and to bear the sunk-costs involved) in cooperation with others agents who would agree, in turn, to specialize in the complementary types of knowledge needed by A (A's knowledge being considered as complementary from the point of view of other agents) seems to be one of the main lines of research that explains the management of knowledge by organizations.

The question of trust in relation to the production of knowledge raises a key topic in the KOP context: the problem of *access to knowledge*. Within networks, there is a considerable amount of negotiation, decentralized to the level of economic units that takes place with the objective of determining the mutual degree of specialization necessary to produce the knowledge and the mutual rights of access to knowledge that are compatible with the system of specialization. The key variable in this perspective is the degree of trust between economic units. This is by nature a very decentralized process; however, a policy that adequately favours an efficient co-determination of specialization and system of rights of access to knowledge in society, would go a long way towards promoting a "climate" of trust between economic units. Networks, therefore, are the institutional framework within which economic units realize a trade-off between property rights and rights of access to knowledge.

The above discussion on trust, in line with the analysis of incentives, leads to the following table, which summarizes four types of cases of the use of property rights instruments, when considered from two dimensions:

trust, and the nature of incentives of economic units. In this table, the first case (absence of trust, appropriability is the only incentive) equates to the traditional context[7].

Table 3.2. Trust, appropriability and incentives

	Absence of Trust in the Production of Knowledge	Existence of Trust in the Production of Knowledge
Appropriability is the only Incentive in the Production Of Knowledge	Use of patents as pure Property Rights	Trade-off between using Pure Property Rights and Rights of Access.
Different Incentives in the Production of Knowledge	Use of patents as Property Rights, Signalling, and Reputation	Complex use of Patents and Rights of Access

Public policy must find a way to build trust between economic units, or at least, acknowledge that this is what some initiatives are about. For example, stimulating the participation of public laboratories, private firms, or public research units in complex networks of innovation (such as the industrial networks financed by the EU programmes) can be considered as a way of increasing the level of trust in the daily functioning of the network. Individual units in such a network of collective knowledge formation can actively build trust by shouldering the risk of high specialization, while remaining open to the wide dissemination of results throughout the network (following the rules of "open science communities"). They can also absorb the sunk costs of building a common architecture of knowledge within these networks, thus facilitating the building of a common trust. These aspects require in-depth investigation to establish an appropriate direction for public policy.

3.5 Selected Conclusions for KOP

The above leads to a renewed framework for designing appropriate KOP initiatives in a knowledge-based context. To a large extent, this framework, which severely questions some of the results of traditional theory, could be seen as complementing the way that policy instruments are conceived within an evolutionary perspective. However, we consider that the KOP vision, based on the concepts of the economics of knowledge, em-

[7] In this figure we only consider distinct economic units, we do not consider members of some specific communities (for whom the incentive mechanisms are driven by the functioning of the community).

phasizes, more than does evolutionary theory, the active units of knowledge (communities), which constitute a nucleus for the formation of competencies in the economy. While evolutionary theory focuses on the interplay between routines (and derives from it some fundamental results for technology policy, such as the formation of standards), the KOP vision focuses on what happens "behind the scenes" where, through their everyday practice, members of cognitive communities shape the domains of knowledge. We propose in the following paragraphs to investigate some of the main KOP initiatives that can be derived from the theoretical framework that has been presented.

3.5.1 Patents Revisited in a KOP Perspective

In the new knowledge context, patents retain their essential feature of protecting the innovator, but the discussion in this chapter reveals new roles linked to the practices of the active communities that develop the new knowledge. Patents begin to play a more important strategic role in negotiations and often are the first signs of a cooperation or knowledge exchange. In this capacity, they determine the balance of power between the members of the network. Patents may also be used by a community to signal the existence of a given competence, in order to enter complex negotiations related to building networks[8]. These multiple potential uses of patents can be linked to the overlapping interactions that lead to innovation. The property rights nature of the patent can be tied to the "innovator", traditionally viewed as the embodiment of "possessed knowledge". The strategic role in negotiation can be recognized through the organization that

[8] As Foray (2002, p. 3) notes, the intensity of signalling depends on the institutional context: "the institutional articulation of [patents] that can vary a great deal across countries. For example, the information disclosure rules matter: The Japanese system is effective for sending signals and placing a large amount of information in the public domain, thus contributing to the essential objective of 'collective invention'. While the European system tends also to have an effective signalling function (though less powerful), the U.S. system, until recently, was not effective in terms of signalling. Minor institutional differences are important to explain the disparities of the value of patents as a source of information and, thus, as a mechanism for efficient co-ordination. When information is properly disseminated (as in the Japanese system) and when the nature of the protection granted is specified in ways that encourage patentees to make their innovations available for use by others at reasonably modest costs (narrow patent as well as weak degree of novelty are crucial in this way), the patent system becomes a vehicle for co-ordination in expanding informational spillovers, rather than for the capture of monopoly rents".

owns the patent (e.g. a patent can be an important asset for a start-up company when dealing with financial institutions, or seed capital companies). The signalling nature of patents can be acknowledged through a recognition of the specific community involved in the innovation, which needs protection for its competence in order to enter into collaborative ventures with other communities.

The need for appropriation might even be marginal when incentives to build knowledge within a given community are already strong (e.g. the desire to belong to specific communities of practice). In this case, at least temporarily, members of the community might voluntarily focus on building a common architecture of knowledge, without any explicit regard for appropriation. The "free software" development by Linux is a good example. The participants in this network had a strong conviction that property rights were a threat to user freedom and the dynamics of innovation in industry. Linux adopted the "copyleft" (as opposed to copyright) system in order to allow free use of their software.

Communities, thus, can be seen as semi-public entities, holding something in common, but something, which is not available to all. This is a key aspect of the production of knowledge that public policy needs to grasp. The growth of copyleft systems, and also the increase in many industries of "knowledge platforms", show that in the emergent stages of innovation, when the boundaries and definitions of products and objects are still fuzzy, there is a strong need for a common platform of knowledge, from which interfaces, standards, and designs will progressively emerge. Thus, at first glance, certain aspects of the behaviour of communities, in particular their contribution freely of knowledge as a *semi-public good*, contradict or reinterpret standard micro-economic rules.

All these arguments force a fundamental reconsideration of the traditional view of patents, which historically have been criticized for hampering the diffusion of innovation owing to their emphasis on individual appropriation. What the above discussion suggests is that strong patents can also hamper the *production* of knowledge. As Foray (2002, p. 3) remarks:

"an excess of privatization relates to excessive fragmentation of the knowledge base, linked to intellectual property rights on parcels and fragments of knowledge that do not correspond to an industrial application. This situation is described by the concept of an anti-commons regime and illustrated with the case of biotechnology: when private rights are granted to fragments of a gene, before the corresponding product is identified, nobody is in a position to group the rights (i.e. to have all the licenses) and the product is not developed".

Box 1 summarizes Foray's suggestions on property rights policies in the new knowledge context. In the same vein, what the example of Linux suggests is that, in certain contexts, when some emerging technological developments need common cognitive platforms of knowledge, the "price free" development by communities could be an efficient solution to the production of new knowledge.

Box 3.1 Intellectual Property Rights Policies in The Knowledge-based Economy

Extract from Foray (2002) 'Intellectual Property and Innovation in the Knowledge-based Economy' *Canadian Journal of Policy Research* (Isuma), vol 3, no1: www.isuma.net, Spring , pp 1-12.

In the knowledge economy: "good fences do not make good neighbours".

As Paul David (2001) claims, good fences probably make good neighbours when the resource is land, or any other kind of exhaustible resources. But, simple considerations of the "public good" nature of knowledge suggest that this is not the case when the resource considered is knowledge. Knowledge is not like forage, depleted by use for consumption; data-sets are not subject to being "over-grazed" but, instead, are likely to be enriched and rendered more accurate, and more fully documented, the more that researchers are allowed to comb through them.

Thus, the shift toward a new policy mix is raising many problems and may lead ultimately to major social losses. In most research fields, "creative discovery comes from an unlikely journey through the information space". If too many property rights are assigned to the micro-components of the information space, travelling through it proves to be extremely costly, even impossible, because at every point the traveller must negotiate and buy access rights. We are facing here a great paradox that IPRs [intellectual property rights], which are traditionally used to support the exploitation of knowledge, are ultimately acting to shrink the knowledge base.

Of course, the new system of knowledge production generates its own regulation, which can bring about a certain equilibrium in some instances. We can list four classes of solutions, dealing with the various problems developed below.

Mechanisms are devised to support, in certain circumstances or for certain classes of economic agents, the fast dissemination and free exploitation of private knowledge. There are three main mechanisms:

Compulsory licensing (compulsory diffusion of private knowledge for the general interest).

The state or international foundations buy patents to put them back in the public domain. To illustrate this mechanism Kremer (1997) uses the historical case of Daguerre, the inventor of photography who neither exploited his invention, nor sold it for the price he wanted. In 1839 the French government purchased the patent and put the rights to Daguerre's invention in the public domain. The invention was developed very rapidly!

Ramsey pricing rule suggests price discrimination between users whose demands are inelastic and those for whom the quantity purchased is extremely price-sensitive. The former class of buyer, therefore, will bear high prices without curtailing the quantity purchased of the goods in question, whereas the low prices offered to those in the second category (e.g. scholars and university-based researchers) will spare them the burden of economic welfare reducing cutbacks in their use of the good (David 2002).

Granting non-exclusive licences, presumably with minimal diligence or exclusive licences with diligence, offers a partial solution to the problem of licensing knowledge produced by publicly funded research programmes in universities.

Cross-licensing mechanisms may be a way out of the anti-commons trap. Transactions costs can be reduced through mutual concessions and through the trading of rights (for example, within a consortium). However, this is a solution that can only work with a small number of companies. Thus, the rapid growth of new kinds of firms does caution against over-confidence that anti-commons problems can be surmounted. For example, the computer hardware industry had few problems with its cross-licensing arrangements, until new kinds of semi-conductor companies appeared.

There is a great deal to be done in terms of the ways in which patent offices enforce patent requirements (i.e. make their assessments of utility requirement, non-obviousness, patent scope). One should note however that hybrid and complex objects — such as genes, DNA sequences, software, databases — generate a lot of uncertainty about what IPR policy is appropriate, making the tasks of patent offices very difficult. It is difficult to provide non-ambiguous and clear answers to the question of whether these new objects should be privately appropriated; and, if yes, what class of IPR should be used.

3.5.2 KOP Within a Decentralized Innovation Policy Model

We have already seen that one tool is new usages of patents, for example, use by a community that has produced a new piece of knowledge to convince others of the value of the novelty. Patents can be viewed as visible artifacts designed to command the attention of others and become their frame of reference too. But, for those communities that agree to participate in an "expanding community of interaction"(Nonaka and Takeuchi 1995, p. 59), which leads the innovative idea to the market, patents have a limited role to play in coordinating the interactions of knowledge intensive communities. Other means and mechanisms are necessary, notably the construction of a cognitive web that allows the different communities to communicate effectively.

Such a cognitive web is composed of material mediaries (currency, material goods, books, articles, patents) as well as intangible mediaries (crystallized collective beliefs, negotiated and accepted conventions, internalized values). These mediaries underpin the efforts of a community to interest others in their knowledge activities, by forcing engagement, adherence, a common language, common beliefs, and an alignment of interests in general (Callon 1999a). The procedural construction of a cognitive web between communities makes it possible to code the experiences and the history of the network and, therefore, to give the sense, ex post, of the construction and stabilization of a common vision or culture assuring the global consistency of the distributed venture.

In such a perspective, how can public policies aid in the translation and alignment process? Through state funded intermediaries? Through incentives to help build meta-narratives? Through funds to help communities sell their ideas? The answer, to a large degree, depends on the nature of the community-based projects, since each reveals its own idiosyncrasies and challenges, which can only be addressed in specific ways (as illustrated in Box 2). But, in general terms, policy recognition might learn to accept the inadequacy of actions centred on individual elements of a string of knowledge (e.g. particular technologies or particular know-how), and appreciate the centrality of the varied and often unpredictable mediaries – human and non-human – that hold networks together, as well as the significance of enrolling others into a knowledge network to make it effective (e.g. through publicity, political influence, indispensability, cultural dominance).

Box 3.2 Examples of Community-Based Research Projects

Extract from the appendix to the Executive Summary of the report 'Community-based Research in the United States', by Richard E. Sclove, Madeleine L. Scammell, and Breena Holland, The Loka Institute, Amherst, Massachusetts, USA, July 1998 (www.loka.org).

Harvard School of Public Health, Boston: Helping citizens link leukemia to industrially contaminated wells: During the 1970s, parents in Woburn, Massachusetts noticed an alarming pattern of leukemia, urinary tract and respiratory disease, and miscarriages in their town, and wondered if the water supply was contaminated. State officials told them the water was safe. With the help of scientists at the Harvard School of Public Health, they initiated their own epidemiological research and identified industrial carcinogens in the town's well water. Their civil suit resulted in an $8 million out-of-court settlement (detailed in the best-selling book and Hollywood movie, *A Civil Action*) and provided major impetus for Congressional action to reauthorize federal Superfund legislation.

Neighborhood Planning for Community Revitalization: Minneapolis: Planning to revitalize an industrial area: Residents and business owners in the South East Industrial Area (SEIA), just outside Minneapolis, were concerned that their area's viability was threatened by increasing pollution, over-strict zoning, crime, and the lack of sidewalks, bike paths, and park space. In addition, various groups affected by the SEIA had a contentious history and had not worked together for years. The SEIA community appealed to Neighborhood Planning for Community Revitalization (NPCR) for assistance. NPCR facilitates collaborative research between universities and local community-based organizations. Researchers working jointly through NPCR and the SEIA community members conducted a research project which established that an urban area can compete with the suburbs and still retain industrial and heavy commercial business. As a result, the city, county, and state agencies formed the South East Economic Development Steering Committee, charging it to prepare a master development plan for the area. This project was funded by NPCR and involved 960 hours of time contributed by graduate student researchers.

Center for Neighborhood Technology, Chicago: Maintaining jobs and environmental standards in the metalworking industry: In Chicago, metal finishing provides many jobs in low income neighborhoods. During the 1970s and 1980s, two waves of environmental regulation caused the immediate loss of 2,500 metal finishing jobs when non-complying plants were forced to shut down. It became clear that environmental regulations threatened this key industry and thousands of related jobs. The non-profit Center for Neighborhood Technology (CNT) collaborated with industrial development organizations to conduct an in-depth study of options for bringing Chicago's remaining metal finishers into regulatory compliance. CNT helped the groups identify the problems facing metal finishers, access free environmental audits of their plants, investigate alternative technologies for compliance, determine criteria for a centralized approach that would offer economies of scale, and secure financing for implementation. This effort represented a remarkable collaboration between manufacturers and environmentalists.

Crucially, however, what needs attending will vary according to the character of the network. In the context of supporting the creation and translation of community-based knowledge, we can consider, along with Callon (1999a), two extreme situations.

The first is a situation of "emergent relations", corresponding to cases where the process of creation is at an embryonic stage, so that only a community (usually epistemic) has experimented and validated the creative idea. The problem here is to design incentive schemes for other communities, as well the means of translating emergent creativity. The degree of uncertainty is so high that agents cannot anticipate the behaviour of others. Agent behaviour remains largely opportunistic. It is deliberately procedural: through processes of negotiation and continuous sets of feedback, the community initiates a process of convergence and formation of collective beliefs to reach stabilization. This period of convergence (of elimination of uncertainty) is a fertile period in the process of formation of collective beliefs. The collective beliefs are more likely to converge if a "metacode" has been established between communities, helping to develop "compromises between the need to make knowledge more explicit and the need to avoid excessive technicalities and local jargon" (OECD 2000, p. 27).

In the context of emergent relations, public policy could respond, firstly, by directly financing communities. The recent policy adopted in the 6th Framework Programme of the EU, in the domain of science and technology, is a significant example of the shift in focus of European public policy from financing individuals or institutions towards directly stimulating the functioning of communities. This programme is designed to directly finance large communities of scientists who will have the opportunity to decide among themselves the type of collective work to be done. Secondly, public policy could steer meta-codes and collective beliefs, in order to support a discourse of emergence through community, inter-community collaboration, and meta-code construction. The development of reference standards by public bodies could also serve to simplify communication between communities and to increase compatibility of systems designed in different countries by reducing non-strategic varieties in design. Thirdly, public policy could facilitate the "distributive power of knowledge" (David 1995) by encouraging the development of modern information infrastructures, such as uniform protocols and formal standards, alongside providing opportunities to researchers and scientists to build careers across national systems (allowing them to go back and forth without having to build up nationally-based accreditation).

Also, there could be a situation of "consolidated relations", in which the existence of common codes and languages allows the different communities to share their respective knowledge on a particular innovation domain, and to interact by continuous feedback to improve the creative principles. Here, the degree of uncertainty is considerably reduced: the individual agents act based on their past experience. Therefore, when interactions between communities are consolidated, agent behavior is conditioned by history and past experience. Thus, individual behaviour converges more and more towards a pattern of substantive behavior: through a set of tacit or casual elements translating the simplified diagrams that one develops to interpret experiences (routines, conventions, heuristics, etc.), the environment becomes consolidated. Thus, the stabilization of interactions between communities plays an important role in stabilizing the collective beliefs of the agents and in sense making.

In this context, in order to stabilize the different interactions in a community of communities, the role of public policy could be to set standards and norms of quality, to enact the efficient practices and routines by diffusing them, to recognize the outcomes of the complex process of interactions that leads to innovation. In particular, as we have seen, patents in this process are important signalling devices. Thus, an instrument, originally designed for its property rights characteristics, has assumed a new role. However, an important question for public policy is whether there are less costly instruments than patents that could be used, reserving patents for their original purpose? For example, in other domains of the economy, there are such things as "public or community utility merits" applying to certain non-profit organizations, or "certificat d'appelation contrôlée" applying to wines, and so on. These types of measures help to underpin the work of communities and protect them from the erosion of diversity that tends to accompany the development of market mechanisms.

Perhaps the most crucial aspect of policy, though, lies in achieving full public recognition that innovation is an emergent process based on the gradual introduction of interactions that link previously unconnected agents, knowledge, and goods in order to produce a relationship of interdependence: the network, in its formal dimension, is a powerful tool for making these connections and for describing the forms that they take. What produces innovation is the alchemy of combining heterogeneous ingredients within a process that cuts across institutions, forges complex and unusual relations between different spheres of activity, and, at the same time, draws on inter-personal relations, the market, the law, and science and technology. Given all of this, the aims of public policy must be modest in terms of the effectiveness of top-down interventions; there needs to be a move away from detached and pre-conceived science and technology programmes towards a hermeneutic approach based on providing nodal support for existing and emergent networks.

Finally, turning to firm-specific policies, perhaps a prime general principle should be the reversal of a public policy culture of market-driven or efficiency-driven restructuring programmes that have reduced redundancy, 'slack', and memory in the pursuit of the lean organization, and maximized labour flexibility, and short-term profit. That employees – especially in the lower tiers – should enjoy some social interaction, should be content to interact without utilitarian gain, should be given autonomy, and allowed to develop their own creativity, have come to be seen as counter-efficient and wasteful. Our analysis points to the reasonableness and benefits of having public incentives in support of industrial democracy, employment security, cultural development in firms, knowledge "vacations", competence-enriching mobility, and grass-roots experimentalism; in other words, for all those things that have become taboos in the contemporary repertoire of corporate efficiency and competitiveness. This is not an argument between the deficiencies of the short-term obsession with "leanness and fitness" and the virtues of a longer-term horizon. It is primarily about recognizing that a distributed system of knowledge production requires inputs from those engaged in the everyday practice of doing.

The above examples suggest that some of the elements that are frequently found in well-managed communities might be adopted by a public policy aimed at enhancing the desirable aspects of community governance. As Bowles and Gintis (2000, p. 16) argue, some major shifts need to be made. The first, strongly supported by experimental evidence, is that members of the community should own the fruits of their success or failure in solving the collective problems they face (this is consistent with our suggestions regarding the use of patents). Second, well-functioning communities require a conducive legal and institutional environment. It is widely recognized that at times government intervention has destroyed community governance capacities. Third, it seems clear that the ability of communities to solve problems can be impeded by hierarchical divisions and economic inequalities among its members. Thus, an institutional environment that complements the distinctive governance abilities of a community and underpins a distribution of property rights such that its members can become the beneficiaries of community success, seem key aspects of policies designed to foster community-based problem-solving.

3.5.3 KOP Initiatives to Bridge Between "Expert" and "Lay" Knowledge

The implication, then, is that public policy will have to take account of these aspects. Callon et al. (2001, p.140) emphasize this growing potential to associate scientific (in their terms "confined" research) with lay (or "profane") research:

"The main weakness of confined research does not reside in the risk of being in a total isolation, though this risk should not be underestimated. It resides essentially in the great difficulty that this type of science faces when it has to reduce the world, and then to reconstitute it. A laboratory, even if well connected to the outside world, as well researchers, even if fully convinced that they alone can achieve the translations that allow them to work efficiently, face insurmountable obstacles if they refuse to build and co-operate with those with profane knowledge".

Accordingly, Callon and colleagues call for *hybrid forums* (with the support of public agencies) to bring together, in innovative ways, the insights of the scientific and the lay communities, leading to the encapsulation of useful pieces of practised knowledge in day to day activities. For instance, it is accepted that some disease-specific patient organizations often know more than do the doctors about the specific traits of a disease, and thus produce knowledge that complements the theoretical and practical knowledge held by the physicians. To a large extent, the establishment of hybrid forums could be at the level of society, mirroring the "modular platforms" that are developed in some industries (such as in the automobile industry) to cope with the generation of complex projects. In terms of public policy, this move has important implications: hybrid forums, like modular platforms, do not just bring communities together in the hope that some positive outcome will result from the matching of active knowledge units. Their success requires a cognitive architecture to be built between the particular communities, the implementation of which involves sunk costs and requires time. Hybrid forums must have established procedures, a common grammar and rules, and specific interfaces between the common platform and each community. To achieve this there must be strong public policy willingness to modify the nature of the knowledge architectures of the "expert" and "lay" communities of knowledge.

Lay knowledge across the ever-widening realms of society where it is developed and held – in the workplace, in associations, in interest groups – must, therefore, be recognized, in terms of public policy, as being one of the mainstays of a knowledge society and its innovative arenas. This imperative, plus the various new policy orientations outlined, that follow from serious acknowledgement of the powers and potential of learning in doing through community forms of social organization, will radically alter the work of government in the knowledge economy. Science, education, and technology policies will have to focus more and more on the social foundations of learning and creativity, the task of joining up and aligning distributed pieces of knowledge, and eliminating the historical hierarchy between expert and lay knowledge. They will have to accept the centrality of the democracy of the commons and of grounded practical knowledge, for survival in the knowledge economy.

3.5.4 KOP and Access to Knowledge and Co-Evolution of Emittive and Absorptive Capacities: Technology Transfer Revisited

A re-thinking of recent trends in technology transfer policies will be at the top of the agendas of policy-makers in relation to KOP initiatives. As an example, in a case study of the locality of Brescia in Italy, Lissoni (2001) rejects the typical description of (Italian) clusters of small and medium sized enterprises (SMEs) (especially industrial districts) as homogeneous cultural settings, wherein technological findings are quasi-public goods. Rather than flowing freely within the cluster boundaries, knowledge related to technological findings is shown to circulate within a few smaller "epistemic communities", centred around the machinery producers that the research chose to investigate, but often extending outside the cluster's geographical boundaries. We can see therefore that:

- such communities are better seen as being composed of people, linked by personal ties of trust and reputation, rather than of inter-firm arrangements, although they arise out of successful commercial partnerships and deals, and respect firms' appropriation strategies;

- the localization of members of the epistemic communities is affected by the frequency of contacts required for transmitting information effectively, as well as by the size of the members' companies;
- public laboratories and universities seem to be almost totally absent from those communities.

From this it follows that epistemic communities might be a better policy target than either firms or specific geographical units, which require specific policy actions: new firms may be established as a result of community members seizing a technological opportunity, which was the basis for many SME clusters. Allowing community members to access knowledge from other sectors or from academic research may help, although policy measures in this direction may contrast sharply with the appropriability measures and staff management practices of employees of the epistemic community members. It should be noted that many technology transfer actions that currently target existing SMEs as potential innovators, could instead be directed towards giving members of local epistemic communities the chance to found their own start-up.

More generally, within a KOP concept, it is necessary to distinguish between traditional technology transfer and knowledge transfer. The target groups for technology transfer are mainly users and practical communities - generally epistemic communities. Technology transfer policies, which focus on specific sectors and locations, but do not arise out of an agreement with local members of the existing epistemic communities, will likely result in very generic, and possibly irrelevant services being offered (as many assessments of technology transfer policies have demonstrated). Since knowledge circulates within a number of relatively close networks, policy initiatives have to focus on access to knowledge and inter-personal networks, the degree of geographical dispersion of the relevant epistemic community, and the extent to which knowledge can be considered as "public" (i.e. shared by different communities) or "semi-public" (i.e. circulating within only one community). Some of the links between SMEs and larger firms, which many technology transfer policies try to enable, are already in place within the existing epistemic communities.

Amesse and Cohendet (2001) view the process of technology transfer as one that depends on the ways firms and other institutions deal with knowledge. On the one hand, they underline the role of absorptive capacities as essentially active within the perspective suggested by Cohen and Levinthal. They show that the more groups, teams, and communities within the firm are receptive to new ideas, the higher are the chances of an efficient absorption of technologies from outside. On the other hand, the quality of the process of technology transfer is fundamentally dependent on the firm's capabilities to emit knowledge beyond its frontier. When firms provide significant assistance to their strategic partners, through multiplying functional interfaces, and investment in knowledge sharing routines, for instance, they in fact are deliberately contributing to enhancing the absorptive and emittive capacities of their key suppliers. These authors also show that, when negotiating within networks rights of access to the complementary forms of knowledge that they need, firms are making a detailed assessment of the absorptive and emitting capacities of the other members of the network. In other words, the management of the technology transfer process is essentially bi-directional. What matters is the co-evolution of the mutual absorptive and emitting capacities between partners, rather than the mere observation of the technology flow between an emitter and a receiver.

Similar conclusions can be drawn about university/industry linkages, and the dense interactions between knowledge intensive services and their customers (Cohendet and Meyer-Krahmer 2001). D'Adderio's (2001) study of software development demonstrated that standardized, "coded" procedures and models are of little use unless they are locally appropriated and effectively transformed into actional routines and prototypes. Diffusion of standardized practices, models and methodologies runs the risk of seriously misrepresenting the organizational costs and productivity effects of software adoption processes. As a consequence, her study shows that software producers need to build greater flexibility and potential for customization into their systems in order to facilitate the process of adaptation of generic systems to local, context specific, circumstances and requirements. These emittive and absorptive capabilities lead to specific requirements, to dynamic learning, to translation routines, etc. All these cases demonstrate that policy needs to go beyond just R&D and to focus more on competencies.

3.5.5 KOP Initiatives for SMEs: Shifting from R&D to Competencies

Within their analysis of knowledge intensive businesses (KIBS), Muller and Zenker (2001) underline that, compared to medium-sized manufacturers, small enterprises are involved in a lower level of knowledge-intensive interactions. This means that generally small firms acquire less innovation-related information from competitors, suppliers, and research institutes and thus have more limited access to external knowledge than do large firms. The consequences of this are twofold: first small firms have less knowledge to draw on for innovation projects, and, second small firms have fewer opportunities to improve their absorptive capacities. These aspects and their small number of personnel, especially marketing experts, are the major obstacles to innovation for small manufacturing firms. Small firms carry out R&D on an occasional basis, which provides fewer opportunities to codify the knowledge produced, whereas large manufacturing firms, that are permanently engaged in development activities, are able to codify most of the knowledge produced in the frame of these activities. Through the development of routines, it can be assumed that large firms will find it easier to codify their knowledge. Therefore, the strategic aims for policy in relation to small firms should be: to raise awareness about the significance of knowledge and learning, and to compensate for their relative weaknesses in terms of knowledge codification.

One important way to raise awareness about knowledge codification, and innovation issues in general, is to provide manufacturing firms with information on innovation projects and on the importance of knowledge. This includes the introduction of routines, such as knowledge monitoring tools in the firm, involving organizational, technical, financial, and human factors. In order to distinguish their competencies, firms must identify their specific strengths and define the contents of their knowledge bases. This is an important process, which requires in-depth examination of firm activities with equipment and capital goods suppliers and with staff. Firms must find a way to organize their knowledge flows and to manage their knowledge base. These activities will need to be supported by available tools such as computer networks, or specific software tools, and require a certain technical standard. All these aspects mainly refer to codified knowledge rather than to technical, social, or organizational knowledge.

Policy measures in this direction would include incentives and project support for introducing knowledge monitoring routines in small and medium-sized firms, which lack the (financial) means to purchase equipment and to train their employees. Innovation policy should also support "knowledge managers", i.e. persons that visit firms and raise firms' awareness about the importance of these aspects. In addition, the development of – at least partly – standardized knowledge monitoring tools would help firms, since the "barriers" to firms developing their own measurement tools are high. Organizational skills and appropriately qualified personnel are necessarily accompanied by the technical skills and experience that are important for research and development activities, for conducting successful innovative projects, and for the absorptive capacity of firms. Some recent studies have talked about the phenomenon of "innovation without research" emphasizing firm networks as a knowledge source. This means that individual firms are, to a lesser extent, being seen as research performers , but the innovation networks in which firms act and interact are emphasized as being a pre-requisite for innovative activities. Nevertheless, a "non-researching, but innovating" firm still has to acquire external knowledge and apply it to its individual problems, which means that the firm must have knowledge appropriating capability. Therefore, the absorptive capacity of firms' employees, i.e. the capacity to "know-what", to become familiar with external knowledge, and to be able to apply it internally, are of crucial importance. Also, a certain level of skills and knowledge is necessary for employees to acquire new knowledge. Thus, political measures should increasingly include support for fostering skill levels in firms, for a qualified human capital base in firms, and for human capital mobility, for instance, through exchange programmes with research organizations.

3.5.6 KOP and New Agents of Knowledge

The new agents of knowledge are not really "new". What is new is that as economic actors they play a far greater role than in the past. The new actors are mostly small firms characterized by skilled knowledge workers. Creplet *et al.* (2001) analyze in more detail the role of two specific new agents of knowledge from a cognitive point of view. Muller and Zenker (2001) focus on KIBS in general, and on their characteristics in creating, reengineering, and diffusing knowledge. The core characteristics of these new agents are a very high level of interactions with customers, a deep access to the knowledge structure of these customers, and significant capabilities in knowledge re-engineering. It is essential that policy takes account of the very different characteristics of their target group compared with the traditional target groups, such as R&D performing firms within manufacturing industries.

Creplet et al. demonstrate the differences between consultants and experts from a cognitive dimension: a simple definition is that consultants contribute to the problem-solving process of their customers by applying standardized methods, routines, and processes, and through their knowledge of best practice. The development of their competencies is mostly based on links with communities of practice. Experts, on the other hand, mainly intervene in complex situations and create and operate relatively new knowledge. The development of their learning process is mostly based on links with epistemic communities. This study demonstrates how differently the new agents of knowledge behave due to their different roles and roots in epistemic communities or communities of practice. It is not possible, therefore, to draw simple policy conclusions. Nevertheless, it is obvious that several policy shifts will be needed. The

KOP must put more emphasis on skills, competencies, and personnel (rather than R&D), on soft factors, such as management organization and training (rather than hard factors), on changes of behaviour (rather than achieving technological advances), and on knowledge management (rather than R&D projects).

In relation to KIBS, Muller and Zenker reach the following conclusions: KIBS perform knowledge processing, re-engineering, and diffusing for innovation. In order to fulfil this function they must have the capability to achieve knowledge transformation from its generation to its application in client firms. KIBS thus act as go-betweens between research organizations that produce scientific results and firms that use and apply this new knowledge. Since firms generally are unable to directly apply new knowledge, and since KIBS are familiar with firm-internal processes and demands, they process the "new" knowledge, diffuse it among their clients, and support its application in firm innovation processes. This knowledge transformation process consists to a large extent in a modification and re-engineering of the codified knowledge made public by research organizations, into tacit (or specific) knowledge that is communicated to firms and can be applied by them. Interactions between KIBS and SMEs produce a circle based on the exchange of knowledge in both directions. This fosters innovations in both types of firms and can be described as the mutual activation of knowledge resources. Cooperating manufacturing SMEs and KIBS treat certain problems in the same way and participate in shared learning processes. Very close interaction, a wide access to the knowledge structures of their customers, and re-engineering of knowledge are the specific characteristics of KIBS as the new agents of knowledge. KIBS may also be able to compensate for regional weaknesses in the research infrastructure since they appropriate scientific results and make them accessible for application in manufacturing firms. As a consequence, one strategic aim will be to support the expansion of the KIBS sector in Europe, and to acknowledge their contribution to "boosting" innovation, both internally and among their clients.

This has two implications for innovation policy: on the one hand, policy should pay more attention to these new agents of knowledge as a new target group. On the other hand, in order to stimulate cooperations between KIBS and other types of firms, the visibility of the former firm type should be raised, especially for small and medium-sized firms, which often lack information concerning cooperation partners. One way to achieve this visibility would be some type of certification for KIBS that indicated their competencies. This would help KIBS to market themselves while at the same time providing information to manufacturing SMEs about what particular KIBS could offer them. Furthermore, innovation policy could increase collaboration by providing incentives for exploiting KIBS services. The benefits of this kind of support would be seen in the innovation activities of manufacturing SMEs, and also in the internal innovations of KIBS, "nurtured" by the knowledge acquired through cooperation with manufacturing partners. The emittive capacities of KIBS, the absorptive capacities of SMEs, and the level of interaction between both types of firms must be the main targets of policy.

3.6 Conclusion

The state of the art in KOP is far from being comprehensive, coherent, or mature. This is mainly because this field of research is at an early stage. Therefore, our policy conclusions are selective, preliminary, and inevitably lack very specific proposals within detailed policy actions. Nevertheless, we believe that we could develop some elements of a conceptual framework, which would bring to the relevant actors, their processes and contexts, a broad and rich set of empirical evidence.

As emphasized in the above, the conceptualization of knowledge production based on the notion of community, suggests an entirely different set of policy principles to those applicable to the traditional theoretical context, where the agent at centre stage in the knowledge production process is the individual.

Acknowledging the key role of communities in the production of knowledge indicates the need for institutional norms that can support communities. Public policy instruments are still very far from achieving this aim. In many research settings (including the academies), the style of public incentives is still to focus on individual publications and publicity, which, in our view, is a major barrier to the diffusion of the community-based knowledge economy. But some change is discernible. For example, a very different approach to incentives has been adopted recently in the UK, where the "laboratory" is increasingly being taken in science and technology policy as the unit of reference for reputation-based and other non-financial incentives. A focus on communities suggests the desirability of extending incentives beyond the boundaries of the laboratory, to reward the *network* of research centres and laboratories that produce new knowledge[9]. How far incentives can move in this direction will be crucial in de-

[9] A revealing example cited by Joly (1997) is an article in *Nature* on the sequence of chromosome III in yeast that was signed by 147 researchers from 40 different research institutions - one among many signs of the need for research and science policies to acknowledge and reward knowledge chains.

termining whether the trend of indisputable achievements in production in research.

From a theoretical point of view, work still needs to be done in order to clearly distinguish the KOP vision from other theoretical visions, for instance, the policy instruments arising out of evolutionary theory. Our belief is that the KOP vision usefully complements the results of the evolutionary approach in terms of policy instruments, but, to be confirmed, this aspect requires further and in-depth analyses.

3.7 References

Abramowitz M, David P (1996) Technological change, intangible investments and growth in the knowledge-based economy: the US historical experience. In: Foray D and Lundvall B-A (eds.), Employment and growth in the knowledge-based economy. OECD, Paris.

Adler E, Haas P (1992) Conclusion: epistemic communities, world order, and the creation of a reflective research programme. International Organization 46(1): 367-390.

Amesse F, Cohendet P (2001) Technology transfer revisited, in the perspective of the knowledge based economy. Research Policy 30(9): 1459-79.

Amin A (1999) The Emilian model: institutional challenges. European Planning Studies 7(4): 389-405.

Amin A, Cohendet P (2004) Architectures of knowledge: firms, capabilities and communities Oxford University Press, Oxford.

Arrow K (1962) Economic welfare and the allocation of resources for invention. In: The rate and direction of inventive activity. Princeton University Press, Princeton, pp. 609-25.

Bessy C, Brousseau E (1998) Technological licensing contracts: features and diversity, International Review of Law and Economics, 18(December): 451-489.

Bowles S, Gintis H (2000) Social capital and community governance. Working Paper 01/01/003, Santa Fe Institute, available at www.santafe.edu /sfi/ publications / Working-Papers/01-01-003.pdf

Callon M (1994) Is science a public good? Science Technology and Human Values 19(4): 395-425.

Callon M (1999a) Le réseau comme forme émergente et comme modalité de coordination. In : Callon M, Cohendet P, Curien N, Dalle J-M, Eymard-Duvernay F, Foray, Schenk E, Réseau et Coordination. Economica, Paris, pp 13-64.

Callon M (1999b) The role of lay people in the production and dissemination of scientific knowledge. Science, Technology and Society 4(1): 81-94.

Callon M, Cohendet P, Curien N, Dalle J-M, Eymard-Duvernay F, Foray D, Schenk E (1999) Réseau et Coordination. Economica, Paris.

Callon M, Lascoumes P, Barthe Y (2001) Agir dans un monde incertain: Essai sur la démocratie technique. Paris: Seuil.

Callon M, Latour B (1991) La science telle qu'elle se fait. Anthologie de la sociologie des sciences de langue anglaise. La Découverte, Paris.

Cassier M (1995) Les contrats de recherche entre l'université et l'industrie. Thèse de doctorat, Ecole des Mines, Paris.

Ciborra C, Andreu R (2002) Knowledge across boundaries: managing knowledge in distributed organizations. In: Choo CW, Bontis N (eds), The strategic management of intellectual capital and organizational knowledge. Oxford University Press, New York, pp. 575-86.

Cohen MD, Burkhart R, Dosi G, Egigi M, Marengo L, Warglien M, Winter S (1996) Routines and other recurring action patterns of organizations: contemporary research issues. Industrial and Corporate Change 5(3): 653-98.

Cohen WH, Levinthal D (1989) Innovation and learning: the two faces of R&D. The Economic Journal 99: 569-596.

Cohen WH, Levinthal D (1990) Absorptive capacity: a new perspective on learning and innovation. Administrative Science Quarterly 35: 128-152

Cohendet P, Creplet F, Dupouët O (2000) Organizational innovation, communities of practice and epistemic communities: the case of Linux. In: Kirman A, Zimmermann JB (eds.) Economics with heterogeneous interacting agents. Springer, Berlin, pp. 303-326.

Cohendet P, Diani M (2003) L'organisation comme une communauté de communautés: croyances collectives et culture d'entreprise. Revue d'économie politique septembre-octobre : 697-719.

Cohendet P, Meyer-Krahmer F (2001) The theoretical and policy implications of knowledge codification Research Policy 30(9): 1563-92.

Cook SDN, Brown JS (1999) Bridging epistemologies: the generative dance between organizational knowledge and organizational knowing. Organization Science 10(4): 381-400.

Cowan R, Foray D (1997) The economics of codification and the diffusion of knowledge. Industrial and Corporate Change 9(2): 211-53.

Cowan R, David P, Foray D (2000) The explicit economics of knowledge codification and tacitness. Industrial and Corporate Change 9(2): 212-53.

Cowan R, Jonard N (2001) The workings of scientific communities. MERIT-Infonomics Research Memorandum series WP n° 2001-031.

Créplet F, Dupouët O, Kern F, Mehmanpazir B, Munier F (2001) Consultants and experts in management creating companies. Research Policy 30: 1517-35.

D'Adderio L (2001) Crafting the virtual prototype: how firms integrate knowledge and capabilities across organisation boundaries. Research Policy 30: 1409-24.

Dasgupta P, David PA (1994) Towards a new economics of science. Research Policy 23: 487-521.

David PA (1992) Knowledge property and the system dynamics of technical change. In: Proceeding of the World Bank Annual Conference on Development Economics, 1992, Washington DC, The World Bank.

David, P A (1995) Standardization policies for network technologies: the flux between freedom and order revisited. In: Hawkins R, Mansell R, Skea J (eds.)

Standards, innovation and competitiveness: the politics and economics of standards in natural and technical environments, Edward Elgar, Aldershot, UK, pp: 15-35.
David PA (2000) The digital technology boomerang: new intellectual property rights threaten global 'Open Science'. Stanford University, WP n° 00-016, October 2000.
David PA (2001) Digital technologies, research collaborations and the extension of protection for intellectual property in science: will building 'Good Fences' really make 'Good Neighbours'?; STRATA-ETAN Workshop.
David PA, Foray D (1995) Accessing and expanding the science and technology knowledge base. STI Review, n°16, OECD, Paris: 13-68.
David P, Foray D (1996) Information distribution and the growth of economically valuable knowledge: a rationale for technological infrastructure policies. In: Teubal M, Foray D, Justman M, Zuscovitch E (eds.) Technological infrastructure policy: an international perspective. Kluwer Academic Publishers, Dordrecht and London, pp. 87-116.
Dosi G (1988) The nature of the innovative process. In: Dosi G, Freeman C, Nelson RR, Silverberg G, Soete L (eds.). Technical change and economic theory. Pinter, London, pp. 221-238.
Foray D (2002) Intellectual property and innovation in the knowledge-based economy. Isuma, 3(1): 1-12, available at: www.isuma.net.
Foss N (1998) Firm and coordination of knowledge: some Austrian insights. Working Paper, DRUID, Copenhagen Business School.
Foss N (1999) Understanding leadership: a coordination theory. Working Paper DRUID, Copenhagen Business School.
Gibbons M, Limoges C, Nowotny H, Schwartzman S, Scott P, Trow M (1994) The new production of knowledge. Sage, London.
Hicks D (1995) Published papers, tacit competencies and corporate management of the public/private character of knowledge. Industrial and Corporate Change 4(2): 401-24.
Joly PB (1997) Chercheurs et laboratoires dans la nouvelle économie de la science. Revue d'Économie Industrielle 79: 77-94.
Joly PB, Mangematin V (1996) Profile of public laboratories, industrial partnerships and organization of R&D. Research Policy 25: 900-22.
Knorr-Cetina K (1981) The manufacture of Knowledge. Pergamon Press, Oxford.
Knorr-Cetina K (1999) Epistemic cultures: how the sciences make sense. Chicago University Press, Chicago.
Kogut B (2000) The network as knowledge: generative rules and the emergence of structure. Strategic Management Journal 21(March): 405-25.
Kogut B, Zander U (1992) Knowledge of the firm, combinative capabilities, and the replication of technology. Organization Science 3: 383-97.
Kogut B, Zander U (1996) What firms do? Coordination, identity and learning. Organization Science 7: 502-18.
Kremer M (1997) Patents buy-outs. A mechanism for encouraging innovation. NBER Working Paper no. 6304.

Latour, B (1986a) Visualization and cognition: thinking with eyes and hands. Knowledge and Society 6: 1-40.

Latour, B (1986b) Science in action: how to follow scientists and engineers through society, Harvard University Press, Cambridge MA.

Lave J (1988) Cognition in Practice. Cambridge University Press, Cambridge.

Lissoni F (2001) Knowledge codification and the geography of innovation: the case of Brescia mechanical cluster. Research Policy 30(9): 1479-1550.

Meyer-Krahmer F (1997) Public research/industry linkages revisited. In: Barré R, Gibbons M, Maddox J, Martin B, Papon P (eds.) Science in tomorrow's Europe. Economica International, Paris, pp. 153-73.

Milgrom P, Roberts J (1988) Economic theories of the firm: past, present, future. Canadian Journal of Economics 21: 444-58.

Muller E, Zenker A (2001) Business services as actors of knowledge: the role of KIBS in regional and national innovation systems. Research Policy 30: 1501-16.

Nelson RR (1959) The simple economics of basic scientific research. Journal of Political Economy 67: 297-306.

Nelson RR, Winter S (1982) An evolutionary view of economic change. Belknap Press of Harvard University Press, Cambridge MA.

Nonaka I, Takeuchi H (1995) The knowledge-creating company: how the Japanese companies create the dynamic of innovation. Oxford University Press, New York.

Nowotny H, Scott P, Gibbons M (2001) Re-Thinking Science. Polity Press, Cambridge.

OECD (2000) Knowledge management in the learning economy: education and skills. Organisation for Economic Cooperation and Development, Paris.

Orlean A (ed.) (1994) Analyse économique des conventions. Presses Universitaires de France, Paris.

Pavitt K (1984) Sectoral patterns of technological change: towards a taxonomy and theory. Research Policy 13(6): 343-73.

Polanyi M (1958) Personal knowledge. Routledge and Kegan Paul, London

Prahalad CK, Hamel G (1990) The core competence of the corporation, Harvard Business Review 68(May/June): 79-91.

Rabeharisoa V, Callon M (2002) The involvement of patients in research activities supported by the French Muscular Dystrophy Association. In: Jasanoff S (ed.) States of knowledge: science, power and political culture. University of Chicago Press, Chicago.

Romer P (1993) Implementing a national technology strategy with self-organizing industry boards. Brookings Papers Microeconomics. 2: 345-99.

Saxenian A (1996) Regional advantage: culture and competition in Silicon Valley and Route 128. Harvard University Press, Cambridge MA.

Schumpeter J (1942) Capitalism, socialism and democracy. Harper, New York.

Stephan P, Levin S (1997) The critical importance of careers in collaborative scientific research. Revue d'économie industrielle 79: 45-61.

Von Hippel E (1993) Trading in trade secrets. Harvard Business Review March/April: 56-64.

Wenger E (1998) Communities of practice: learning as a social system. Systems Thinker.

Wenger E (1998) Communities of practice: learning, meaning, and identity. Cambridge University Press, Cambridge.

Wenger E, Mcdermott W, Snyder M (2002) Cultivating communities of practice. Harvard Business School Press, Boston.

Wenger E, Snyder WM (2000) Communities of practice: the organizational frontier. Harvard Business Review Jan-Feb: 139-45.

Zuscovitch E (1998) Networks, specialization and trust. In: Cohendet P, Llerena P, Stahn H, Umbhauer G (eds.) The economics of networks. Springer Verlag, Berlin, pp. 243-64.

Part II New Technology Procurement: Knowledge Creation, Diffusion and Coordination

4 Technology Policy and A-Synchronic Technologies: The Case of German High-Speed Trains[1]

Patrick Llerena and Eric Schenk

BETA, Strasbourg, E-mail: pllerena@cournot.u-strasbg.fr
LICIA, Strasbourg, E-mail: eric.schenk@insa-strasbourg.fr

4.1 Introduction

Public support for research and development (R&D) can be oriented towards various objectives: at early stages of the innovation process, exploration of technological opportunities is sought. Indeed, short run, profit oriented research strategies might lead to too early a focus and to lock-in to an inferior solution (Cowan 1991). At later stages, public support often seeks to foster the adoption of the new technology. There are situations where private incentives lead to under-adoption of the new technology (Farrell and Saloner 1986). Even though these objectives may be distinct, they can overlap, for instance when several technologies are supported simultaneously. The purpose of this chapter is to shed some light on the difficulties that could be encountered in such situations.

This we do by studying the case of the German high-speed train programme. Several stages have been identified since the launch of this programme in the early 1970s: in the first (1971–1977), innovations in the Magnetic Levitation (at that time a very “un-mature” technology) and Wheel/Rail technologies were pursued under the sponsorship of the Federal Ministry for Research and Technology (BMFT). The splitting of the

[1] We wish to thank Dominique Foray, Frieder Meyer-Krahmer and participants in the ‘Investment decisions in technological breakthough projects’ seminar (IMRI, Université Paris Dauphine) for their helpful comments, and Stéphanie Danner-Petey and Arman Avadikyan for their research assistance. This research was supported by a grant from the Ministère de l’Equipement et des Transports (PREDIT 1996–2000 n°98 MT 07). Part of this research was carried out while Eric Schenk was visiting the Ecole Nationale Supérieure des Télécommunications de Bretagne, Brest (France).

"generic" programme into two separate projects took place in 1977. The BMFT was responsible for the further development of the Magnetic Levitation technology, with a short term marketing objective, while the Federal Ministry of Transports (BMV) took responsibility for the development of a more traditional Wheel/Rail system. We interpret this bifurcation as institutional specialization of the innovation oriented research ministry and the diffusion oriented transport ministry.

From that time, the two projects followed separate paths: the Wheel/Rail technology was marketed under the ICE label; the Magnetic Levitation (MagLev) technology became stable and incremental improvements were embodied in the various Transrapid versions. However, at the end of 2000, despite the maturity of the technology, the Transrapid was not adopted for the Hamburg–Berlin line. Some of the reasons given were the high costs of the technology, its small performance advantage over the existing ICE, and demand uncertainty. An alternative outlet for this technology, namely the 31.5 km Chinese project linking Pudong airport to the Long Yang road-station in Shanghai, was found only recently.

In our view, the difficulties encountered by the Transrapid are associated with the type of policy that was followed. While it may seem a "natural" way to cope with technology evolution, the German policy of providing parallel support for "a-synchronic" technologies (Wheel/Rail being is seen as the "old" technology and MagLev as the "new") raises several non-trivial issues. First, evaluation of the merits of the respective technologies must be conducted at "comparable levels of knowledge". At some point, acquisition of new knowledge requires commercial exploitation beyond the laboratory. Second, implementation of a transport system required high investment in network infrastructures. The need for compatibility with existing infrastructures heavily influences the operator's choice of a technology. Finally, we would argue that delays in technology adoption could have irreversible consequences, as (i) improvements to the unadopted technology do not occur and (ii) the "window of opportunity" for its diffusion might be missed.

4.2 The German High-Speed Train Programmes

The Wheel/Rail technology (presently marketed under the name ICE) followed an incremental development path with the primary consideration being compatibility with the existing rail infrastructure. To a large extent, innovations took place within a pre-defined framework. In contrast, MagLev was a radical innovation, at both system and component levels. According to Büllingen (1997), the MagLev technology emerged from an innovation process which sequentially followed fairly well defined stages: invention (1922–1940), innovation (1960–1967), consolidation (1968–1978), and, finally, implementation (1979–present).

Public support was important in converting what was primarily a technological challenge into an economic one. We therefore look first in our historical analysis at the implications of involvement of public institutions. Two main periods can be identified in the history of the German high-speed train: in the first (1971–1977), MagLev and Wheel/Rail technologies were developed within a global programme. The second period began in 1977 when the programme was split into two separate projects. MagLev was seen as a technological breakthrough project, and Wheel/Rail was considered to be a project of incremental innovation.

4.2.1 The Generic High-Speed Train Programme

The initial German high-speed train programme was launched after a study commissioned by BMV, which identified a need for high-speed guided transports.

4.2.1.1 The HSB Study (1969 – 1971)

The time that public authorities became involved in high-speed guided transports is clearly identifiable. In 1969, the HSB group (HSB is the German acronym for High-Speed Trains), which had been established two years earlier by Bölkow, Krauss-Maffei (KM) and the Deutsche Bundesbahn (DB), was commissioned by BMV to conduct a study with the objective of reducing the gap between the speeds of land and air transport.

The final report of the HSB group was delivered in 1971. Parallel development of the Wheel/Rail and the MagLev technologies was advocated. This raises several points. At that time, it was considered that due to its intrinsic characteristics (and especially the physical contact between wheels and rails), Wheel/Rail technology would not allow a commercial operating speed exceeding 300 km/h. The MagLev system (which had entered the consolidation phase, see Büllingen 1997) offered the possibility of higher commercial speeds (500 km/h was considered feasible). Despite this, both these technologies were seen as being possible substitutes for air transport for distances of less than 500 km. It should also be noted that the HSB study was based on the Hamburg–Köln–Stuttgart–Munich corridor (known as the "C line", and which would later have a connection to Frankfurt).

The HSB group's recommendations had one major consequence, namely the involvement of the BMFT in a high-speed train research programme. In addition, the DB launched a programme for modernization of the rail infrastructure (this programme was known as the "Ausbauprogramm 1970"). The modernized network was designed to support speeds up to 300 km/h.

4.2.1.2 The "Technologies for Transport Systems" Programme

The purpose of the BMFT funded research programme was to find medium and long term answers to the problems raised by the increasing demand for transport. Based on the recommendations of the HSB group, the programme had two essential components:

- developing the MagLev technology up to technical maturity (from 1970 onwards);
- identifying the technical and economic limits of the Wheel/Rail technology (from 1972 onwards).

Five research stages were scheduled for each technology:

- Conceptual study
- Components study
- System development and experimentation
- Exhibitions under commercial conditions
- System validation by trials in "reality-like" environments.

The funding scheme adopted by the BMFT was the following:

- Financing of all research concerning the MagLev technology: this was justified by the high immaturity (at that time) of the technology, and the (commercial and technical) risks associated with it;
- Financing of academic research: the argument was that academic institutions had *a priori* no financial interest in either of the projects;
- Financing of 50% of the research undertaken by the private sector into Wheel/Rail technology, which had short term commercial perspectives.

The overall BMFT funding for the 1970-1991 period amounted to 1.56 milliard DM (approximately 780 million Euros) for the MagLev technology and 0.64 milliard DM (approximately 320 million Euros) for the Wheel/Rail.

The BMFT programme enabled the construction of a dual-purpose trial circuit in Donauried. The construction of a Wheel/Rail prototype was scheduled for 1977, with a target speed of 400 km/h. The MagLev technology remained unchanged at a speed of 500 km/h, establishing a trend that would be reconfirmed over time, namely a reduction in the (perceived) speed gap between the MagLev and the Wheel/Rail technologies. Paradoxically, this did not translate into the DB policy. In the mid 1970s, the high-speed ambitions of DB were revised: instead of the initially planned speed of 300 km/h, the redesigned network was only capable of supporting speeds up to 250 km/h. This conservative policy had dramatic consequences for the development of the high-speed Wheel/Rail system. At that time, the conventional E120 locomotive was able to achieve 200 km/h. Thus, the need for a breakthrough technological solution decreased as the revised target speed became achievable through incremental innovations.

The Donauried trial circuit project was abandoned in 1977. This can be explained in part by certain exogenous factors: low social acceptance of the project, a cut in the public budget, etc. However, it can also be seen as a willingness on the part of BMFT, as the main financial contributor to the project, to focus on breakthrough technologies. Although BMFT's financing of the Wheel/Rail research continued to increase up to 1980, institutional specialization had begun in 1977: BMFT increased its commitment to the MagLev technology, while the conventional Wheel/Rail players adopted an incremental approach to innovation.

In addition, the diffusion and the rapid growth of air transportation (increased number of airlines and routes and significant decreases in fares) were having an effect. This increased the pressure to develop high-speed trains, with the focus being on the competing alternative high-speed train technologies. Thus, high-speed train technologies were seen as defining a new generation of land transportation, the various alternatives being regarded as competitors of, but not exactly substitutes for, air transport.

4.2.2 High-Speed Trains in an Institutional Specialisation Context

From 1977 onwards, the MagLev and the Wheel/Rail technology projects followed different paths. A decision to build a 31,5 km long MagLev circuit in Emsland was made in 1977. In 1978, the DB provided what was at the time an unused line (the 23 km long Rhein–Fehre section) for the construction of a Wheel/Rail trial line.

4.2.2.1 Incremental Innovations to the Wheel/Rail Technology

The purpose of this trial line was to improve the knowledge about the effects of certain parameters (ground stability, rail inclination, etc.) on the Wheel/Rail system and to check the operational character of new information and guidance systems. The decision to interrupt the construction of the Rheine–Fehre section was quickly taken by BMFT. But when the French TGV came into service (in 1981), the DB launched the "Hochgeschwindigkeits Verkehr" programme, out of which was born the ICE project (1982). In 1991 (*i.e.,* 10 years later), the ICE train was put on the market on the Hannover–Würzburg and Mannheim–Stuttgart lines.

The ICE demonstrated the willingness of the DB to benefit from a solution, which was compatible with the existing infrastructure. This led to a partial disengagement of the BMFT. To an extent, the ICE development was seen as the answer to international competition constraints.

4.2.2.2 Emergence of the Transrapid: Elimination of Options and Implementation

In 1970, and following the HSB group's recommendations, the BMFT launched a research programme aimed at supporting development of the MagLev technology. The BMFT policy had two stages. The first involved preservation of the technological options. The second was characterized by a focus on two specific solutions, namely the Electro Magnetic System (EMS) technology (supported by the so-called Transrapid EMS consortium and by Thyssen-Henschel) and the Electro Dynamic System (EDS) technology (supported by the AEG-BBC-Siemens consortium). The main principles of the EMS and EDS technologies are depicted in Figure 4.1.

The decision to build a specific MagLev trial circuit in Emsland came about because of the adoption of the Thyssen-Henschel EMS technology. This final reduction of the "technology space" was justified by cost minimization considerations. Also, it could be argued that the exploration period had yielded "sufficient" knowledge concerning the comparative advantages of the competing technologies. Finally, in 1977, the BMFT expressed its desire to accelerate the pace of development of the technology in order to achieve rapid commercialization of the MagLev. Therefore, the decision favoured the least risky, most economic, technology, which, it was considered, could be implemented in the short term.

It is interesting to draw a parallel between what happened in Germany and the choice made by the Japanese in favour of the EDS technology (embodied in the MLX prototype). The difference can be explained by such factors as the lower sensitivity of EDS to earth tremors and Japan's larger potential market. However, it can also be seen as the result of a divergence in terms of the willingness to develop a breakthrough technology: the EMS had always been a more mature technology than EDS, and was generally seen as a "low breakthrough" technology[2].

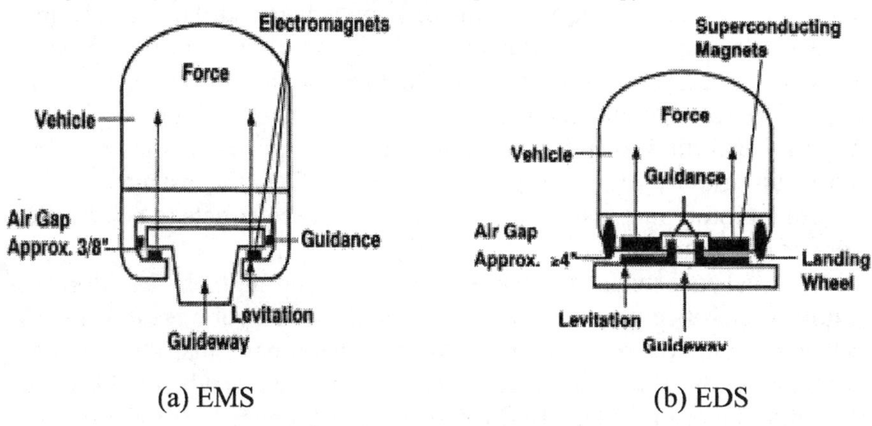

(a) EMS (b) EDS

Source : http://inventors.about.com/library/inventors/blrailroad3.htm

Fig. 4.1. MagLev technologies

The BMFT decision gave rise to the emergence of the Transrapid International[3] (TRI) consortium in 1982. The MagLev technology then entered into the maturation stage characterized by a sequence of incremental innovations. This period officially ended in 1991 when the Deutsche Bahn confirmed the maturity of the technology.

4.3 The Role of Institutions in the Management of Options

Institutional specialization between the Ministry of Transport and the Ministry of Research allowed various options to be retained. Option preservation is particularly relevant in mission-oriented projects, which are subject to a conflict between increasing information and the degree of freedom concerning the future course of the project.

[2] The EDS system relies on superconducting magnets, while the EMS employs electromagnets. A brief description of MagLev technologies can be found at <http://inventors.about.com/library/inventors/blrailroad3.htm>.

[3] Composed of Krauss-Maffei, Messerschmitt –Bölkow-Blohm and Thyssen.

4.3.1 The Importance of Maintaining Options

Experience reported by ECOSIP (1993) show that the dynamics of a project are constrained by a conflict between the willingness to reduce uncertainty (*i.e.*, to acquire various forms of knowledge) and the desire to preserve a sufficient degree of flexibility (or freedom) concerning the future course of the project. Those strategies that reduce the conflict between the level of knowledge and flexibility either delay the "freezing" of designs (*i.e.*, irreversibility), thereby maintaining options, or enable a faster reduction of uncertainty regarding possible options. ICTs, for instance, the use of virtual prototyping, can be seen as allowing these targets to be achieved more or less simultaneously.

The usual investment evaluation methods (such as those based on investment rate-of-return) are not suited to analysis of option preserving projects, since they do not account for the fact that a particular decision may shape the set of future opportunities. One method that can be used is to apply option theory (Kester 1984; Cohendet and Llerena 1989). On this basis, an investment will be considered if future options are given a high enough (subjective) value. Conversely, abandoning a particular technological option should be seen as a reduction in the future opportunity space, and evaluated as such. Unfortunately, difficulties in parameter measurement make the application of option theory problematic[4], but we consider that the mode of reasoning it involves is crucial for understanding the German policies under consideration. We argue that the option preservation policy should be linked to the institutional framework that surrounded high-speed train developments in Germany.

4.3.2 The Differentiated Role of Institutions

In the early phases of the projects (and essentially during the 1970s) neither the BMV nor the DB played an active role in the development of high-speed train technologies. All the projects we have mentioned were dependent on financial support from BMFT. It is fairly clear that BMFT's aim was primarily to promote exploration of different technological options (including the Wheel/Rail technology).

Basically, BMFT wanted to maintain all the technological options. In particular in the 1970–1977 period, BMFT's policy reflected a willingness to preserve all the options associated with the MagLev. Keeping all the op-

[4] Bowe and Lee (2004) apply a real option methodology to evaluate the Taiwan high-speed rail project (THSRC).

tions meant that a final decision could be postponed allowing the projects to profit from new knowledge. Such a strategy can be particularly relevant in the case of breakthrough (and immature) technologies. Indeed, in this case, preserving all the options enabled the acquisition of knowledge relating to the various MagLev solutions. In 1977, the BMFT decided to put its main focus on the EMS solution. However, the BMFT continued to actively support the MagLev project (*e.g.*, by financing the Emslang trial circuit), even after the ICE project was launched in 1982. The BMFT had always been optimistic about the opportunities that a MagLev technology would open up, even though some economic and technological uncertainties persisted. Supporting the MagLev was seen as maintaining a short term option to make it possible to switch to a new trajectory were it to prove viable and profitable.

When the international competition and the technological trials showed that the high-speed Wheel/Rail system was workable and even economically interesting, the BMV and the DB adopted leading roles.

In this historical process, the BMFT was the manager of options and the BMV/DB jointly acted as the "adopting institutions". The BMFT was responsible for keeping the MagLev option alive until a higher level of maturity was reached. We would contend that this specialization by the different public institutions involved in the high-speed train projects favoured the preservation of technological options.

However, the necessity for options to be preserved only exists if there is the expectation of adding knowledge in the future, i.e. to the expected learning processes, either through continued research or through experimentations and/or commercialization.

4.4 Why and how Learning is Done ?

Learning appears to be a central element of technological evolution, and even more so in the context of breakthrough technologies. Whether done consciously or not, learning may serve several purposes. The first is to acquire information as to the approximate performance, and the potential of alternative technologies. We define this type of learning as "exploration". The second is to enhance the performance of a particular technology. We define this as "exploitation" (the distinction between exploration and exploitation was developed by James March (1991) in an organizational context). Whatever its general aim, learning may occur through several modes. For our purpose, we adopt a classification of learning modes based on their degree of "representativeness" of the real environment. Representativeness is the extent to which experiments are conducted in conditions that mirror the real life environment (Pisano 1996). How representative a learning mode should be is, in turn, related to the existence (or not) of the relevant scientific knowledge (Pisano 1996).

4.4.1 Exploration vs. Exploitation

By definition, exploration requires that a diversity of options prevails. Conversely, exploitation is the outcome of focused learning within a reduced set of options. Exploration and exploitation are usually considered as being sequential (exploitation follows exploration). This raises the issue of timing: when should exploration be stopped? From a decision theory point of view, the situation can be modelled as a stopping problem, the point being to identify the time when "enough" information (*e.g.*, as to the merits of the technologies) has been acquired. Formally, this issue may be solved by means of bandit theory (see Cowan (1991) for an application to the Technology Policy dilemma).

The question of timing is crucial for several reasons. First, it is argued that, eventually, both types of learning are subject to diminishing returns. Moving from exploration to exploitation learning is one way to overcome decreasing returns and follow an "optimal" learning curve. The distinction between exploration and exploitation learning is not sufficiently fine, however, to allow an analysis of how learning takes place in a breakthrough technology context. In the following, we look at the environment in which learning takes place.

4.4.2 The Learning Environment

Exploration and exploitation learning may occur in various environments. Following Pisano (1996), we focus on the ability of various environments to represent "reality". Table 4.1 presents a classification of learning environments in terms of their representativeness. The efficiency of a particular learning context depends on the knowledge structure that characterizes the sector being considered. In sectors where a strong base of scientific and organizational knowledge exists, problem identification and problem solving are likely to be conducted "in the laboratory". Conversely, in emerging sectors, characterized by a low level of relevant knowledge, problem identification and problem solving are likely to require commercialization.

Relevant elements of a sector knowledge structure include the theoretical understanding of fundamental processes, the ability to fully characterize intermediary and final products, and knowledge concerning possible scale and second order effects (Pisano 1996).

Table 4.1. Representativeness of learning environments (based on Pisano 1996)

Representativeness	Learning environment
High ↑	
	Commercial exploitation
	Experimental running on "production" site
	Experimental running on R&D site
	Laboratory experiments
	Computer aided simulations
Low	

The choice of a specific learning environment can be trivial, as in the case of very mature technologies where low representativeness environments yield interesting outcomes. However, it can be a strategic decision, which may involve several trade-offs. Especially relevant for our study are the trade-offs between

- cost of experimentation and representativeness of its results;
- flexibility (due to, *e.g.*, technological or investment irreversibility) and representativeness of the learning environment.

As a further step in our analysis, we recapitulate some elements of the role of "doing" in the learning process.

4.4.3 Learning-by-Doing

The issue here is to what extent learning requires some form of "doing". Following Rosenberg (1982) or Habermeier (1990), it is commonly accepted that practice is an essential element of learning, since interactions between products and their use environments are often too complex to be predicted. Von Hippel and Tyre (1995) propose a further development of this proposition by analyzing the role of doing in processes of problem discovery and problem solving. Indeed, they argue that doing entails the juxtaposition of two complex elements (*e.g.*, a machine-tool and a factory environment). Doing provokes the "precipitation of symptoms" (*e.g.*, a weak performance), which, in turn, reveal unexpected interferences between the product and its use environment. It should be noted that this argument is far from being obvious: as doing implies increased complexity, it might well reduce one's ability to identify problems.

Following Von Hippel and Tyre (1995), we argue that in stable environments (*i.e.* environments that are under the control of the decision maker under either perfect expectations or knowledge of the probabilistic distribution of events), learning-without-doing can be achieved, provided a sufficient number of possible interactions within the system is investigated. Although not a necessary step for learning, doing remains a candidate learning device as it can reduce learning time and/or monetary costs. This is probably even more so in the case of highly complex systems (or technologies). However, the importance of doing seems much higher in unstable environments (*i.e.* environments that are out of the decision maker's control, where events are unexpected, and there is high non-probabilistic uncertainty). Here, symptoms emerge as the outcome of an endogenous conflict between the system (or the technology) and what Von Hippel and Tyre (1995) define as "autonomous problem solvers". This conflict gives rise to problems that are difficult (if not impossible) to anticipate and generates sets of solutions that are *a priori* non-predictable from the developers' standpoint.

In turn, we are led to conclude that assessing the weight of "doing" in the learning process requires an evaluation of the level of scientific knowledge and of the stability of the system environment.

We have argued that the timing of the switch from exploration to exploitation could have an influence on the competition between technologies. We can now push this argument further: considering complex systems (technologies), which benefit from a limited level of scientific knowledge, and which are developed within an unstable environment, an anticipated switch to practice can be considered as a source of competitive advantage as it can speed up the learning process. And, if this is the case, it means that the institutional specialization mentioned earlier also becomes crucial, because it is related to the nature of the learning processes.

4.5 Technology Competition

While it may seem a "natural" way to cope with technology evolution, the German policy of giving parallel support to "a-synchronic" technologies (Wheel/Rail is seen as the "old" technology and MagLev as the "new") raises several issues. First, evaluation of the merits of the respective technologies must be conducted at "comparable levels of knowledge". At some point, acquisition of new knowledge requires commercial exploitation beyond the laboratory. Second, the implementation of a transport system requires heavy investment in the network infrastructures. The operator's choice of technology is inevitably influenced by the need for compatibility with existing infrastructures. Finally, delays in adoption of new technology might have irreversible consequences, as (i) improvements to the unadopted technology will not occur and (ii) the "window of opportunity" for its diffusion might be missed.

Even though the public authorities (BMV and BMFT) did not initially consider them as such, the Wheel/Rail and MagLev technologies should be seen as virtual competitors in the market for guided transport. Therefore, we would argue that the timing of the projects, especially in terms of commercial exploitation, has an influence on the eventual outcome of the competition between these technologies.

4.5.1 The Role of Learning

The description of the German experience in section 4.2 and the theoretical framework presented above are a first step in the appraisal of the German high-speed train technology policy.

Following the earlier arguments, we present a (very) schematic representation of the learning experienced for the systems considered in Figure 4.2. The first step refers to the simultaneous development of candidate solutions (exploration period), mostly "in the laboratory". After a certain amount of information has been acquired concerning the merits of the technologies, elimination of candidate technologies enables acceleration of the learning process. This might be due to a concentration of financial efforts. A few technical solutions are first tested experimentally on a trial circuit. Finally, commercial exploitation enables learning on the basis of "real experience".

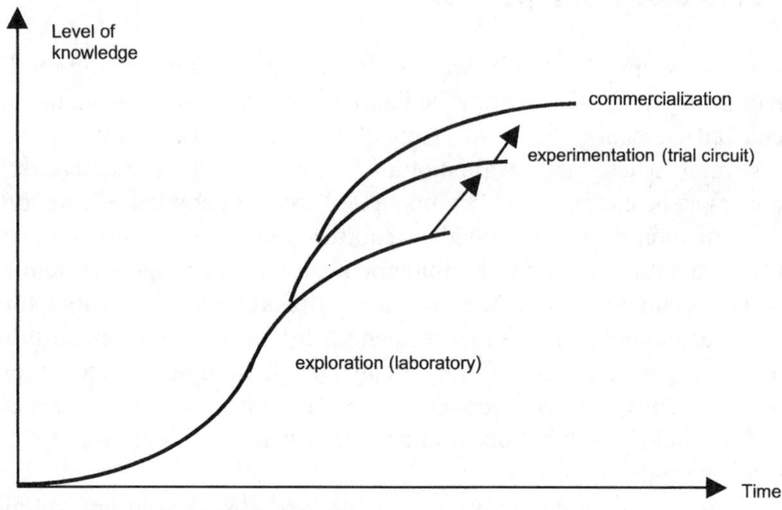

Fig. 4.2. Successive learning steps

Next we consider the parallel development of the Wheel/Rail and the MagLev technology as represented in Figure 4.3. Both projects were launched in 1971 as part of a global research venture, but, due to accumulated knowledge in similar applications, the Wheel/Rail technology underwent a rapid learning curve. The "Hochgeschwindigkeits Verkehr" programme launched in 1982 gave rise to the ICE project and the ICE train was put into service on the Hannover–Würzburg and Mannheim–Stuttgart lines in 1991. This meant that diminishing returns from the learning period were minimized.

The most important decisions (focusing on the EMS technology and construction of the Emsland trial circuit) concerning the MagLev technology were taken in 1977. We contend that commercial exploitation of the Transrapid would facilitate the acquisition of new knowledge. As long as there is no "real scale" development and commercialization, the MagLev technology will not embark on a new "learning curve".

The consequences of delaying the implementation of the Transrapid are manifold. First, it is still not possible to evaluate the merits of the MagLev and Wheel/Rail technologies on the basis of "common experiences". Second, after more than 500,000 km of cumulative trials, the MagLev displayed low returns to outlays on experimentation, while the ICE has remained the subject of (incremental) improvements (Jänsch and Keil 1999). This is likely to result in a "learning gap", which may have consequences for the opportunities of diffusion of the Transrapid, in particular in those markets where the technologies are in direct competition.

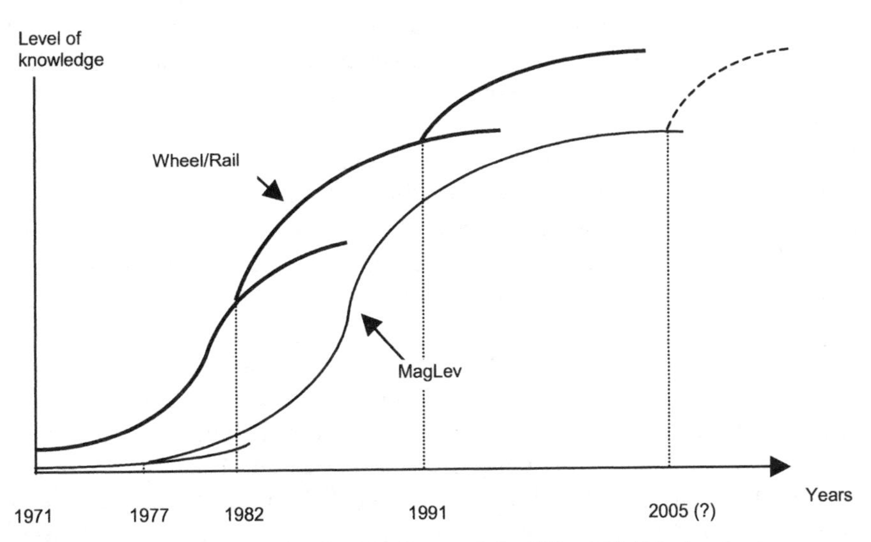

Fig. 4.3. Learning curves for the MagLev and the Wheel/Rail technologies

The German case demonstrates a contradiction, which can be observed more broadly at the European level. On the one hand, Wheel/Rail technologies, such as the French TGV and pendular systems in France, Italy, or Sweden, benefit from similarities with existing systems. In these cases, innovations raised some specific (technical or organizational) problems, but they were developed within (more or less) established boundaries. Therefore there was a degree of predictability, giving greater opportunity for learning-without-doing (see section 4.4). On the other hand, MagLev technologies, such as employed in the German Transrapid and the Swiss Swissmetro, did not benefit from previously acquired knowledge[5].

The effects of these differences are far-reaching. First, identification of the "relevant" MagLev technologies required the exploration of numerous *a priori* feasible options. This led to a costly and lengthy exploration period. Second, learning in a context where little previous knowledge exists requires highly "representative" learning environments. In addition to cost considerations, the timing of the learning sequences becomes more important. Finally, we would argue that there are differences between breakthrough and continuity oriented technologies in terms of the need to carry out a "full sequence" of learning steps.

Due to the structure of the knowledge and considerations of cost, learning in the German MagLev technology was conducted through a complete

---

[5] See in particular Foray (2001) for an analysis of these cases.

sequence of steps. For instance, the propelling technology a set of rules concerning the infrastructure and control electronics. The necessity for learning to follow the full sequence is lessened in relation to Wheel/Rail systems since spill-overs from other similar projects can be expected. This was the situation that occurred in the late 1990s: before the French pendular system was developed, there had been several trials conducted in France (*e.g.*, Lyon–Modena) based on the Italian ETR460 pendular system (*cf.* Saubesty and Vernimmen 1999). In Germany, the various MagLev technologies did not benefit from cross-learning because the systems were at a very early stage of maturity and embodied few common characteristics.

This section has introduced some of the elements that can be used as the basis for an appraisal of the role of "doing" in learning processes. We argue that this is of particular significance in the switching time from pre-commercial exploitation to commercial exploitation. The arguments developed by Von Hippel and Tyre (1995) and Pisano (1996) suggest that doing plays a crucial role in the learning process involved in "breakthrough" technologies. But, the necessity for "pre-doing" learning should not be underestimated. First, evaluation of the various learning environments must account for the (direct or indirect) costs entailed. Further, in the context of high-speed train technologies, factors such as "acceptable risk" must be considered. More precisely, the possibility of "catastrophes" tends to reduce the decision makers' flexibility in determining the duration of the pre-doing learning process. Whether the development of ICTs will enhance the efficiency of the learning process substantially (for example, through the use of virtual prototyping and computer simulations) is an open question worthy of further investigation (see, *e.g.*, Nightingale 2000).

Doing and access to a representative learning environment are strongly correlated with the existence of a "lead user" (Von Hippel 1988). In this respect, the contexts within which the two alternative technologies were developed were different.

In the case of the ICE, the DB had a greater commitment after 1977 to the development of the technology. Under pressure from BMV, the DB participated directly in the design and experimentation phases of the ICE, mostly as a reaction to the development of the TGV in France. In this respect, the experimental platform (Rheine–Fehre section) and later the first commercialization on the Hannover–Würzburg and Mannheim–Stuttgart lines, were critical for the fine tuning of the technical and operational solutions. In addition, DB's support was essential in providing commercial credibility and increasing take-up by other users.

The Transrapid case stands in stark contrast. The project has never been able to attract or integrate a significant "lead user". The position of DB in relation to this project has always been ambiguous, probably because the MagLev technology was judged to be too "disruptive". Moreover, the Lufthansa airline company, which was involved in the process (especially in the latest phase), was unable to provide the necessary knowledge and support (in terms of infrastructure, for instance) to enable improvements and the fine tuning of the technology to user needs. Partially as a consequence of this, the Transrapid, up to the time of writing, has never been able to demonstrate its feasibility and economic viability, and thus has lacked credibility.

In this section, we have focused on the learning processes supporting the development of technologies. We consider that the "efficiency" of a learning process is closely related to the "correct" management of learning environments, which, in turn, is related to commercial exploitation.

However, such commercial success of a technology is also dependent on effective diffusion opportunities. This issue is particularly relevant to the Transrapid case: even though competition between the Transrapid and the ICE is officially precluded, the established ICE network is likely to have an effect on the diffusion opportunities for Transrapid, particularly because of the increasing returns to adoption that occur in network technologies (*cf.* Arthur 1989).

4.5.2 Network Competition

Guided ground transport systems comprise complementary elements, such as infrastructures, rolling stock, etc. In this context, the geographic diffusion of a particular technical system in part defines its economic value: a better diffused system enables the junction between more sites, which would be expected to increase demand. As a corollary to this, the economic value of any technical system depends on its ability to become integrated into established ones. The deciding factor is the compatibility between established and new systems (see, *e.g.*, Cohendet and Schenk 1999), a characteristic of systems that exhibit network effects: the "value" of a technology increases with the size of the associated network. The "associated network" can be a technology-specific network when there is incompatibility with existing technologies, or a shared network when there is compatibility with the existing technologies.

As we have seen, the solutions adopted for the ICE favoured continuity (and compatibility) with traditional Wheel/Rail systems. Thus, the high-speed network that was developed during the 1970s provided the base for the initial commercial exploitation of ICE, with specific parts being developed incrementally. The rationale for this approach was the low deployment costs and the speed of diffusion it allowed. Moreover, the compatibility enabled its speedy implementation on the high-traffic lines, and especially along the Köln–Rhein–Main corridor.

There are consequences in terms of the opportunities for diffusion of the Transrapid in Germany. For both economic and "political"[6] reasons, in the short run, the opportunities for diffusion of the Transrapid lie in connections where the ICE does not operate: the fall of the Berlin Wall produced some good prospects since at the time there were no high-speed connections between the former East and West Germany. A Hamburg–Berlin Transrapid project was launched in 1992, but was abandoned in 2000 after re-evaluation.

The ICE has the benefit of "first mover advantage" over the Transrapid, which considerably reduces the latter's scope for diffusion in Germany. However, although the incompatibility of the Transrapid with established systems is a drawback from a network effect standpoint, it does have some positive effects. For the ICE, compatibility generates incentives to exploit the existing network. In other words, there is little motivation for DB to develop a specific ICE network. Therefore, we would argue that the potential of ICE is not being fully exploited: heterogeneity in terms of network quality, high and low speed trains using the same tracks, and the constraints imposed by the frequency of stops, are all handicaps to higher transportation speeds. Such constraints would not apply to the Transrapid, since its incompatibility would necessitate development of a specific network[7].

Since the abandonment of the Hamburg–Berlin project, export has been seen as a credible alternative to national implementation of the Transrapid. China is constructing a 30 km Transrapid connection between Pudong airport and the Shanghai Lujiazui financial district, and this should become commercialy operational in 2004. Other projects in the United States and in the Netherlands have been positively evaluated.

The Chinese Transrapid project demonstrates a change to the initially perceived opportunities for MagLev technology: short distance connections (*e.g.*, airport–city connection) are now being given deeper consideration. Such a "re-encoding" is linked to a re-evaluation of the competitive advantage of MagLev: flexibility, space saving and ecological aspects (low

[6] Several actors (such as Siemens, Adtrans and the DB) are engaged in both the Transrapid and the ICE.

[7] As an instance, the transportation time on the actual ICE Hanover-Frankfurt line (339 km) is over 2 hours, while the Transrapid Hamburg-Berlin connection (292 km) would take 1 hour.

energy consumption and minimal noise) and not just commercial speed are being classed as essentials.

Our analysis suggests that, in the short term, implementing the Transrapid was of particular importance since its commercial exploitation acts as a "technological display" and enables new technological learning. In the longer term, the successful introduction in China of the first Transrapid connection could open the way for its further exploitation, and once again, particularly in China where many high-speed connections are still needed.

4.6 Conclusion

The history of the German high-speed train offers rich opportunities for analyses. Observation of the development of "a-synchronic", but potentially competing technologies led to our focus on the interplay between learning and adoption processes. The main point that this chapter has tried to emphasize, is that the timetable of these projects, especially in terms of commercial exploitation, had an influence on the competition between these technologies.

On the one hand, delaying the commercial implementation of the Transrapid has prevented "real scale learning" from taking place and may lead to a learning gap between the MagLev and the Wheel/Rail technologies. On the other hand, the Wheel/Rail technology benefited from increased network advantage, which made the adoption of the MagLev even less likely. Excessive specialization between research oriented and implementation oriented institutions and difficulties in "relay transmission" makes the management of these aspects even more intricate.

Thus, designing policy recommendations in this context is a very complex business. The optimal decision sequence is determined by the value that public decision makers (and, by extension, society) attach to the preservation and the eventual exercise of options. This study could be helpful for anticipating the difficulties that might be encountered in innovation processes.

4.7 References

Arthur WB (1989) Competing technologies, increasing returns and lock-in by small historical events. Economic Journal 99: 116-131.

Bowe M, Lee DL (2004) Project evaluation in the presence of multiple embedded real options: evidence from the Taiwan high-speed rail project. Journal of Asian Economics 15: 71-98.

Büllingen F (1997) Die Genese der Magnetbahn Transrapid. Deutscher Universitäts Verlag, Wiesbaden.
Cohendet P, Llerena P (1989) Flexilibité, information et décision. Economica, Paris.
Cohendet P, Schenk E (1999) Irréversibilités, compatibilité et concurrence entre standards technologiques. In: Callon M et al. (eds.) Réseau et coordination. Economica, Paris: 65-109.
Cowan R (1991) Tortoises and hares: choice among technologies of unknown merit. Economic Journal 101: 801-814.
ECOSIP (1993) Pilotages de projet et entreprises: diversités et convergences. Economica, Paris.
Farrell J, Saloner G (1986) Installed base and compatibility:innovation, product preannouncements, and predation. American Economic Review 76: 940-955.
Foray D (2001) Choix technologiques dans les projets de ruptures technologiques. Rapport final pour le PREDIT. IMRI, Université Paris Dauphine.
Habermeier KF (1990) Product use and product improvement. Research Policy 19: 271-283.
HSB mbH. (1971a) Studie über ein Schnellverkehrsystem (Kurzbericht: Systemanalyse und Ergebnisse). Schriftreihe des Bundesministers für Verkehr.
HSB mbH. (1971b) Ein neuartiges Verkehrssystem für die Bundesrepublik Deutschland. Schriftreihe des Bundesministers für Verkehr.
Jänsch E, Keil J (1999) ICE und Transrapid: Konzeptionnelle Planungen für Elemente im HGV-Angebot der DB AG. Bahn Report 99, ETR Edition.
Kester C (1984) Today's options for tomorrow's growth. Harvard Business Review 62: 153-160.
March JG (1991) Exploration and exploitation in organizational learning. Organization Science 2: 71-87.
Münchschwander P (Ed.) (1989) Schienenschnellverkehr, vol. 1-4. R.v. Decker's Verlag, G. Schenck, Heidelberg.
MVP- Versuchs und Planungsgesellschaft für Magnetbahnsysteme mbH. (1988) Einsatzfelder neurer Schnellbahnsysteme, vol. 1-4.
Nightingale P (2000) Economies of scale in pharmaceutical experimentation. Industrial and Corporate Change 9: 315-359.
Pisano GP (1996) Learning-before-doing in the development of new process technologies. Research Policy 25: 1097-1119.
Rosenberg N (1982) Inside the black box: Technology and economics. Cambridge University Press, Cambridge.
Saubesty C, Vernimmen S (1999) Incrémentalisme et rupture technologique: étude de cas à la SNCF. IMRI, Université Paris Dauphine.
Von Hippel E (1988) The sources of innovation. Oxford University Press, Oxford.
Von Hippel E, Tyre M (1995) How 'learning by doing' is done: problem identification in novel process equipment. Research Policy 24: 1-12.

5 Institutional Arrangements of Technology Policy and Management of Diversity: the Case of Digital Switching System in France and in Italy[1]

Patrick Llerena, Mireille Matt and Stefania Trenti

BETA, Strasbourg, E-mail: pllerena@cournot.u-strasbg.fr
BETA, Strasbourg, E-mail: matt@cournot.u-strasbg.fr
Banca intesa, Milano, E-mail: stefania.trenti@bancaintesa.it

5.1 Introduction

The importance of the development of a Digital Switching System (DSS) lies in its positive effect on the entire telecommunication network. DSS has provided greater reliability and speed, and enabled the introduction of new value added services, all of which have benefited the entire economic system. This chapter focuses on the development of DSS in France and Italy. We analyze how the 'organization' of a public research and development (R&D) programme can influence the relative success of a policy. By organization we mean coordination of the actors involved, and also the different technological options involved in innovation in telecommunications, which emerge, and should be publicly supported. Based on previous theoretical development, we state that the organization of a mission-oriented programme (such as the DSS) depends mainly on the learning ability of the policy maker and its proximity to the participating firms and institutions. Moreover, the management of technological diversity may have important impacts in terms of timing, costs, competition, technological diffusion and lock-in phenomena. The purpose of this chapter is to use this

[1] The empirical research had financial support from the EU-TSER Programme, under the project "Innovation Systems and European Integration", coordinated by C. Edquist. We would like to thank all participants in the project for helpful comments. We also profited from discussions with Dominique Foray and Patrick Cohendet. Earlier versions of the paper were presented at the Interdisciplinary Seminar at Compiègne University, 1998; and at the Santa Anna School of Advanced Studies, Pisa, 1998.

dynamic framework and evolutionary concepts to assess the relative success of the French and Italian DSS public programmes.

To understand the importance of DSS development and, thus, the critical role of technology policy, we will briefly set it within the broader context of technological developments in telecommunication, i.e. the paradigm shift from electromechanical to digital technologies. Between the mid 1950s and the end of the 1980s, the telecommunication industry was at the centre of a technological revolution with the shift from the electromechanical to the digital paradigm. Electronic devices were gradually introduced in the three main sub-sectors of the telecommunication industry: transmission, switching and terminals. This chapter focuses essentially on the switching part of the network.

The possibility of introducing electronics into the telecommunication network was first conceived in the 1940s. But it was not until after the Second World War that the Bell Laboratories in the US and the research laboratory of the UK Post Office began to work on an electronic switching system. The discovery of the transistor by the Bell Laboratories in 1948, as a by-product of this research, was a fundamental step. However, it was over 20 years before a completely digital switching system was realized. The main problem was the reliability of the electronic components. In England, the first ambitious attempt in 1962 to introduce a totally electronic switching system failed. The research path followed by the US Bell Labs was more successful. Instead of being oriented towards a completely digital switching system, Bell Labs incrementally introduced new sophisticated devices in a traditional electromechanical switching system: the first prototype of a semi-electronic switch (ESS 1) was installed in 1965.

To understand the technological choices made by Italy and France it is worth describing briefly the two main trajectories within the digital paradigm: the space division and the time division trajectories. A space division switching system is characterized by a physical connection between the entry and the exit of the signal. It is thus possible to follow the path of the signal through the space by the means of contacts that are generally electromechanical, but may also be electronic. This was the technological option adopted by the US and Germany. In a time division DSS the signal is "translated" into a digital code and then transmitted through purely electronic devices. In France and in Italy, time division was the preferred trajectory, but was arrived at by different routes (Libois 1983). The aim of this chapter is to compare the technological developments in France and Italy.

The evolution of the technology related to the introduction of digital switching had various consequences for both the service provider and the manufacturer. For the telecommunication service provider, the technological shift meant, first of all, a reduction in costs: these were the result of overall size, maintenance and numbers of personnel employed on installed lines[2]. This shift allowed improvements in the capacity of the switching systems and offered greater reliability and improved quality[3]. Finally, it opened up the opportunity for the introduction of new services and offered the possibility for digitized transmission of data and images in addition to voice (i.e., an ISDN - Integrated Services Digital Network). For the service providers it can be claimed that: "the advantages of electronic systems towards electromechanical ones undoubtedly exceed the risks connected to the management and organization of the technological conversion and the costs connected to the qualification of the personnel" (Bragho 1988).

For the manufacturing firms, this technological evolution meant a shift from a *labour intensive* technology (electromechanical) to a *competence intensive* one (digital). This shift involved a change in the number and type of people employed, the need for them to accumulate new competencies, and increased economies of scales, especially in R&D (Zanfei 1990). The progression from the electromechanical to the electronic paradigm involved a radical process of restructuring of the workforce, an increase in fixed capital investment, and the introduction of new flexible production technologies. Moreover, manufacturing firms were forced to accumulate upstream competencies in new and different fields, such as microelectronics and software, in order to develop and incrementally improve new products. Finally, the economies of scales in R&D increased. R&D investments need to be high in terms of the minimum efficient threshold, and they tend to remain high throughout the entire product life cycle, which eventually becomes shorter.

Thus, it can be seen that technology policies have a tremendous impact on the further development not only of an industry – in this case telecommunication – but also of the whole economy.

The two cases described in this chapter are the development of the E10 in France and the Proteo/UT family in Italy. The time period spans the 1950s to the 1980s for France, and the 1960s to the end of the 1980s for Italy. Based both on theoretical hypothesis and on empirical facts, we show that the relative success of the efforts in France can be explained by the ability of policy makers to appropriately coordinate the actors in the telecommunication innovation system, and the various technological options.

---

[2] For example, the passage from the Cross-Bar switch to the ESS 4 in 1976 in Chicago led to a saving of 25% in size and more then 30% in energy and maintenance (Libois 1983).

[3] The already cited substitution in Chicago leads to a five-fold increase in capacity.

However, political decisions related to competition entailed some failures, such as the demise of one national company. The Italian R&D programme, on the other hand, suffered from a lack of coordination between the actors involved. After a long period of experimentation, national resources were finally pooled and Italian firms cooperated with foreign companies. The delay in this collaboration induced high costs and significant delays in the development of the technology, thus explaining its relative lesser success, compared to France.

The first part of the chapter focuses on the institutional set up of the telecommunication systems in each country. The analysis mainly compares the characteristics of the two information structures and highlights the coordination mechanisms used to develop the DSS. The specific institutional arrangements had some consequences for the dynamics of DSS diffusion. We then go on to analyze the policies in terms of coordination failures, diversity exploitation, and lock-in and diffusion effects. This allows us to assess the relative success of each country's policies.

5.2 Institutional Arrangements, Information Structure and Coordination

A technology policy is embedded in institutional arrangements, specific to each country, resulting from the particularities of its institutional history. One way to deal with the problems associated with these specificities is to compare (with some kind of implicit "ceteris paribus" assumptions) some "parameters" of the institutional structure. Following on from previous work (Foray and Llerena 1996), we have chosen to focus on the information structure and the coordination between the different actors in the national telecommunication system. Our purpose is not to achieve a precise description of the institutional mechanisms supporting technology policy; but to suggest that there is a clear link between the informational structure of, and the coordination mechanisms implemented by, a policy, and its degree of success. We first present the analytical framework used to define the relevant parameters, and then apply it to each case.

5.2.1 Information Structure and Coordination in Technology Policies: Analytical Framework

The aim of this section is to establish a link between the informational structure of the technology policy system and the mode of coordination of activities, in order to analyze the coherent and incoherent elements of a technology policy. For this purpose, we use the model developed by Foray and Llerena (1996), which restates Aoki's (1986) results, and compare two types of technology policy organization.

5.2.1.1 Informational Capacities of the Policy Maker

Foray and Llerena (1996) state that the policy-maker's informational capacity is determined *de facto* by the nature of the technology policy. Using the classification developed by Ergas (1987), they distinguish between mission-oriented and diffusion-oriented policies.

Mission-oriented policies correspond to radical innovation projects. Their main characteristics are the centralization of the decision making process and the pursuit of goals involving the implementation of complex systems. Frequently, one particular public agency is in charge of the programme. By their very nature, missions concern strategic technologies. The number of projects is limited, as is the number of actors involved. Mission-oriented policies often imply the creation of a new technology, a specific (i.e. large technical systems) or a new generic technology, without any *a priori* specification of the modes of use.

Diffusion-oriented policies are characterized by decentralization. Public agencies have a restricted role, entrusting responsibilities to professional bodies, or to cooperative research organizations. Public resources are widely spread throughout the systems, so as to reach small and medium sized enterprises (SMEs), the main goal being to diffuse the technology, and to ensure adaptation to the local and specific needs of individual firms.

In both cases, the *informational capacities* of the policy maker need to be differentiated. It can be supposed that in the case of diffusion-oriented policies, *de facto*, the policy maker has a very imperfect perception of the precise needs and characteristics of firms, and that there is probably a long delay between perception of a possible need and implementation of the relevant solution (i.e. the policy maker's reaction time). In the case of a mission-oriented policy, however, the policy maker has a clear 'vision' of what should be done (it is clearly a centralized design and decision process, with a precise definition of the technological goals to be achieved) and there is strong institutional proximity between the policy maker and the chosen firms ("National Champions").

5.2.1.2 Learning Capacities of the Policy Maker, the Firms and the Intermediary Institutions

The main learning characteristics of the local elements of the system (i.e. firms and intermediary institutions, such as technology centres) are represented by the initial ability of these elements to identify their needs, their learning rate, and the efficiency of horizontal coordination between them. More precisely, the ability to identify needs could be interpreted as an in-house research capability, relevant for defining the absorptive capacity of firms. The coordination skill refers more specifically to a firm's and institution's ability to slot into cooperation networks. These parameters help to determine the appropriate mode of coordination of diffusion-oriented policies, since the informational capacities of the policy maker are *de facto* low in these types of policies.

The learning ability of the policy maker is its ability to accumulate past experience and know-how. The rate at which it takes into account historical events demonstrates capacity to define a coherent system within a mission-oriented policy (i.e. coherence between the policy and the mode of coordination). The learning ability of the policy maker will induce the relevant coordination mode since the selected firms are characterized by significant learning abilities and the informational capacity of the policy maker is given (i.e. high).

The Aoki model allows the relevant coordination mode (vertical or horizontal) to be defined for a given set of values of parameters characterizing the institutional arrangements of the technology policy. The main results of the model can be summarized as follows (see Aoki 1986; Foray and Llerena 1996).

In the case of a *diffusion-oriented policy*, since the informational capabilities of the policy maker are by definition low, horizontal coordination will be more effective than vertical when the learning rate, the initial knowledge and the coordination skill of the firms are high. "In other terms, a diffusion-oriented policy with horizontal coordination needs a minimal technological capability or learning potential on the side of potential users" (Foray and Llerena 1996, p. 164). If the converse is true, then vertical coordination will be more appropriate.

In the case of *mission-oriented policies*, the policy maker has, by definition, a high information capability, and firms are characterized by a high learning rate, and significant initial knowledge and coordination skills, since the policy maker can choose the most relevant firms and research institutions. Which coordination mode is the most appropriate will crucially depend on the capacity of the policy maker to accumulate knowledge based on past experiences. If the technological competences of the policy maker are high, then the preferred mode of coordination would be vertical; if not, then horizontal coordination would be better.

Our purpose is to use and develop this analytical framework in the case of technology policies implemented in France and Italy. The DSS case is particularly interesting because this technology played a major role in defining the relative competitive advantage of firms in the telecommunication industry. In other words, the DSS programme can clearly be said to be a mission-oriented policy, with the explicit goal of developing a new device for the switching system based on new technological principles.

5.2.2 Relevance of Coordination Modes in France and in Italy

France and Italy have different institutional arrangements in relation to their technology policies, and especially for DSS. In France, there is a powerful specialist centre (CNET, National Centre of Telecommunication Studies), which is close to the policy maker, and which is able to capitalize on knowledge, know-how, and political expertise. According to our analytical framework (i.e. the ability of the policy maker to accumulate knowledge is high), vertical coordination seems to be appropriate. In contrast, in Italy there is no institution with the capabilities to develop such competences. Therefore, horizontal coordination should have been effective, but did not really emerge. Table 5.1 presents the technological evolution of DSS in each country.

5.2.2.1 The French Case: a Mission-Oriented Policy with Centralized Coordination

In France, especially during the first phase[4] (1958-1974), the development of DSS was marked by the creation of a "specific organizational device" (Quelin 1992), with CNET, the research laboratory of the French PTT, playing a central role. In 1958, after the opening of a new switching department in Lannion, CNET formed an alliance with Socotel, the pool of the French manufacturers of switching equipment. Under this arrangement the two French subsidiaries of ITT (Compagnie Générale de Constructions Téléphoniques – CGCT – and Le Materiel Téléphonique – LMT), the French subsidiary of Ericsson (Société Française de Téléphones Ericsson-SFTE) and the two French manufacturers, AIOP and CIT-Alcatel, collaborated to conduct research on the new technological paradigm of electronics.

Two projects were initiated. The first, Socrates, involved Socotel and CNET. It aimed at the development of a digital switching system following the space-division trajectory (incremental innovation). The second,

[4] This first period ended in 1974 with the election of V.G. D'Estaing to the presidency of the French Republic, whose government recognized that important changes were occurring in the telecommunication industry and intervened to weaken the role of CNET.

Aristote, involved only CNET and Société Lannionnaise d'Electronique (SLE), a new company opened in Lannion by Cit-Alcatel. The Aristote project chose to start directly on the time-division trajectory (radical innovation).

Table 5.1. Technological evolution of the Italian and the French digital switching systems

Year	Technological evolution of the Italian digital switching system	Technological evolution of the French digital switching system
1958		Creation of the Switching Department of CNET Creation of Socotel Project Socrate and Aristote
1967		Platon and Pericles
1970	Beginning of Proteo	First E10 installed
1972	Proteo CTA installed	E10-A (improvement in program capacity and components)
1979	Proteo CT-2 (improvements in memory) Beginning of UT	
1980	TN-16 and TN-5	E10-B (new circuits, modular architecture, improvements in program capacity)
1981	UT10/3 prototype	
1983	UT10/3 industrialized	
1985		ALCATEL E10 (new access unit for network compatibility - ISDN level)
1987		First ISDN trial
1989	UT100/60	
1992	First ISDN trial	

Source: Our elaboration on Chapuis Joel (1990)

The reasons for this division of labour were twofold. One was technological: the technological advance of manufacturing firms in the space-division technology. Given their greater experience in terms of the electromechanical paradigm, these firms were thought to be in a better position to gradually introduce electronic control and management systems within a crossbar switch. On the other hand, CNET, in the previous ten years, had accumulated vast experience doing basic research on electronics, especially transmission, and was thus better suited to developing a new elec-

technology contributed to the objective of building French independence in this strategic field. It was clear that if the time-division project proved successful, French manufacturing industry would be in a monopoly position vis-à-vis foreign subsidiaries. In fact, the group of researchers working on Platon (the successor to the Aristote programme on the time-division solution) had the explicit task to "realise the prototype of totally electronic switch and then to pass immediately to the industrial phase" as quickly as possible (Libois 1983, p.158). The consequence of the "exclusive" French efforts on time-division technology, if successful, would be to push ITT and Ericsson out of the market.

Socrates was followed by a new project, Pericles, while Aristote was followed by Platon, the project that finally led to the installation of the world's first time-division switch. The Platon prototype, later known as E10, was installed in Lannion in 1970 followed six months later by a new and bigger prototype. Between 1970 and 1972, work continued in Lannion to progress from a prototype to an industrial product. In 1972, the E10A was ready to be produced and sold commercially. During the inauguration of the time-division switch in 1972, the Ministry of PTT confirmed the importance of the new technology for the modernization of the French network in announcing that the E10 would cover 2% of the French switching market by 1973, and 10% by 1975 (Libois 1983). Moreover, in 1973, the new Ministry of PTT confirmed support for the time-division technique, predicting further development of the system in order to serve bigger towns (Libois 1983).

It is important to underline the crucial role played by CNET in this first phase. CNET had two main responsibilities: R&D and control over equipment. This dual function gave CNET the advantage of being able to interact both with the service supplier and with manufacturers with relative autonomy. Also, during this period CNET had a very charismatic leader: Pierre Marzin (Griset 1995). Through a combination of personal contacts and trust, CNET developed a dense network of relationships with industry (facilitating technological transfers), with policy makers (accelerating the funding of projects), and with academia (strengthening the flows and exchange of knowledge and personnel).

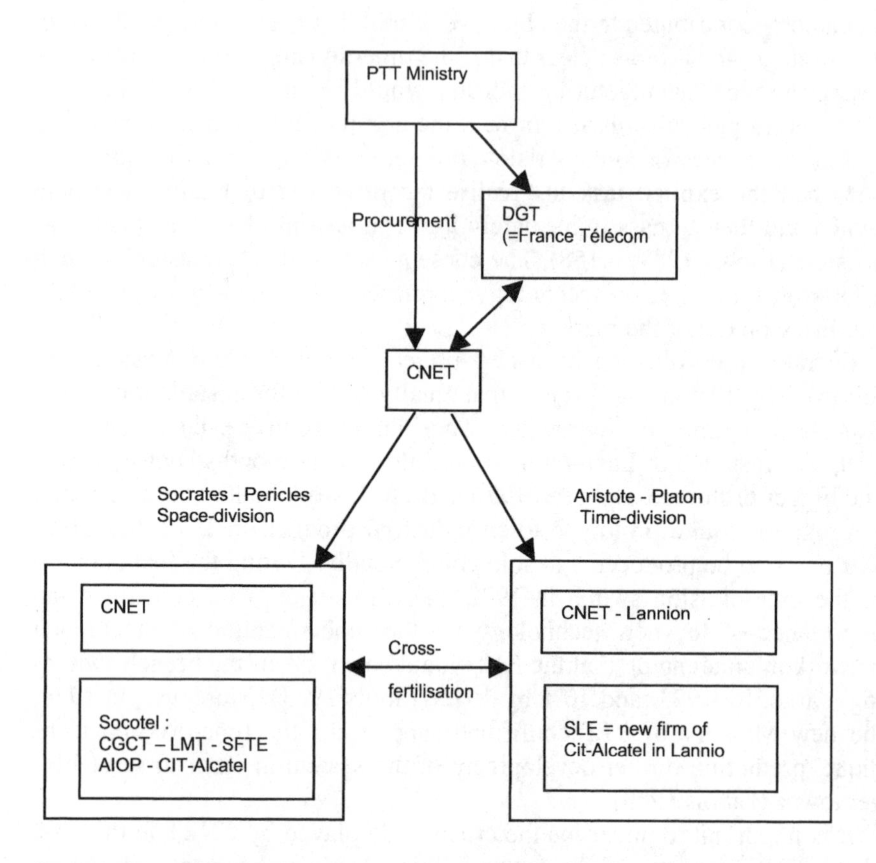

Fig. 5.1. The Organisational Design of the DSS Policy in France

A crucial aspect of the technological developments was the involvement of all the suppliers acting in the French market and the division of labour between the different members of Socotel. One of the bonuses of this division of labour in the two series of projects was the opportunity for the two groups to interact, given the presence of CNET researchers in both projects. For example, the manufacturers' experience in relation to the reliability of the switch promoted adoption, in the Socrates prototype, of the principle of "load sharing" i.e. the simultaneous use of two (or more) parallel computers to control the switch. This principle was later adopted in the design of the DSS architecture in the Platon project.

5.2.2.2 The Italian Case: a "Mission-Oriented" Policy with a Lack of Coordination

In Italy, there was no such centralized process or closeness to the policy makers' decision process, and horizontal coordination (between sub-sets of firms and institutions) emerged. "Competing" solutions, promoted by competing "networks of firms" or "competing actors horizontally coordinated", were produced; but indecision about the adoption of early technology and industrial choices resulted in loss of opportunities and a lag in the adoption and diffusion of DSS.

The first phase of development of DSS spanned the period between 1960 and 1980. In the course of these twenty years, three main actors (Sit-Siemens, Telettra and CSELT) were involved in research on digital switches. Despite some attempts to reach agreement (between Sit-Siemens and Telettra and between Sit-Siemens and CSELT inside the STET[5] group), and despite the strong complementarities that existed among the different players, most of the research was isolated. We briefly describe below the three axes of this research.

In the early 1960s, a small group of researchers in Sit-Siemens started working on a prototype for an electronic PBX (Private Branch Exchange). At that time Sit-Siemens was producing the majority of its equipment under licence from the German company, Siemens. Based on this first prototype, a project for the development of an electronic public switching system was conceived. The research project was conducted completely *intra-muros*[6]. The top management in STET were part of the decision process, but no participation from the vertically integrated service provider was planned. In 1969, a first draft of the project was presented to SIP, in order to obtain the approval of the service provider concerning the architecture and the main characteristics of the switch. SIP's R&D department approved the main characteristics[7].

Even though no formal arrangements were put in place in relation to the future market, the vertical integration within the STET group, and the historical dominance of SIT-Siemens in the Italian market for switching

[5] STET is the Italian holding company for telecommunications. It includes the service providers (SIP for local calls, Italcable for international calls, and Telespazio for satellite), the national research centre (CSELT), and the main national manufacturing firm (SIT-Siemens).

[6] An early participant in the research asserted that the project was conducted almost in secrecy for fear of reaction from the German licensor. Interview with a Sit-Siemens engineer - January 1992 (Trenti 1992)

[7] Interview with the General Director of CSELT, working at that time at the SIP R&D Department (Torino, 4/7/97)

equipment, gave Sit-Siemens a great advantage. The decision to directly "jump" into the time division trajectory was taken on the basis of its initial success in France (Llerena, Matt and Trenti 2000a,b). The project was extremely ambitious for a firm with no autonomous technological capability and no experience in the new electronic paradigm. Sit-Siemens specialism was in switching systems and not transmission, which introduced electronics into telecommunication equipment. The new digital system presented a specific architecture formed by a central transit switch and linked with peripheral small switches. The full design was never realized. Instead, one small prototype of the peripheral switch was installed in Milan in 1972. The service provider was involved only in the installation phase. The prototype did not become operative until two years later. During the 1970s, other prototypes were installed (in Rome, Pordenone, Florence and Messina in 1975), but were not activated until the end of the decade.

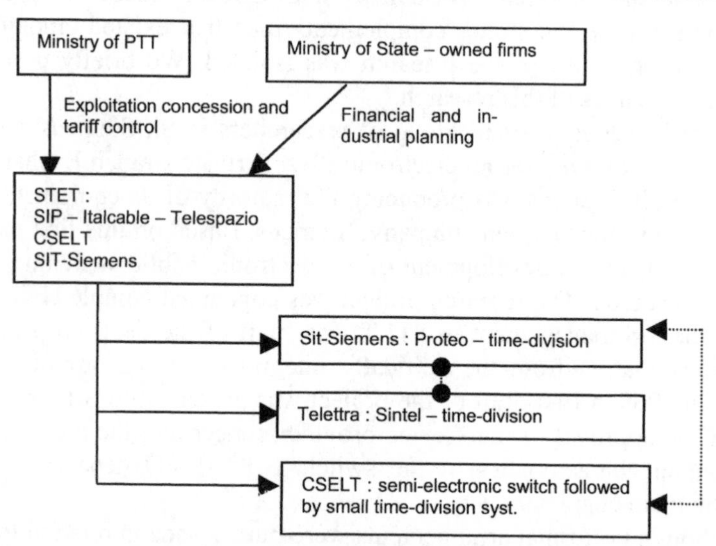

Fig. 5.2. The Organisational Design of the DSS policy in Italy

From the early 1960s, the second Italian supplier of telecommunications equipment, Telettra, historically specialized in transmission technology, had been conducting research on electronic switching (Sintel project). Telettra was convinced about the inevitability of the microelectronics paradigm infiltrating the whole range of telecommunications equipment. Therefore, Telettra researchers "felt the hazard of being excluded from the world of future communications" (Bellman 1976). Unlike Sit-Siemens, Telettra had already accumulated competencies in electronics thanks to their historical specialization in transmission technology. However, Telettra had no practical experience in electromechanical switching and did not enjoy the historical market share in the switching sector of Sit-Siemens. During the first decade of research, efforts were focused on acquisition of the knowledge needed to solve switching problems, and on training up researchers in the new paradigm, working on small time-division switches and on hardware and software equipment[8]. In the early 1970s, a second phase of research began aimed at the production of marketable switching products (Bellman 1976). The initial Sintel project on development of a complete time-division switch was considered to be too ambitious. Research focused on two main products: the DST1, a space-division switch with an electronic cross-point network and a distributed control; and the DTN1, a transit switch with a time-division switching network and stored-program control, an outcome of earlier research on the Sintel system. The DST1 was first installed in the Italian telephone network in 1975 while a bid to provide the international switch in Verona was awarded to the DTN1. The DTN1 was due to be installed in 1977, but installation was postponed and the switch was not finally installed until 1979[9]. The Telettra digital switching equipment was later modified (AFDT1) to accommodate the voice and data transmission network of SIP.

In the meantime, CSELT was conducting research on electronic switches. The aim was primarily to acquire a competence in the new paradigm and to respond to the needs of the service provider. Various products were developed. In 1967 a small semi-electronic switch was successfully put into service. During the 1960s, answering a specific need of SIP concerning the documentation of traffic, a prototype for a small time-division switch (TECA) was developed. Then, in 1971, the prototype of a small time-division switch and stored-program control with advanced feature (GS - Gruppi Speciali) was installed in Mestre (Venice) just one year after the successful Platon product in France. Further improvements to the GS switch were developed and installed during the 1970s (CSELT 1994).

This brief description of the research carried out in the 1960s and 1970s highlights the lack of coordination among the three actors involved. Given the shortage of high skilled personnel in the Italian National System of Innovation (NSI), the novelty of the technology, and the complementarities between the three actors, this lack of coordination is particularly signifi-

---

[8] One of the first versions of the system (Sintel III) was presented in an international symposium in Boston in 1972. The Sintel III featured two main characteristics (a time-division switching and a stored programme control) in order to be economically attractive and technologically valid (Bellman 1972).

[9] Interview with the Vice-President of Italtel, at that time R&D Manager at Teletra (Milano, 25/6/97).

cant. During the 1970s (in 1972, 1976 and 1978) Telettra and Sit-Siemens tried, without success, to coordinate their R&D activities and the development of the digital systems[10]. Moreover, even inside the same group in STET, coordination between Sit-Siemens and CSELT was poor. CSELT was only involved in the improved version of a Proteo switch (CT-2) installed in 1982 for which its Turin laboratory provided an advanced component (Chapuis and Joel 1990). Taking the French case as a benchmark, where the R&D laboratory of the service provider was given the task of coordinating national research, it is suggested that CSELT had the technical expertise to become the catalyser of the Italian efforts, as demonstrated by its technical success in the digital switching. However, CSELT was never given a clear commitment to develop a national technology and transfer it to national firms (lack of political decision).

5.3 Coordination and the Management of Diversity

Based on these cases it is interesting to see how a given coordination mode, or a coordination failure, can influence the mechanisms of selection of technologies. This section will analyze in more detail some dynamic properties of coordination modes. In the French case, vertical coordination induced an internal exploitation of controlled development of the technological alternatives (diversity). In the Italian case, lack of coordination produced a tendency to postpone final choice, leaving the diversity of solutions unexploited. Finally, policy makers chose to exploit external diversity, by employing a process of national coordination and cooperation with foreign actors to achieve catch-up. As a result there was delayed diffusion of the new technology in Italy, i.e. a relative failure of the policies and policy organizations implemented in this country, compared to France.

We first look at the contradictions faced by policy makers in their decisions about the appropriate technological option in a pre-paradigmatic phase. We then describe how French and Italian policy makers managed the technological diversity. Finally, we assess the relative successes of both policies, based on analytical elements and some simple statistics.

5.3.1 Analytical Elements

Each policy faces paradoxes, especially in the early phases of the new paradigm. David (1987) has highlighted three difficulties encountered by

[10] In 1976, Telettra passed into the control of FIAT.

policy makers in technology selection: (1) "*the narrow policy window paradox*" underlines the limited time period in which public intervention is possible; (2) "*the blind giant quandary*" shows that the possibility of intervention is necessarily localized at the beginning of the development process - the period in which the policy maker has little information concerning the relative efficiency of competing emerging technologies; (3) "*the angry orphan*" underlines the problems related to early adopters that are finally eliminated by the policy maker choices.

In other words, once the policy maker selects a technology, there is greater opportunity for the characteristics of this technology to be demonstrated and greater chance of its being adopted. If an inferior technology is chosen, the system is very likely to become locked into this technology. The probability of this phenomenon is amplified by the existence of increasing returns. To avoid this, the policy maker should subsidize different options until there is sufficient information available for a decision to be made as to which is the best. This, however, is not a fool-proof solution since technology improves as the result of diffusion, and trajectories are difficult to assess *ex-ante* Thus, the technology chosen might ultimately be the most inferior. For practical and financial reasons, the policy maker will have to put an end to experimentation at some point and back the technology that is believed to be the best at the time.

According to Cowan (1991), these three paradoxical situations evolve in different ways depending on whether the focus is on the degree of increasing returns, which shortens the competition period, or on the level of uncertainty, which extends this period. Increasing returns will tend to limit the window of opportunity and increase the cost of interventions to keep it open. But decreasing the period preceding lock-in will limit the number of orphans. A certain level of uncertainty will encourage policy maker to extend the experimentation period, which will increase the number of orphans. It will also heighten the "blind giants quandary", because the characteristics of the technologies and their future performance are ignored.

To manage these problems it is necessary "to build organizational systems, involving a coordinated set of decentralized experimental projects, mechanisms and procedures for exchanging and distributing information produced in the course of these projects, and a centralized procedure of assessment to decide the timing for switch (to the standardization phase)" (Foray and Llerena 1996, p.171). Our aim is to illustrate via the case studies how such coordination could influence the selection mechanisms and deal with the paradoxes.

5.3.2 The French Case: a Successful Technology, but a National Orphan

During the second period of development (1975-1983), the French situa-tion was characterized by important political changes. After the election o V.G. d'Estaing to the Presidency of the Republic, the role of CNET ir French industrial policy was called into question. Criticisms of CNET con-cerned its near monopoly of the financial means and decision function which reduced the flexibility of the telecommunication industry. More-over, the technological leadership of CNET was considered to be a majo factor in explaining the high dependence and lack of entrepreneurshi among French manufacturers (Griset 1995). Thus, in 1975 a change wa made in the "leadership" of the telecommunication sector from CNET t DGT (i.e. the French Ministry in charge of telecommunication). The ain was to modernize the network and to obtain lower equipment prices, intro ducing more competition and control in a market dominated b SOCOTEL, considered by DGT to be a cartel. This change in policy ha important consequences for digital switch development.

In 1975, DGT decided to publish an international tender to equip the Pa risian network. DGT, after a long and difficult decision process, decide not to equip the entire Paris network with time-division technology. Th decision was made to share the market between two space-division sys tems (the Metaconta of ITT and the AXE of Ericsson) and one time divi sion system: the E10. At the same time, Thomson created a "secon French pole" for switching equipment (space-division) with the acquisitio of LMT (ITT) and SFTE (Ericsson). Space-division technology was give priority in the planned modernization of the French network, while time division technology was relegated to the ancillary role of equipping the ru ral and low-density parts of the network.

Three years later, in 1978, the choice of space-division technology wa itself called into question. Use of time-division technology was seen as th only way to achieve the objective of modernizing the French network, an to sustain the planned evolution towards telematics. Thomson, which wa supplying the space-division system, was no longer able to sustain compe tition with CGE (later Cit-Alcatel), which had already developed the tech nology with the fundamental help of CNET. In 1983 (after the election o François Mitterand, a socialist president, in 1981), the telecommunicatio department of Thomson was sold to CGE, which had been recently nation alized.

In other words, the announced strategy of the 1960s to push the foreig subsidiaries out of the French market, was suddenly halted by the 1975 de cision to opt for space-division technology. There are two possible inter pretations of this change in policy. Each differently assesses the role of government technology procurement in the development of a national technology.

On the one hand, the choice of space-division technology can be seen as being completely incoherent. According to Griset (1995), "this strategy was a major failure and delayed the international development of the French industry". The decision to give Thomson the leadership for the space-division technology represented the starting of a "French-French competition", producing losses in terms of years of development and financially to several billions of francs. The positive climate created by CNET, and the success of the technology transfer strategy, was disrupted by the intervention of government procurement, aimed mainly at creating a competitive environment[11].

On the other hand, the choice of the space-division technology can be seen as facilitating a gradual modernization of the network. From this perspective, the time-division technology was not sufficiently well developed to be used to equip bigger towns (Libois 1983). Moreover, it was costly and risky. The "dilemma" between protecting the "national champion" and gaining from competition with other suppliers while waiting for the mature phase of the national technology, was resolved with the creation of a second "national champion" for space-division technology. In this sense, we could say that government procurement was used to enhance variety, allowing competition between two different technological trajectories at an early phase of development (Cohendet and Llerena 1997). The three year delay in the full adoption of the E10 in the French network, coupled with its installation in rural networks, might have helped CNET, which was still responsible for technological improvement[12], to gradually improve the performance and the reliability of the system. In this interpretation, the decision to sustain a "bridge technology" (the space-division trajectory) was not as totally incoherent as in the previous interpretation. The knowledge about time-division technology, totally new to the market, was insufficient to allow DGT to follow this trajectory from the beginning. At that time, no other country had attempted mass introduction of time-division switching in the network. Even if, inside CNET, there was great optimism about the

[11] According to an observer at the time (*Le Monde,* July 30, 1976) quoted by Griset (1993), government intervention should have been suspended and reconsidered, because "competition is a good thing if it does not turn into anarchy".

[12] For the whole of 1970s, Cit-Alcatel was responsible only for the production and the marketing of the new system, while technological improvements were still conducted under the umbrella of CNET.

development of microelectronics and semiconductors, DGT had no guarantees that the E10 would have the capacity to handle high traffic nodes.

However, with the support for time-division technology in 1978, CNET's plans were finally realized. Government procurements started to diffuse the new technology, which was ready to be sold and installed. According to Libois (1983), after 1978, the situation became clearer and the superiority of the solution provided by CNET was proved by the prodigious development of semi-conductors and by the introduction of microprocessors. The French Director General of Telecommunications (DGT) confirmed, at a conference held in Paris in 1979, that "the development of time-division switching has become a reality demonstrated by the already taken or imminent decisions in most of countries" (our translation) (Libois 1983). In 1980, time-division switches accounted for 70% of the procurement of DGT. In 1982, the orders for E10 amounted to 8 million lines, of which 2.5 million had already been installed.

The role of the national market was fundamental for Cit-Alcatel to enhance performance of the system. New versions of the E10 were developed in 1980 and 1985. The last version (Alcatel E10) was at the basis of the first ISDN trial in 1987. The success of the E10 in the national market had given Cit-Alcatel an important competitive advantage in the field of electronic switching. The process of digitalization of the French network was a key way to prove to foreign clients the reliability of the system. At the end of the 1980s, the E10 was in use in 40 international markets.

5.3.3 The Italian Case: Late Coordination and Delayed Choice of Technology

In Italy during the 1970s, the two service providers, SIP and ASST, continued to buy and install electromechanical switches. Only ASST chose to buy semi-electronic switches while SIP decided to skip the semi-electronics phase and wait for fully electronic switches. SIP's decision was based on an evaluation that showed that electronic switches were better suited to equip bigger transit nodes than the semi-electronic switches in the ASST network. Moreover, the move to an electronic switching system was seen by SIP as a way of reducing the number of suppliers. Given the rigidity in the supply of electromechanical switches, it was very difficult for SIP to change the historical shares in the switching market[13]. The Italian switching system was initially seen as a way to simplify the market and

[13] Interview with the General Director of CSELT, working at that time at the SIP R&D Department (Torino, 4/7/97)

sustain the vertical integrated "national champion". The development process was long and difficult, highlighting the problems related to the Proteo system. The presence of a second Italian supplier, Telettra, was not considered as a possible alternative (only ASST attempted to install a Telettra switch, which experienced a two year delay).

The result was that SIP was "waiting for the Proteo" (Morganti 1980). The Telettra system was chosen only to equip the national voice and data-transmission network, pushing Telettra to develop a specific switch for this purpose. The first trial for the voice and data-transmission network (Rete Fonia Dati - RFD) started in 1976. At the end of the 1970s, the network was ready for service. However, the RFD was not activated until 1984, because of prolonged battles between the Ministry, SIP and ASST over tariffs and charges.

At the end of the 1970s, and after the late and the partial failure of the Proteo system, SIP was aware that achievement of a "unique system" would be difficult and was still a long way off. For this reason, SIP decided to establish, enshrined in specific documents, the characteristics of the switching systems to be purchased in the future. These specifications were laid down as a way to allow the installation of different systems in the Italian network[14].

The role of policy in this period must be interpreted with caution. Two particular issues had been the object of criticism and debate: to pass directly to the fully electronic switch, or to wait for the Proteo.

The suggestion to leapfrog the semi-electronic phase was criticized for the reason that trialling the semi-electronic systems could have enhanced the capacity of SIP in handling the new microelectronics paradigm (Pontarollo 1989). However, as in other countries (for example, Canada), technological leapfrogging was the path chosen.

The "waiting for Proteo" strategy might be considered as being "providential" for the vertical integrated manufacturer (Cozzi and Zanfei 1996). The Proteo I experienced many technical problems during the 1970s, and it was thus considered too risky for SIP to install. This delay in modernization of the network gave Sit-Siemens the chance to develop a new product (see below). However, the decision to push Telettra to develop a specialized switch for the voice and data-transmission network, together with the difficulties encountered by Telettra in installing its digital switches, show that government procurement was not used in this period as an instrument to coordinate the already scarce national technological resources.

[14] Interview with the General Director of CSELT, working at that time in the SIP R&D Department (Torino, 4/7/97)

At the end of the 1970s, while Telettra was successfully equipping the voice and data-transmission network, Proteo finally entered the industrial phase, notwithstanding certain technical difficulties. A group of engineers from Sit-Siemens was sent to Dallas to cooperate with a US company, the Advanced Business Communication (ABC), in order to develop a new generation switch. The decision to collaborate with a foreign partner was motivated by the difficulty of hiring skilled personnel in Italy, especially for software (Chapuis and Joel 1990). Progress in the next two years was rapid: the first prototype of the new system (Proteo II) was installed in Milan in 1981.

One year later, an agreement was signed between Telettra, GTE and Italtel to build a complete switching system, finally pooling national resources. The distribution of roles inside the system saw Telettra supplying a peripheral switch, GTE a transit switch, while Italtel (the new name for Sit-Siemens since 1981) supplied the new switches, which became known as UT10/3. In the same period, responsibility for the switching system passed from Telettra's R&D department to Italtel. Industrial production began in 1983 and production of the old system was terminated in 1984. The UT10/3 evolved very rapidly in this period from an isolated switch to a family based on a modular architecture (Linea UT). The initial architecture of UT10/3, proposed by ABC, was completely reviewed to reach a truly modular solution that allowed low cost and an almost flat curve for the cost per subscriber (Bellman 1987).

The Proteo/Linea UT system development, which began as an isolated project, was strongly marked in this period, by one major organizational device: the search for external technological competencies, both at national (the Telettra R&D) and at international levels (ABC). After the difficulties encountered during the 1970s, the UT was developed in a relatively short time. The system was ready to be sold in 1984, but full development to satisfy the demands of the Italian market was not achieved until 1987 when "Piano Europa" was finally approved[15]. There was thus a three-year delay in the launch of the national procurement plan . This delay could be interpreted as an inability to sustain "the national government procurement" in line with the new system, the Linea UT, which was by far "more reliable and better performing" than the Proteo (Cozzi and Zanfei 1996).

The digital switching system evolved rapidly. In 1987, a new and bigger switch (UT100), the upper end of the modular architecture, came on line. The industrial phase of UT100 began in 1989, the year of the Joint Venture between Italtel and AT&T, introducing new features and improvements to

[15] Despite SIP's low level of investment and procurement, in 1986, 335 switches were delivered and 294 were already in service.

the Linea UT. At the same time, the investment plan for the modernization of the network was finally approved. Starting in 1988, the level of investments steadily grew. Around 30% of investment was devoted to digital switches. The number of digital lines installed in the Italian network increased dramatically from 1,5 million in 1986 to 8,5 in 1990. In relative terms, digital lines as a percentage of total lines increased from 11,8% in 1986, to 33% in 1990. The diffusion of the Linea UT followed a similar exponential path.

5.4 Assessment of the Policies and Conclusion

In the French case, during the first phase (1958–1974), the existence of CNET, a strong and competent centralized research centre, promoted vertical coordination. CNET participated in the development of two different technological options: space-division technology, which involved mainly a pool of foreign subsidiaries; and time-division, a trajectory that involved a new company being established by the French Cit-Alcatel. The presence of CNET in both experimental projects favoured cross-fertilization, and the centralization of information, and allowed the policy maker to take appropriate decisions. The time-division technology was well supported and in 1972, it (E10 from CIT-Alcatel) was ready to be commercially produced and sold. In this first phase, appropriate coordination helped to decrease the "blind giant quandary" and to increase the probability of the most efficient technology being selected. During the second phase (1975–1983), political changes induced some strategic modifications. CNET lost its leadership in the telecommunication sector. The government wanted to introduce more competition in the French telecommunication sector and, in 1975, launched an international tender to equip the Paris network. The market was finally shared between space-division (ITT and Ericsson) and time-division (E10) switches. Space division was given priority and Thomson absorbed the foreign subsidiaries to form a second French pole. In 1978, time-division switches were considered to be the only solution to modernize the French network. Thomson could not sustain the competition and its telecommunication department was sold to Alcatel in 1983. The second period could be described as follows. The high level of uncertainty, especially in relation to the ability of time-division technology to equip large towns, induced the government to extend the experimentation period and to sustain the alternative technology, with the risk of generating an orphan. Finally, the superiority of the time-division switches and the French-French competition pushed Thomson out of the market (national orphan).

The political decision to extend the experimentation period brought increased costs and longer development time. These disadvantages could perhaps have been reduced if CNET had been given similar leadership and if a truly competitive situation in relation to foreign firms had emerged. In other words, the organizational system was *a priori* coherent to deal with some of the paradoxes during the first phase, but the choices made during the second phase entailed some shortages. Despite these limitations, France was the first country to introduce time-division switches in a substantial way into its network.

In the Italian case, during the first phase (1960–1980) horizontal coordination between competing suppliers would have been appropriate due to the absence of a skilled central research centre. Three competing projects (mainly time-division systems and a semi-electronic one) emerged. The lack of real coordination between the actors seems particularly disadvantageous given the shortage in high skilled personnel, and the complementarities of the actors. Moreover, the procurement policy did not work as an instrument to coordinate existing national experience. In other words, there was no coordination between the decentralized projects and no centralized procedure favouring public decision making. During the 1970s, prototypes were installed on local basis. The experimentation period was lengthy, the "blind giant quandary" was not reduced, and the risk of a national orphan was high. At the beginning of the second phase, Telettra successfully equipped a special switch for the voice and data-transmission network and Proteo (Sit-Siemens) entered the industrial phase. The development of the technology and its late success can be explained by two crucial events: the collaboration of Sit-Siemens with an American company (ABC) to develop a new generation of DSS, and the pooling of national resources (Telettra, GTE and Sit-Siemens signed an agreement to build a complete DSS). This new organization of competencies induced the emergence of a switching system family based on a modular architecture: Linea UT. This second phase clearly shows the positive impact of the horizontal coordination: it decreased the "blind giant quandary" by allowing firms to develop a new performing switching system, without a national orphan. But it did not compensate for the coordination failures of the first period and the negative impacts on the development of the DSS: very high costs of development and a tremendous loss of time. This organizational failure induced a less successful development than the French case.

One way to measure the relative success of each policy is to compare the digitalization of the telecommunication networks. Table 5.2 presents data concerning the digitalization of networks in the main industrialized countries from 1988 to 1993. France, thanks to the massive installation of the national switching system that began at the end of the 1970s, appears as the leading country: it had the highest percentage of digital lines over the period considered. The Italian data confirm the delayed diffusion, and the importance of the most recent investments and the efforts of SIP to modernize the network.

Table 5.2. Digitalization of the network 1988-1993 (% of digital lines on total lines)

	1988	1989	1990	1991	1992	1993
France	60.5	65.8	69.1	73.5	78.1	83.0
Italy	20.1	24.7	32.6	39.9	47.4	54.5
USA	35.4	41.6	48.1	54.4	60.9	68.1
Canada	40.0	45.0	50.0	55.0	62.0	68.0
Japan	17.4	24.9	31.9	38.9	46.4	53.3
Gemany	5.0	7.3	10.5	15.0	19.4	23.7
UK	23.5	31.3	39.7	41.1	53.6	59.4

Source: Zanfei (1990) from NBI and SIP Development Plan

This chapter has focused on the technological evolution and, in particular, on the introduction, of DSS in France and Italy. We elaborated on the institutional framework of the policies involved in terms of information and coordination structures, and also in terms of lock-in and diversity management. We proposed for both dimensions an analytical framework, which allowed us to analyze the elements of both histories and to assess the relative successes of the two experiences, which differed in nature and timing. Further research should be devoted to examining whether these two rather specific histories had some impact on the introduction of subsequent generations of technologies and products, such as Minitel and/or Internet, in these two countries.

5.5 References

Aoki M (1986) Horizontal versus vertical information structures of the firm. American Economic Review 76: 971-983.

Bellmann A (1972) An integrated time division PCM exchange. Paper presented at the International Switching Symposium, Boston, May.

Bellmann A (1976) Syntel system: evolution, development, feature. ISS, Kyoto.

Bellmann A (1987) The linea UT concept: architecture, technology, evolution. Paper presented at the Italtel Telecommunications for Today and Tomorrow Conference. Geneva, October.

Bragho' A (1988) La concorrenza globale nel settore della commutazione pubblica Economia e Politica IndUStriale n. 58

Chapuis R, Joel AE (1990) Electronics, computers and telephone switching: a book of technological history. In: North Holland Studies in telecommunication, 100 years of telephone switching, Volume 2 : 1960-1985. Amsterdam, New York, Oxford, North Holland Publishing Company.

Cohendet P, Llerena P (1997) Learning, technical change and public policy: how to create and exploit diversity. In: C Edquist (ed.) Systems of innovation – technologies, institutions and organisations. Pinter Publishers, London and New York, pp. 223-241.

Cowan R (1991) Rendements croissants d'adoption et politique technologique. In : J De Bandt, D Foray (eds.) L'évaluation économique de la recherche et du changement technique, Edition CNRS, Paris, pp. 381-398.

Cozzi G, Zanfei A (1996) La commutazione elettronica in Italia: il mancato decollo di un'innovazione di frontiera. In: C Bussolati, F Malerba, S Torrisi (eds.) L'evoluzionz delle industrie ad altra tecnologia in Italia : Entrata tempestiva, declino ed opportunità di recupero. Il Mulino, Bologna.

David PA (1987) Some new standards for the economics of standardization in the information age, in P Dasgupta, P Stoneman (eds.) Economic policy and technological performance, CUP, Cambridge, pp. 207-240.

CSELT (1994) CSELT-Trent'anni. L Bentovoglia (ed.). CSELT, Torino.

Ergas H (1987) The importance of technology policy. In: P Dasgupta, P Stoneman (eds.), Economic policy and technological performance. CUP, Cambridge, pp. 50-96.

Foray D, Llerena P (1996) Information structure and coordination in technology policy: a theoretical model and two case studies. Journal of Evolutionary Economics 6(2): 157-174.

Griset P (1993) The Centre National d'Etude des Télécommunications and the competitiveness of the French telephone industry, 1945-1980. In: W Aspray (ed.). Technological competitiveness: contemporary and historical perspectives on the electrical, electronics, and computer industries. IEEE Press, New York, pp. 176-187

Griset P (1995) Innover comment? Innover pourquoi? Cinquante ans de recherche au Centre National d'Etudes des Télécommunications. Les cahiers – Télécommunication, Histoire et Société, n°2. France Télécom, pp. 18-35

Quelin B (1992) Trajectoires technologiques et diffusion de l'innovation : l'exemple des équipements de télécommunication. Revue d'économie industrielle 62(4): 83-105.

Libois LJ (1983) Genèse et croissance des télécommunications. Masson, Paris.

Llerena P, Matt M, Trenti S (2000a) Public policy procurement: the case of digital switching systems in France. In: C Edquist, L Hommen, L Tsipouri (eds.) Public technology procurement and innovation. Kluwer Academic Publishers, Dordrecht, pp. 197-216.

Llerena P, Matt M, Trenti S (2000b) Public policy procurement: the case of digital switching systems in Italy. In: C Edquist, L Hommen, L Tsipouri (eds.) Public technology procurement and innovation. Kluwer Academic Publishers, Dordrecht, pp. 217-239.

Morganti F (1980) Le prospettive nel settore elettronico. In: Ministero PPSS. Rapporto sulle partecipazioni Statali (Annex 6). Franco Angeli, Milano.

Pontarollo E (1989) Domanda pubblica e politica industriale: Fs, Sip, Enel Venezia, Marsilio.

Trenti S (1992) Acquisti pubblici e innovazione sul mercato internazionale delle telecomunicazioni e del termoelettromeccanico. Il caso italiano. Tesi di Laurea, Università Bocconi, Milano, March.

Zanfei A (1990) Complessita e crescit esterna nell'industria delle telecomunicazioni. Franco Angeli, Milano.

6 A Study of Military Innovation Diffusion Based on Two Case Studies[1]

Arman Avadikyan, Patrick Cohendet and Olivier Dupouët

BETA, Srasbourg, E-mail : avady@cournot.u-strasbg.fr
BETA, Strasbourg and HEC, Montréal : E-mail : cohendet@cournot.u-strasbg.fr
BETA, Strasbourg, E-mail : dupouet@cournot.u-strasbg.fr

6.1 Introduction

The diffusion of military innovations, and how this compares with the diffusion of civil sector innovations, has long been the object of debate in the literature. At the risk of oversimplifying, most of the contributions are in agreement that there are two radically different periods involved:

- The first phase, between the end of WW II and the mid-1980s is characterized by the spin-off paradigm in the sense of Alic et al. (1992). The thesis for this period is that military expenditure, the importance and the growth of which was naturally justified by the Cold War context, created significant technological spin-offs into the civil sector. As Sachwald (1999) underlines, this view can be justified from a combination of military and industrial perspectives. The examples given by Sachwald are familiar: the development of semiconductors, of telecommunication satellites, of civil launchers, of aircraft, and of composite materials and other technologies owe a lot to the research and development (R&D) efforts expended within military "big programmes".
- The second phase, which emerged progressively from the mid-1980s, can be provocatively interpreted as resting on a quite different mechanism: military expenditure decreased in response to the ending of the Cold War; meanwhile the innovation dynamics shifted towards the civil sector, thereby reversing the direction of technological flow, from the civil to the military sectors (some authors describe this transformation as the *spin-in* paradigm).

---

[1] This work was conducted for a study supported by the *Observatoire Economique de la Défense* (OED). The authors are responsible for any errors or omissions.

This hypothesis is in many respects excessive. Accepting it without qualification runs the risk of erroneous interpretations in terms of industrial policy. For example, within the first phase, it is important to stress that military technologies *per se* were not always at the origin of spin-offs. Such spin-offs were based essentially on the enabling technologies developed for military use (e.g. information and material technologies) and on some of the generic industrial organizational and complex management methods practised within military projects. However, it is clear that within the second phase military projects continued to influence the major technological trajectories (as in the cases of the Internet and Global Positioning Systems (GPS)). The recent successes of high-speed trains and of the Airbus show also that the concept of big technological programmes continues to hold its relevance.

Nevertheless, this idea constitutes a major feature of the relationships between civil and military technological diffusion processes by highlighting their contextual dimension. At a given point in time, these relationships depend on the maturity of generic technologies, of the innovation system in place, and on the perception, according to which military and civil efforts are considered to be complements or substitutes.

The above remarks lead to a better understanding of the nature of the diffusion process for military innovations and allow its main characteristics to be highlighted. The analysis in this chapter is based on a series of interviews conducted with people responsible for the management and coordination of military radar and missile projects. Our work also benefited from the experience accumulated at BETA from studies conducted in the space sector. However, this chapter mainly focuses on two case studies and should be considered as exploratory work. Our contribution aims simply to highlight some traits, which we consider to be important in relation to the process of innovation diffusion from the military to the civil sector.

The chapter is organized in three parts. The first part analyses the contextual characteristics to allow greater understanding of the specificity of military diffusion processes. The second part summarizes the results of our field studies in two firms involved in complex military projects (Thomson-CSF Airsys[2] and Matra BAe Dynamics[3]), which help to illustrate the speci-

[2] Airsys pools the air defence, ground-to-air missile and air control activities of Thomson-CSF. In 2000, Thomson-CSF and Airsys were renamed respectively Thales Group and Thales Air Defence, following the acquisition of Racal Electronics Ltd.

[3] Matra BAe Dynamics (MBD) was created in 1996 by the merger between Matra Defence (Lagardère Group) and the missile subsidiary of BAe, BAe Dynamics. In 2001, MBD was extended to Alenia Marconi Systems (Finamecannica and

ficity of the innovation process based on military projects. Finally, we give some brief details about the firms studied concerning the factors that might be relevant for – and have an influence on – the innovation diffusion process.

6.2 Major Characteristics of Military Innovation Diffusion : Context Matters

The theoretical developments concerning the diffusion of innovations emphasize a number of determinants of the quality and the intensity of the diffusion process. These include the nature of the technologies (particularly their more or less generic character), the nature of industrial organizations that develop and experience them, and the nature of user or adoption networks. In this section, we try to specify the diffusion characteristics of innovations stemming from military projects in relation to these criteria.

6.2.1 Nature of Technologies

Military projects aim at developing products and systems responding to specific military needs and require generally high technological performance. They are often defined as *complex product systems* (CoPS) (Hobday 1997). The notion of complexity refers to the following features: 1) the variety of pieces of technological knowledge that must be combined; 2) the competencies that must be mobilized at the technological frontier[4] in order

Bae Systems) and Aerospatiale Matra Missiles (EADS) and was renamed MBDA, in a bid to establish an integrated missile company at the European level.

[4] Military innovations can be conceptualized in terms of core, peripheral, and linking technologies (Prencipe 1997). Behind these technologies are the competencies needed for their production. Although the technologies and the competencies necessary to develop them are not the same (Brusoni et al. 2000; Chiesa and Manzini 1998) the links between them are close. Thus, the competencies that support technologies can be categorized according to the above distinction. Prahalad and Hamel (1990) showed that the competitive advantage of a firm depends principally on the identification, maintenance, and improvement of its core competencies. What is important is to identify which competencies the firm should definitely maintain in-house. Brusoni et al. (2000) analyse the firm by linking its organizational dimension to its competencies based on two criteria: responsiveness and distinctiveness. An organization is said to be responsive if it closely controls the different technological components and has a strong integra-

to develop each of the components and products; 3) the necessary competencies for integrating these diverse technologies; and 4) the high cost constraints that generally inhibit the possibility of developing military products and systems through standard industrial methods. CoPS are generally supplied by a single entity (a firm, a production unit, or a temporary organization, constituted around an identified target and involving several firms). CoPS are typically purchased by a single user, usually through a formal contract defining a precise project.

Examining the technological features of CoPS helps *a priori* to situate the outline of the debate on the diffusion of innovations stemming from military projects.

First, most enabling technologies have a dual character, in that they can find applications both in military and civil products. What differentiates military technologies from their civil counterparts is their application conditions (in extreme environments) and their functionality (speed, reliability, discreteness, etc.). From this perspective, it is easy to see why spin-in (improvement and adaptation by the military, of technologies developed in the civil sector) is a natural diffusion path from the civil to the military sectors. This approach favours the generation of incremental innovations with respect to the knowledge available at a particular time. By relying on civil technologies military projects can diversify their innovation sources and benefit from scale and speed economies. The spin-in process can also create new innovation opportunities for civil products, thus favouring a virtuous cycle of innovation stemming from close interaction between the two sectors.

Second, in the event that the technologies necessary for the realization of military projects are not available, the capacity for military resources to be mobilized to develop these technologies and stimulate research capabilities, can produce radical innovations. This is what happened, for instance, at the beginning of the computer era, and more recently with the GPS conception phase, where the capability of military programmes to accumulate cognitive resources and research competencies has initiated major technological breakthroughs. This capacity of military projects to generate technological ruptures depends, of course, on the level of military expenditure to stimulate basic research. The diffusion potential of such technologies beyond the military domain is linked, on the one hand to the more or less generic nature of these technologies, and on the other to their degree of maturity. The more mature a technology (implying a high speci-

tion capability. Distinctiveness refers to the variety of competencies coordinated by the firm, and the relative independence of the organizational units developing them with respect to the control centre.

ficity with regard to military needs), the more difficult will be its diffusion to other than military applications.

Third, the ability within military projects to integrate diverse technologies, rooted in very different knowledge bases, in order to develop new products or systems, is one of the main sources of innovation diffusion. The diffusion process may take two principal forms. Having been tested and developed within military projects, new organizational forms (project management, risk management methods, etc.) may be disseminated and transferred to other domains. Also, products and systems realized within military projects may be adapted and redesigned for civil applications (e.g. aircraft, launchers, radar). Naturally, this form of diffusion will only occur in the case of civil applications that are technologically close to military needs (e.g. space and aeronautics technologies).

6.2.2 The Nature of the Organization

The diffusion of military technologies will only be effective if the organizational forms of the industries conducting the military projects promote the circulation of knowledge flows. Thus, the internal organization of, and the nature of the relationships between the firms participating in the projects (system integrators, sub-contractors, partners, research networks, etc.), are critical.

The specific nature of military projects leads to organizational forms, internal as well as external, that are not *a priori* suited to encouraging or speeding up the diffusion process. Internally, the performance of military projects generally depends on big project management firms, organized as multidivisional structures, or business units. As shown by Marengo (1993), such structures favour knowledge specialization, but have a tendency to slow down the circulation of knowledge and the cross-fertilization between individual departments. In addition, the projects being extremely complex, firms must be able to make fast modifications to their behaviour, to reduce uncertainties, to communicate on a regular basis, and finally to generate positive externalities (Hobday 1997). This implies that there must be a high degree of *responsiveness*, i.e. tight control over the different components. Therefore, the network of firms involved in the elaboration of a military product is generally hierarchical and limited to a small number of firms: a network structure that leads to a high integration of competencies such as is rarely observed in the civil sector. This integration carries the risk that external relationships become under-valued, since almost all the necessary competencies exist internally. The knowledge and competencies developed inside the firm cannot circulate outside the firm, which is, in itself, an obstacle to the diffusion process. It is clear, however, that the recent moves to create a more cooperative framework between the European defence industries, which led to an increased number of alliances and mergers/acquisitions, are providing a real opportunity to work within structures that are more open than in the past.

Another obstacle to knowledge diffusion from the military sector is the obvious need for secrecy and the need to control the diffusion of certain strategic technologies. This is inherent in all military projects, and the challenge for the firms managing these projects is to find a compromise between controlling the communication of certain types of knowledge and the desire to communicate as much know-how as possible beyond the frontiers of the military domain. This is achieved by favouring the capabilities that facilitate the differentiation between technologies or systems that are more or less strategic, while at the same time trying to avoid a culture of excessive secrecy, which would hamper knowledge communication, and the adaptation of technologies developed with military effort, for the civil sector.

6.2.3 The Relationships with Users

The importance of interactions with users through «*learning by interacting*» in the development of new products has been underlined by several economists (Lundvall and Johnson 1994; von Hippel 1988). A rich interface with users leads to *feedback* processes that contribute to continuous improvements of products and technologies. What is supplied becomes progressively more and more adapted to user needs. Through this interaction, the user becomes an integral part of the technology creation process.

In most cases military projects concern only one user – the state – so the variety of interactions is necessarily limited. In order to compensate for this lack of variety, and to adjust ongoing programmes in a timely way to emerging problems and opportunities, it is crucial that the relationship between the client (in this case the Délégation Générale à l'Armement (DGA)) and the supplier of military systems is based on rich and intense interactions. The existence of stable and integrated project teams throughout the life cycle of a programme and their effective management (ensuring, for example, that critical information and knowledge are appropriately circulated and integrated into the decision-making process) are all the more important since the technical resources and competencies of the user and the supplier are highly complementary. The user is in a privileged position with operational experience in detecting, defining, testing and validating possible improvements, whereas the supplier is in a privileged position in translating them into technical solutions, integrating them into military systems, and finding potential diffusion paths for the technologies developed.

6.2.4 Context Matters

The arguments developed above are an attempt to characterize the potential for and the constraints to diffusion faced by military projects. Positive factors related to the diffusion process are linked to the capability of military projects to develop radically new technologies by concentrating R&D resources in new fields of knowledge, by developing new organizational competencies, and by producing new complex industrial systems transferable to "neighbouring" civil sectors. Factors hindering the diffusion process are related to the specific organizational structures dedicated to the management of military projects and the lack of variety of interactions with the demand side. These opposing factors can be combined in different ways according to the economic and strategic environments. Bearing in mind the caveats stressed above, it is possible to relate again to the two distinct periods (the spin-off and spin-in paradigms) that are quoted in the literature on the diffusion of innovations developed through military projects.

- The period from the 1960s to the 1980s was marked by the Cold War and a context in which military funds were regularly increasing and significant enough to allow concentration of resources on research into new knowledge fields and the building of new competencies. Military priorities in the conduct of national R&D programmes during this period played a structuring role, not only with respect to applied research, but also with respect to fundamental and curiosity driven research. Furthermore, the technological context lent itself to being stimulated by military projects since this period was characterized by the development of autonomous innovations[5] that were then further recombined. The technological life cycles were long, and the conception time, in particular, was important in the military as well as the civil sectors. In this context military projects have been significant for the creation and diffusion of new technological opportunities in the civil sector.

---

[5] Chesbrough and Teece (1996) identified two types of innovation: autonomous and systemic. Autonomous innovations can be accomplished independently of other innovations. They do not imply a redefinition of the whole system within which they are integrated. Systemic innovations, on the other hand, must be developed jointly with other innovations in order to be viable.

– However, since the 1980s, the context has changed considerably. Technological creation and innovation diffusion mechanisms are increasingly systemic, and each innovative activity depends on the capability to access rapidly the complementary competencies held by other stakeholders (hence the importance of networks and partnerships) in a global environment. The conception and adjustment times of products have become considerably shorter. The civil sector firms have equipped themselves with the competencies necessary to allow them to absorb external innovations. In this context, the factors that favoured the diffusion of knowledge and technologies from military projects became blurred, while with the end of the Cold War, military expenditure began to decrease, which significantly affected the capability to concentrate massive resources around projects with technological rupture potential. The process of diffusion from military projects must therefore be approached differently. As noted above, military projects can still play a significant role in the innovation diffusion process through incremental improvements or spin-ins. But the question is whether there are any other mechanisms that will allow greater leverage of current and future military projects. New approaches to duality management can, in this perspective, be an important determinant of the diffusion and interaction processes between civil and military actors. They need, however, to be considered within a broader perspective than *ex-post* technology transfer schemes and must rely more on organizational and institutional mechanisms creating the appropriate conditions for stimulating interaction, communication, mobility, and exchange of knowledge and information between the civil and military communities on an ongoing basis.

The next section first examines the issue of military innovation diffusion through two specific case studies. Through these two examples we try to show how, in the new industrial context, it is possible to adopt more efficient ways to support the diffusion process of innovations stemming from military projects.

6.3 An Analysis of Diffusion Mechanisms Based on Two Case Studies : the Airsys Radars and the MBD Apache Missile

The case studies were carried out by the BETA in the course of a contract with the Observatoire Economique de la Défense (OED). The firms selected were Airsys, as project manager for radar systems in France, and Matra-BAe Dynamics (MBD) as project manager for the Apache missile.

In both firms, in-depth interviews were conducted with industrial managers in charge of devising and implementing the projects concerned. An analysis based on two case studies cannot pretend to be exhaustive: the aim is rather to underline some significant features that are illustrative of the theoretical developments described above, that should be the subject of subsequent in depth studies.

The information gathered through our interviews allowed us, on the one hand, to situate the positive effects generated by the diffusion process (6.3.1), and, on the other hand, to better explain the diffusion hindering factors (6.3.2) in order to suggest solutions for improving the impact of military projects.

6.3.1 The Positive Economic Effects of the Two Projects

For each of the firms analysed the positive impacts related to the projects considered are linked to precise effects. In the case of the radar systems, the positive effects result from the use of radar systems in civil airports. We can see direct spin-offs from the military to the civil sector that allow important economies in terms of R&D, conception, and adjustment efforts in the civil sector.

In the case of the Apache missile, the principal effect concerns the firm itself and remains in the defence sector (we will therefore avoid the term spin-off in this specific case) since it is probable that had the Apache missile not been developed, MBD would not have engaged in the next generation Scalp G missile.

Both these direct effects[6] (the monetary gains of which are judged as being quite significant compared to the costs) show a particular way of evalu-

[6] BETA has proposed two main criteria for assessing the technological performance of innovation projects (see Chapter 9):
a) The *direct effects* of a R&D programme are those economic effects that are directly linked to the objectives of the project (these objectives are specifically mentioned in the contracts defining the projects). If, for instance, the objective of the project is to develop a new product (a missile), the sales of this product can be considered to be a direct effect.
b) The *indirect effects* are those economic effects that have been realized outside the projects' objectives. They correspond to unintentional effects not anticipated in the contract. The literature has often described such effects as spillovers or spin-offs. These descriptions can be misleading since they narrow the spillover to new products derived from a project. A more in depth analysis shows that these effects can be richer and cover a wider range of phenomena (development of new competencies, organizational learning within the network, commercial

ating the efforts stemming from military projects and the competencies of the contracting firms. In both cases military expenditure allowed the firms to accumulate industrial competencies and to realize them either in civil or military projects.

Nevertheless, despite the importance of these effects, it seems that the two military innovation processes did not generate any other major positive impacts. There have not been varied diffusion forms as might have been expected, if we consider, for instance, equivalent civil projects in the space sector (see in this book Chapter 1 by Bach and Matt). From the set of phenomena that are likely to reflect the existence of positive effects stemming from big projects (e.g. technology diffusion outside the contracting firms, technology transfer, development and dissemination of new industrial methods, diversification processes, constitution of radically new competencies at the origin of new competitive advantages, increased market share, etc.), very few were observed in the projects under examination. We have therefore tried to identify the main factors hindering the diffusion process of innovations stemming from military projects.

6.3.2 Factors Hindering the Innovation Diffusion Process

6.3.2.1 The Reduction of Military Basic Research Expenditure

The direct counterpart to the reduction in the defence basic research budget is the weakening of military projects as a driving force for radical innovations. The effect of this decrease in the budget has been all the stronger since the costs of R&D in all sectors have tended to increase drastically. The contribution of defence expenditure to the development of the technological base of industrialized countries is no longer as great as in the period of the Cold War. Thus there has been a profound change in the diffusion of military innovations.

---

advantages, new markets, etc.). These effects can be classified according to the following typology:

- Technological indirect effects: new products, improvements brought to existing products, processes of diversification, new processes, etc.
- Organizational and methodological effects (quality procedures, complex project management methods).
- Commercial effects resulting from brand image and reputation and leading to increased market share for existing products.
- Effects in terms of competencies. R&D projects often lead to the grouping of a critical mass of highly qualified personnel that disperses after the project to take up other activities in the participating firms.

- The dependence of defence related firms on innovations and technological developments from the civil sector is steadily increasing. This situation increases the risk that, as far as technological innovation is concerned, defence-related firms lose their capabilities to generate radical innovations and increasingly adopt the role of follower. The tendency is towards "upgrading", i.e. the adaptation to military needs of technological bases developed in the civil sector (spin-in), rather than spin-offs from the military to the civil sector. For instance, it is now the case that only the software programs for signal processing and the programming of processors are written by Airsys. Likewise, apart from stealth bomber technologies for the Apache, and image processing technologies for the Scalp, MBD purchases most of the technological components and subsystems it needs from the civil sector.
- The basic research workforce in defence related firms is also declining. However, it is clear that if a major proportion of their basic research is carried out by military research entities, such as ONERA (Office National d'Etudes et de Recherches Aérospatiales), or contracted out to universities, firms will try to maintain a high absorptive capacity for new and innovative ideas coming from the outside. The management of this capacity is key to firms' abilities to integrate generic knowledge bases quickly. As far as research is concerned, the trade-off is never simply between make or buy. The capacity to understand, to accumulate, and to integrate what is performed outside is a critical element for competitiveness. As Brusoni et al. (2000) emphasize, defence related firms, which are generally characterized by hyper-specialized structures, undoubtedly have a strategic advantage, since such specialization facilitates technological monitoring and increases the absorptive capacity of firms. But good absorptive capacity does not automatically confer the capacity to diffuse innovations effectively. The knowledge emission capabilities of firms are also critical, but defence related firms adopt a more cautious position in this regard.
- We can also see an "inversion" in risk taking. When large-scale basic research for the military was financed by public funding, public authorities took the risk of technological development. In the current climate this situation is reversed: industrial actors with the necessary competences are increasingly having to bear the initial risks of technological creation, even if at some later stage public authorities play their part in supporting innovations that turn out to have potential. Nevertheless, on a global scale (taking account of all the individual research budgets), it is not possible to achieve a level of basic research expenditure equivalent to the level that was available from public funding. Industrial firms must

concentrate their efforts on a limited number of specific fields in which they have competencies and resources. The sum of individual R&D efforts, therefore, cannot have the coherence of a global effort guided by the public sector.

In addition to a reduced budget, firms must also be able to cope with the irregularities of military funding. In both of the case study firms, the problems connected to financing of military projects (e.g. delayed or deferred payments) were seen as a major obstacle to innovation. Such irregularities lead firms to plan their research activities according to their own funds. The depth and diversity of basic research is thus curtailed, in depth, since once a product is developed the research field is generally abandoned, and in variety, since firms tend to confine their research activities to specific domains.

For instance, in the case of Airsys, each new radar represented a technological rupture from the previous generation. This phenomenon is ascribed to the irregularity of military funding. Indeed, development of the first project – the TRS 2215 mobile radar – intended for export, was financed through equity and was not finally bought by the state until the final stage of the life cycle. The second project – the TRS 22XX fixed radar – was financed partly through equity and partly by the DGA. In the case of the MASTER category radars, the DGA financed the research on modularity. The situation was similar for MBD: the research involved in the development of the stealth bombers (Apache missile) and image processing (Scalp G) was supported through equity, while the activities related to the guidance system were financed by the DGA.

In short, the cut in military expenditure on basic research had considerable consequences. Some of these were positive: the incentives for military related firms to become more similar to civil firms, to endow themselves with absorption capacities, to take higher technological risks and to develop dual technologies significantly increased. However, if public expenditure on R&D is reduced too drastically, these positive effects will be considerably diluted and could even disappear. For instance, in the case of dual technologies, it is well known that the duality potential is all the more important at the generic or experimental phase of a technology (Cowan and Foray 1995). The more mature a technology, the more civil and military applications diverge, and the more the duality potential diminishes. The development of efficient dual solutions, therefore, requires massive support of basic research. In examining this support, it is helpful to refer to the distinctions proposed by Cowan, Foray and Mohnen (1998) about the different forms of research. These authors distinguish three principal types of research activity: pure basic research (« Bohr » type), finalized basic research (« Pasteur » type) and pure applied research (« Edison » type). They suggest that if public authorities want to promote duality, they must preserve and even increase their support for « Pasteur » type finalized basic research.

6.3.2.2 The Specificity of Military Products

The military requires products with relatively long life cycles (e.g. an average of 20 years for a radar), and places functionality above all other industrial criteria. Thus, in the past, defence requirements have always contributed to distinguishing military production from civil production. However, nowadays the gap between civil and military life cycles has tended to increase. A few decades ago, the life cycle of military products was comparable to that of civil products. With the drastic reductions in the life cycles of civil products (down to a few months for computers), the gap has widened considerably. The innovative capacity of the civil sectors has become more dominant and more systematic. The improvements in civil sector products have been continuous and swift, implying a succession of incremental innovations, whereas the military sector has continued to favour very radical improvements (opening the transition to next generation products). Also, the innovative capacity of the civil sector is all the more significant in that it is punctuated by a competitive environment focused on cost reduction. This new context hampers the diffusion of innovations stemming from military projects for various reasons.

A consequence of the specificities described is that the defence industries must keep within their structures, throughout the life cycle of projects, conception, production, and maintenance capabilities. This constrains the competence management practices of the defence industries, which cannot be as flexible as civil firms. They are sometimes obliged to maintain internally competencies that do not necessarily lead to innovative activities. In the case of Airsys, for instance, although progressive wave tubes (a technology allowing the generation of signals in old radars) are now of little use for the development of new radars, the competencies for developing them have continued to be maintained.

Furthermore, firms in charge of military projects generally are more integrated than civil firms. The number of activities integrated by Airsys and MBD appears to be greater than typical of a civil sector firm. In the case of Airsys, the network participating in the creation of products is practically reduced to the company itself. Although it is considered that Airsys's core competencies reside in system integration, and that it is necessary to acquire some key technologies externally, the company considers that competencies supporting power generation and software creation for signal processing are critical to its activities. This integrated competence management approach is also in accordance with Airsys's strategy: the company aims above all to be a pool of competencies rather than to be constrained by market forces. Nevertheless, the company is more and more being obliged often to take into consideration these constraints, and to modify its approach.

MBD's network for producing missiles is somewhat broader, but also quite limited and very hierarchical. In the case of Apache, the Commissariat à l'Energie Atomique (CEA) and MBD successfully conducted joint research activities on computation codes and stealth materials. Cooperations have also taken place with different sub-contracting firms (in the field of materials, reactors and guidance systems), but always in a hierarchical way. Finally, technologies classified as critical are all developed within the company (e.g. stealth bombers and major components of the guidance system).

One approach that has been adopted by both firms we studied is to introduce modularity[7] allowing the re-use of elementary technological blocks[8]. In order to make cost economies, firms are looking more and more for modularity in the sub-systems of products. Furthermore, they are capitalizing on inter-product technological blocks in order to reduce research

[7] According to Sanchez (1996), modularity refers to a particular product architecture. A modular approach minimizes the interdependence between product components by specifying the interfaces between them. Moreover, modularity is defined at the start of a project in order to offer the product in a variety of forms. A modular architecture creates a flexible platform allowing the product to be designed to accommodate needs expressed. Such an approach has the following benefits. Functionality improvement costs and time are reduced compared to when the architecture must be designed from scratch. Modularity allows also a more reactive response to the market by taking into account demand variety. Finally, economies of scale reduce component or sub-system production costs.

[8] In their innovation approach, defence related firms decompose the system to create elementary "building bricks", which are then developed separately. This decomposition not only has an impact on technology diffusion through competence transfer, but also has an impact in terms of technological artefact transfer, since it leads to a distinction between autonomous and systemic innovations. This latter distinction is also closely related to the notions of *distinctiveness* and *responsiveness*. In fact, in the case of the two firms studied, autonomous innovations favour distinctiveness, while systemic innovations favour responsiveness. This approach is justified by the specificity of military products. The latter are systems integrating complex technological sub-systems. In terms of innovation diffusion, task decomposition can have a positive impact since it allows for the creation of innovative sub-systems and, hence, new sources of potential diffusion.

expenses. In the case of Airsys, the modular approach was explicitly adopted in the case of the MASTERs. Despite the fact that there is still great disparity between products (there are three MASTERs) compared, say, to the automotive industry, the creation of a common platform is considered to be crucial for a given generation of radars. Similarly, MBD has tried to develop technologies that can be used in several products. For instance, the image processing technology in the Scalp G had already been used for the air-air and Apache missiles.

What is even more important is that a modular strategy allows new technological opportunities in CoPS to be integrated in a more reactive way by separating the modular innovation process (increasing in the short term the speed of incremental improvements and thus optimizing the possibilities for a given architecture) from the architectural innovation process in the long term.

However, in terms of technology diffusion, the problems addressed by the modular approach are very similar to those of hyper-specialization of the departments developing the sub-systems. Indeed, organizational modularity induced by product modularity seems to provide for more effective learning processes at each module level, and thus an increase in specialized knowledge. Furthermore, systems integration requires a profound understanding of the product architecture and the building of an informational infrastructure (Sanchez and Mahoney 1996). However, it might be that once such an infrastructure is built, there would be little room for it to evolve. Thus, in spite of a reorganization of the firms' competencies, smooth knowledge communication might not be guaranteed, either within the firm because of standardized interfaces, or with the external environment since military specific factors would not have been removed. One finds here the case developed below where the firm is composed of tacit knowledge pockets, which are not really transferable outside the communities within which they were created and are being used (Pitt and Clarke 1999).

Furthermore, since the module interfaces are standardized only in relation to military systems, they cannot be transferred to the civil sector. It is even possible that these modules, which comply with specific military needs, would be more difficult to transfer than technology systems.

6.3.2.3 Despite Restructuring, Military Activities Remain Isolated

The existence of a very strong local communication culture and the lack of codification (or a common language) to ease the flow of knowledge and information hamper the diffusion process. Within defence related firms, projects are generally divided into clearly identifiable and highly compartmentalized sub-parts, to be integrated at a later stage to form a complex system. Although efforts are being made to integrate these sub-sets earlier in the process, and to codify the knowledge developed by engineers, the segmented approach still dominates. In addition, firms tend to be organized in separate units, responsible for a specific part of the system. Such an organizational structure reinforces the cognitive « insularization » of firms.

Within the firms studied, their organizational forms hindered knowledge communication. Each specialized sub-set was developed by a separate unit. Low turn-over of the workforce increased this specialization and isolated the units from each other. Furthermore, the relationships among units rely on contracts, thus limiting the development of personal links. To sum up, from a cognitive point of view, firms manage pockets of knowledge linked through contracts with low informational content.

Within these pockets of knowledge local codes and communication modes develop, and the knowledge is perceived, by an actor external to the community, as being highly tacit (Cowan et al. 2000) and, thus, difficult to transfer or translate. In fact, the codes are built on shared beliefs and shared values that are specific to the members of each group. From these beliefs and values emerges a collective knowledge system, which conditions the representations of the members of the community (Dibiaggio 1998), and which cannot easily be shared. Firms progressively become composed of fairly separate entities that find it difficult to communicate with each other (Brown and Duguid 1998). Under these conditions, knowledge circulation, as an integral part of the diffusion process, tends to be mismanaged.

A similar situation exists in terms of the relationships between defence related firms and their environment. Due to their high degree of integration, and due to the fact that they develop specific technologies for a limited category of users, the knowledge bases within these firms are rather poorly codified, and therefore cannot be transferred. Consequently, the information network that supports the circulation of codified knowledge between the civil and the military sectors is extremely limited (Cowan and Foray 1995). Since firms are part of a restricted network, the communities belonging to it rarely cross the frontiers. Hence, knowledge – be it tacit or codified – remains confined to the organizational structure.

6.3.2.4 The Commercial Approach of Defence Related Firms

Traditionally defence firms sell complex systems with specific functionality to a small set of actors. The defence market appears to be weakly contestable, and is characterized by high entry barriers. Although there is competition related to technology within defence networks, this is weak compared to the highly competitive environment of a classical market. As a result, the commercial teams in defence sector firms have rather limited competencies. They can deal with state departments, but have little experience of operating within more classical markets. Specificity within defence related firms, therefore, is not only related to products, but also to the network of buyers, to the bargaining modes, and to the legislation in force. The change towards a more commercial environment for defence industries thus raises the following issues.

First, defence firms dealing mostly with governments have not developed commercial networks in the civil sector. This is why they generally lack the competence to create a market for potentially innovative products. Thus, even when there are technological components that could be adapted for the civil sector, their commercialization is poor.

Second, due to the comparative stability of their market, defence firms have developed particular relationships with their clients. The latter become accustomed to the technologies in which they invest through long learning processes such as «learning by using». This learning process is essentially based on the accumulation of tacit knowledge over time. Consequently, the knowledge base necessary for using defence technologies is poorly codified and to train more classical clients would involve high costs in terms of the involvement needed to create a market in the civil sector.

Third, in a military context, firms are not urged to re-deploy their competencies in order to capture new markets, whereas the dynamic management of competencies is a necessary condition of success for innovative firms (Chiesa and Manzini 1998)[9]. Firms engaged in this dynamic management process must necessarily interact with their environment, either to understand it better or to gain access to a knowledge base that they consider necessary. This interaction process favours innovation diffusion since it is based on continuous exchanges with external partners and on the creation of novel networks. The two firms studied, for the reasons we men-

[9] In fact, firms facing a dynamic environment must evolve according to their context. They need to continuously monitor the fit between existing competencies and those required to be successful in the market. Such an approach assumes a close relationship between the strategic and the technological dimensions (Chiesa et al. 1999). Chiesa and Manzini (1998) identify several phases related to the development of competencies in a dynamic environment. These are: competence creation, competence strengthening, competence updating and competence destruction. Firms find themselves within a phase based on their perception of their environment.

tioned above, seem to have found it difficult to re-organize and re-deploy their competencies in a dynamic way in order to access new civil sector markets.

These problems have acted as a severe constraint to innovation diffusion[10]. Although within both firms there was a clear perception of the critical role played by competencies, the management of these competencies is generally organized in a static way (less so in MBD, which increasingly has developed its competencies in line with demand in civil markets). However, even if these firms capitalize on their competencies, they are less likely to investigate fields related to their principal activity. In terms of the evolutionary theory of the firm, this can be interpreted as a deficiency in long term development capabilities, since radical innovation and reorientation strategies are often based on complementary competencies, and, where possible, are conducted in synergy with already existing competencies.

6.3.2.5 The Restructuring of the Defence Industry

While the environment of the early 2000s leads firms more and more to engage in international partnerships and mergers, government policies in response to these changes have evolved more slowly, and consideration of military needs until very recently has imposed severe constraints on the restructuring of the defence industry. This is not without consequences for the innovation diffusion process.

For instance, at the time of our study Airsys was engaged in an international project involving Spain, France, and the UK. Airsys had responded to the call for proposals through a joint-venture between French, German,

---

[10] As we have seen, competence management at Airsys resides essentially in capitalizing on existing competencies. Traditionally the firm has privileged competence maintenance. The orientation of the firm towards external markets is justified by the will to acquire contracts which will allow it to maintain its existing competencies. At the same time, Airsys believes that its ability to obtain contracts largely rests on these competencies. To a lesser degree, this phenomenon can be observed in Matra. The competencies acquired during the Apache project have been maintained within the firm, which looks for opportunities where these competencies might be exploited. For stealth technology, applications have been found in the space sector. Nevertheless, these applications remain marginal compared to the principal application. Another example is image processing for the Scalp G project. For the military project, very powerful real-time image processing software was developed. This software had important civil applications, but it was not Matra-Defence that managed its civil exploitation. The firm continued to give priority to its traditional activities.

British, and North American firms. An international team had been formed and isolated from the rest of the participating firms in order to avoid the risk of industrial and military information leakage. Secrecy concerns in this case added to the problems of cross fertilization and innovation diffusion between firms. The decision to isolate the teams from their parent companies exacerbated the difficulty of bringing back to the firm experience gained during the project (Gibbons *et al.* 1994). This effect was particularly significant in an industrial context in which intra-firm indirect effects prevailed (previous studies conducted by the BETA have shown that this is the case within firms specialized in advanced technologies).

MBD's experience was similar. In 1996, Matra merged its missile activities with BAe to form Matra BAe Dynamics in order to access the UK market. However, national policies did not recognize this change. One problem was that the British part of the new firm benefited from public grants while the French part did not. Moreover, information communication between the two parties was hindered by national defence secrecy aspects. Finally, team working was difficult due to cultural differences and geographical obstacles. Obviously, in such circumstances, articulating competencies becomes very difficult.

6.4 Firm-Specific Choices

The problems outlined above are, in large part, common to the firms we interviewed and result principally from the characteristics of the military context. However, the strategic and specific choices of each firm also have an effect, either by making the problems more pronounced, or through correcting them by the adoption of a more active role in the innovation and diffusion process. In fact, each firm has found specific and original ways to add value to its innovative efforts.

6.4.1 Thomson-CSF/Airsys

6.4.1.1 Organization of Business Unit

In 1992, Airsys initiated an organizational change with the launch of the MASTER missiles. This change was in response to the need to reduce production time. The aim was to reduce average product development time from 8-10 years to 4 years. Prior to 1992, Airsys was organized in technical departments that successively developed radars from the first to the final stages: technical development, conception, industrialization, and production. After 1992, the firm re-organized into *Technical Business Units* (TBU). Thomson-CSF can be seen, along with the automotive industry, as the leader in the adoption of this type of organization.

Each TBU includes several technical sections in charge of a specific field. Each of these sections (antenna, data processing, etc.) has a staff of about 200 and is specialized in and responsible for a complete sub-set process. Integrators are responsible for putting together the different elements. The business unit section within the TBU consists of 60 to 70 people with multiple competencies, and is responsible for defining the product 'policy' of the TBU (e.g. ensuring that the components developed for a particular radar can also be used for others).

The firm admitted that such an organization introduced the risk of compartmentalization between the different units (radar, image processing, etc,). Employees do not rotate much between units. Despite management's efforts to encourage exchanges, this compartmentalization is exacerbated by the geographical distance between sites. Opportunities for cross-fertilization effects, therefore, are limited.

As Hamel and Prahalad (1990) observe a TBU type organization leads to excessive empowerment of individual units, making them reluctant to exchange knowledge and personnel unless it is in their interests. However, there are some informal personal networks that, to an extent, counterbalance the administrative structure and facilitate communication between TBUs. As noted by Pitt and Clarke (1999) TBUs accumulate knowledge bases around already well-identified problems. They are thus well suited to efficient exploitation of existing technological solutions. However, innovation, which consists of identifying and solving novel problems, requires interactions that cross structural barriers, and new combinations of knowledge bases.

Moreover, existing relationships between TBUs are contractual, specifying, for example, the technological characteristics, the costs and the timing of product development. These explicit contracts between units can also restrict the opportunity for experimentation and constrain the necessary ambiguity for solving novel problems (Baumard 1999).

Civil and military radars are developed within the same TBU, accounting for respectively 25% and 75% of turnover. The sectors are seen as complementary. The technological competencies developed for the defence sector are employed in the civil sector. However, even at this level, cross-fertilization is fairly rare since civil requirements are generally less detailed than defence ones. For instance, civil antennae do not integrate all the military functions. However, the data processing and computation systems are very similar. Nevertheless, all things considered, since civil needs differ from defence needs, and since defence system costs are very high, the diffusion of military technologies towards civil markets remains very difficult.

To sum up, the TBU organization seems appropriate for the exploitation of well-defined solutions and for easing access to the market, but does not encourage innovation and even less so its diffusion (Chiesa *et al.* 1999).

6.4.1.2 Knowledge Codification Efforts

Airsys is investing increasingly in knowledge capitalization processes and in the development of an organizational memory. One of the reasons for these efforts is the age structure of the Airsys staff: there are two fairly distinct age categories. There is concern about ensuring that the know-how of the older cohort, nearing retirement age, is transferred to the younger employees. Furthermore, the difficulties encountered by Thomson in past years resulted in large lay-offs of staff and the loss of an important part of its know-how. The need to avoid 're-inventions', and the search for time economies, are two additional reasons justifying knowledge codification efforts.

Concerning competence management, Thomson initiated an internal procedure – MIST (System Engineering Methodology Thomson) – adopted in 1997 by Airsys. The methodology ensures greater rigour in project development and memorization of past projects. Also, this knowledge capitalization aims at avoiding duplication. Efforts are being made to extend this approach to engineering (by developing a data base for specifications). To this must be added the extensive use of information systems. Airsys has been awarded ISO 9000, and since the mid-1990s the firm has been managed at all levels (e.g. human resources management, after sale services), according to codified processes.

All these organizational investments aim at ensuring better knowledge diffusion within the firm. This knowledge codification might, in some respects, counterbalance the disadvantages of the TBU concerning the internal diffusion of technologies. In addition, it is possible that, once developed, this knowledge base could be used outside the firm for training new users. The codification process could also be at the origin of a common meta-language in the firm. Such a language is an advantage in the development of organizational competencies and thus potentially could increase innovation (Pitt and Clark 1999).

6.4.2 MBD

6.4.2.1 Developing Capabilities in Enabling Technologies

MBD's efforts are focused on the systematic integration of a new technology in each of its product generations: the firm tries always to transfe what has been developed at the basic research level to the product projec level. MBD's general policy is to respond globally to the needs of the DGA by referring to what has been developed through basic research activities or within other projects. Stealth technology, for instance, was the key concept developed for the Apache missile, and image processing wa the innovative aspect of Scalp G. In the latter case, the Image Processing Laboratory (LTIS) was created originally for the development of the MICA air-air missile range. The competencies developed for MICA were than transferred to Scalp G in recognition of the specific requirements fo the new missile. The process of identification of strategic technologies is considered by MBD as the key to gaining advantage over its competitors In this perspective the identification of critical technologies is assured through a *«bottom-up»* process of questioning the different parts of the firm. Beginning at the operational level, the questionnaire circulates progressively to the highest hierarchical level where technological orientation choices are finally made according to its results. One of the main reason to involve the operational level in the identification process is to favou communication and exchange of ideas between different technologica communities in the firm and to establish a common vision for the future Involving the operational level in the process brings together a critica mass of actors, legitimizing emerging and critical trajectories, which are than enacted at the strategic management level.

The case of the Scalp G illustrates how MBD manages its technologica bets. In 1992, demand for cruise missiles arose as a result of the first Gul war. This corresponds to the establishment of the LTIS. The Scalp projec began in 1995-96. The first step was the creation of a specialist imag processing, and computer and system guidance team (the three element being closely linked) to manage any problems related to image processing before it was integrated into the missile. The approach taken, therefore was to exploit the firm's well-mastered technologies within the programm in order to meet time constraints. This was only possible because MBI had anticipated which technologies might be critical for future product and had taken the decision to invest significant effort in order to become leader. Once the technology was perceived as mature, MBD decreased it efforts and shelved its corresponding competencies. Presently, MBD merel invests in technology monitoring in stealth bombers and has reduced its other efforts in this specific domain.

The example of Scalp G illustrates the dynamic management process described by Chiesa and Manzini (1998), a process that is based on internal knowledge circulation and continuous interaction with the environment, and which favours innovation diffusion. It also highlights the critical role of functionality competencies (Hamel 1994) to sustain the capabilities to create a new, viable, and dominant technology. Such competencies allow the firm to endow its products with a unique functionality rather than their being based on incremental improvements. Innovation at the functionality level also favours diffusion. For example, stealth technology has been used not only by the air force, but also by the army and the navy. Image processing technologies have found numerous applications in civil sectors (e.g. multimedia, television).

6.4.2.2 Capability to Convince the DGA to Support Basic Research Activities

Another important aspect is the ability of MBD to leverage its own R&D expenditure through public financing (principally from the DGA) of basic research. The case of stealth technology illustrates this ability to call upon DGA's resources. In 1987, when the Apache project was initiated, MBD's efforts in stealth technology were concentrated on winning the Apache contract. MBD had already invested in measuring instruments, computation systems and absorptive materials. After proving its competencies in stealth technology, MBD worked on a "black programme" concerned with next generation stealth technology, financed by the DGA. This programme influenced the development of the Scalp G. Stealth technology was thus financed initially by equity and then by the DGA, once MBD had demonstrated the necessary competencies and proved technology relevance.

In terms of passive stealth technology, i.e. materials, it has reached a certain threshold (although there is a need to improve sensor technologies and to reduce costs). Passive stealth technology is no longer considered to be satisfactory since low-observability is not guaranteed at low frequency. Efforts have thus been focused on active stealth technology. This has led to a new contract with the DGA involving several units of Thomson and a partnership between the company and the CEA[11].

---

[11] The GPS technology is another example of a project to which the DGA gave significant financial support. GPS was however not initiated until 1996/1997 since the DGA refused to use US technology.

As the case of stealth bombers illustrates the role of DGA continues to be critical in financing technological innovation. The research activities conducted in this context are extremely valuable for MBD since the results obtained support a variety of products developed by the firm. Moreover, in the case of stealth technology, its application has been widened to other military sectors. Some of the developments within this framework have also been commercialized in the civil sector (e.g. measuring instruments, computer systems).

6.5 Conclusions

The analysis in this chapter describes the evolution of the defence industries during the 1990s, and attempts to explain why military technology diffusion may differ from the process of diffusion of civil technologies. It is clear that the very mechanisms that characterized the diffusion context up to the 1980s have changed so profoundly that a paradigm shift has occurred (from spin-off to spin-in). Defence related firms have progressively adapted their strategies and structures to the new environment, assisted in this by the strategic vision of the DGA (Helmer 1997). The interviews we conducted show that, if some major obstacles, of which the firms are perfectly aware, are removed, new diffusion opportunities stemming from military projects would be created, and, more generally, these projects would continue to contribute to society's knowledge base.

The following points should be emphasized:

- Defence projects remain particularly relevant in mixing new technologies. An important feature of defence products and systems is that they are often the result of applying and combining technologies from different sectors (aeronautics, space, materials, electronics, optics, etc.). Defence industries have traditionally drawn technologies from various sectors and integrated them into new configurations. The "high-tech" dimension of the defence sector can thus be seen as the result, not so much of the advanced characteristics of a particular technology, as of the novel combination of existing technologies in order to satisfy specific military needs. It is precisely through such novel combinations within defence projects that several process and product innovations have emerged. The higher the technological variety (in the sense of technologies stemming from different sectors), the more important the possibility of generating new technological opportunities seems to be.

- A model oriented towards knowledge exchange (spin-in as well as spin-off) opens up several promising perspectives. The commonly held idea that only spin-off is virtuous, and spin-in merely rests on technology imitation, is misleading. On the one hand, there are numerous cases where spin-off is questionable (as in the case of having tried to justify the Apollo programme efforts from an economic point of view with the example of Teflon®). On the other hand, and, more essentially, from an evolutionary perspective, spin-in has creative power. The adaptation of a technological transfer to a specific application is a key phase within the general innovation creation/diffusion process, revealing new characteristics and usage modes and contributing to society's knowledge base (through new feedbacks or spin-offs). In this sense, spin-in creates a favourable context for bringing the defence and civil sectors closer. Aspiring in an integrated way to knowledge spillovers in both directions should introduce a strong potential for diffusion.
- The diffusion of innovations stemming from defence projects will only occur if some of the functionality developed by defence projects is close to what is needed by the civil sector. The defence sector naturally develops technologies with certain privileged functionality (distance monitoring and maintenance, security, high level of quality and reliability, mastery of highly complex information flows). However, it is important to underline that it is the generic character of the developed functionality rather then the generic character of the technology that conditions the creation of technological externalities and the importance of potential transfers. For instance, a distinctive functionality in defence projects, related to the capability to quickly mobilize a deployment logistic of human and material resources at the theatre of operations, would satisfy civil sector needs in critical situations (e.g. earthquakes, or other major disasters). Defence related competencies in this field are thus a source of diffusion potential towards civil applications.
- The tendency in defence projects to sub-contract work to SMEs, a significant number of which are engaged in important civil activities, could be a source of learning diffusion, the more so since these SMEs benefit from the competencies of big defence firms. The case of some SMEs that participated in R&D consortia (e.g. space projects) shows that they can benefit from important economic advantages, not only by learning new technologies, but also by integrating new methods and by accessing new markets. The participation of SMEs in hierarchical R&D consortia is certainly one of the major justifications for their performance. Within consortia, SMEs are often chosen because of their capability to develop

a specific component. Their contribution to these projects is generally guided by a big firm. This latter allows them, through strong interactions and learning processes, to widen and improve their methods, their market approach, and their brand image. In this context the small firm strengthens its technological base. This learning mode is however not without the risk to SMEs that their technologies will be captured by the big firms (for instance, through absorption or integration of the most strategic parts of the SME). In the US, in order to guarantee the positive effects of such a learning relationship while minimizing the risk of too high a dependence, « mentor-protégée » type arrangements have been developed within several programmes, including some managed by NASA – National Aeronautics and Space Administration – and the US Department of Defense. These arrangements aim at stimulation of learning and development within SMEs as the result of being backed by a big firm and without the risk of plundering. The birth of EADS or Thales as European companies and project managers with diversified activities in the defence and civil sectors should strengthen the possibilities for Europe to develop incentive policies for SMEs willing to participate in defence projects.

However, these opportunities can only be exploited if certain obstacles to the firms operating in the defence sector are removed. The obstacles identified through our analysis are described below.

The risk of too little basic research. The new industrial landscape implies that a significant proportion of basic research is performed outside the defence firms (especially in universities). This does not imply less research, but means better research and better access to basic knowledge. This has two corollaries. On the one hand, within basic research it is important to distinguish between pure and finalized basic research. If more pure basic research can be sub-contracted to universities, it would be in the interests of defence related firms to engage in more finalized basic research in order to benefit from dual technologies (because the duality potential is essentially found at the basic research level). It is certainly important that at this stage public authorities give strong support to private firms, since the knowledge base will have public good properties. The second corollary is related to the fact that what will be critical in the on going changing environment are the options adopted by firms to manage their absorption and emission capabilities. Absorption capabilities are essential for benefiting from outside positive externalities and particularly from the civil sector. Emission capabilities are necessary for signalling the competencies accumulated by defence related firms (contributing to their reputation as pivotal actors in the technological scene, capable of continuously interacting with other sectors). These capabilities should be a determining factor for signalling to the DGA the technological domains perceived as promising by the firms and for justifying further support for private research.

The gap between the life cycles of civil and defence products. One of the key factors hampering innovation diffusion from defence projects is the length of product life cycles in the civil and defence sectors. This difference is becoming more important and forces defence firms to "rigidify" their competencies in order to keep pace with longer technological cycles. This leads to a situation where firms operate according to generational qualitative leaps, and hinders the dynamic redeployment of competencies in order to keep up with the accelerated and incremental innovation rhythm of the civil sector. Shortening the life cycle of military products would be an efficient way to remedy this. However, should this prove too difficult, because of the specificity of defence projects, it is important that incremental innovations be continuously stimulated and realized within existing longer cycles. This would mean projects being conducted through both a "mission-oriented" approach (oriented by the industrial object to be developed) as well as a "diffusion-oriented" approach (oriented by the will to extend the state-of-the-art frontier), according to the distinction introduced by Ergas (1987).

From the preceding arguments it follows that the recent changes in defence related industries requires a re-consideration of the innovation diffusion process of military technologies. These reconsideration possibilities will certainly depend on the evolving relationship between defence related industries and the DGA. The reform of the DGA in 1996 anticipated the principal components of this new relationship by providing for enhanced public-private partnerships. The greater accountability and risk imposed on firms, the “smart buyer” position stressed by the DGA, and the intensive partnership in each phase of the military products life cycle, but also the narrowing down by the DGA of its technological options for strictly military needs, are the main elements of this reform, which will, in the long term, influence the diffusion potential of defence technologies towards the civil sector.

6.6 References

Alic J, Branscomb L, Brooks H, Carter A, Epstein G (1992) Beyond spin-off: Military and commercial technologies in a changing worldl Harvard School Press, Harvard MA.

Baumard P (1999) Tacit knowledge in organizations. Sage, London.

Brown JS, Duguid P (1998) Organizing knowledge. California Management Review 40(3): 90-111.

Brusoni S, Pavitt K, Prencipe A (2000) Knowledge specialisation and the boundaries of the firm : Why do firms know more than they do? Paper presented at the conference on Knowledge Management: Concepts and Controversies, University of Warwick, UK.

Chesbrough W, Teece D (1996) When is virtual virtuous ? Organizing for innovation. Harvard Business Review(January-February): 65-74.

Chiesa V, Manzini R (1998) Towards a framework for dynamic technology strategy. Technology Analysis & Strategic Management 10(1): 111-129.

Chiesa V, Giglioli E, Manzini R (1999) R&D corporate planning : Selecting the core technological competencies. Technology Analysis & Strategic Management 11(2): 255-279.

Cowan R, Foray D (1995) Quandaries in the economics of dual technologies and spillovers from military to civilian research and development. Research Policy 24: 851-868.

Cowan R, David P, Foray D (2000) The explicit economics of knowledge codification and tacitness. Industrial and Corporate Change 9(2): 211-254.

Cowan R, Foray D, Mohnen P (1998) The relationship between military R&D and the civilian innovation system. Working paper IMRI-WP 98/06, Paris Dauphine.

Dibiaggio L (1998) Information, connaissance et organisation. PhD thesis, University of Nice-Sophia Antipolis.

Ergas H (1987) The importance of technological policy. In: P Dasgupta,P Stoneman (eds.) Economic policy and technological performance. Cambridge University Press, Cambridge: 51-96

Gibbons M, Limoges C, Nowotny H, Schwartzman S, Scott P, Trow M (1997) The new production of knowledge: The dynamics of science and research in contemporary societies. Sage Publications, London.

Hamel G, Prahalad CK (1990) The core competence of the corporation. Harvard Business Review (May-June): 79-91.

Hamel G (1994) The concept of core competence. In: G Hamel and A Heene (eds.) Competence-based competition. John Wiley & Sons, NY: 11-33.

Helmer JY (1997) Pourquoi une réforme de la DGA, L'Armement, décembre1996/janvier1997, pp. 5-6.

Hobday M (1996) Product complexity, innovation and industrial organisation. Research Policy, 26: 689-710.

Lundvall B-Å, Johnson B (1994) The learning economy. Journal of Industry Studies 1(2): 23-42.

Marengo L (1993) Knowledge distribution and coordination in organizations : on some social aspects of the exploration vs. exploitation trade-off. Revue Internationale de Systémique 7: 553-571.

Pitt M, Clarke K (1999) Competing on competence : a knowledge perspective on the management of strategic innovation. Technology Analysis & Strategic Management 11(3): 301-316.

Prencipe A (1997) Technological competencies and product's evolutionary dynamics: a case study from the aero-engine industry Research Policy 25: 1261-1276.

Sachwald F (1999) Banalisation et restructuration des industries de défense, Les Notes de l'IFRI, n°15.

Sanchez R, Mahoney JT (1996) Modularity, flexibility, and knowledge management in product and organization design. Strategic Management Journal17: 63-76.

Sanchez R (1996) Strategic product creation : Managing new interactions of technology, markets and organizations. European Management Journal 14(2): 121-138.

von Hippel E (1988) The sources of innovation. Oxford University Press, New York.

Part III Impact of Incentives Tools on Systemic and Learning Failures

7 University-Industry Relationships and Regional Innovation Systems: Analysis of the French Procedure Cifre

Jean-Alain Héraud and Rachel Lévy

BETA, Strasbourg, E-mail: heraud@cournot.u-strasbg.fr
BETA, Strasbourg, E-mail: levy@cournot.u-strasbg.fr

7.1 Introduction

This chapter aims at understanding the role of universities at the level of territory, or "region", that is, as a sub-national entity. A considerable amount of the economic literature and a number of policy-oriented papers have been devoted to university-industry relationships and regional innovation systems. However, little has been done on looking at the links between university and the regional industrial fabric. We address this gap drawing on a database of contractual PhD research projects involving private firms and public laboratories.

Since 1981, there has been a system in operation in France, under the auspices of the French government, which enables doctoral research students to conduct their research partly in a public research laboratory and partly in a firm. This collaborative arrangement, called Cifre (Convention industrielle de formation par la recherche), is a public-private research training agreement. The PhD student's time is split between the laboratory and the firm. The students are recruited by firms, which receive a subsidy from the public agency that oversees the Cifre arrangements, the Association Nationale de la Recherche Technique (ANRT).

Thus, the student becomes involved in both the industry and scientific communities. He/she could be seen as a "cognitive platform" facilitating the creation and transfer of knowledge between science and industry. This role is particularly important in relation to small firms for which working with a Cifre sponsored PhD student is often their first contact with academia. If the experience is a good one there is the possibility that the relationship with the academic world will continue. One of the objectives of this system is to bridge between the scientific and industrial spheres, and to build durable networks involving business and academic institutions. In relation to the students involved the objective is to encourage and facilitate their integration into the labour market. It has been shown that the transition from being at a university to getting a job has been significantly easier for Cifre trainees.

In this chapter we will compare the regional distribution of the laboratories and firms involved in the Cifre scheme to see to what extent this transfer of knowledge between universities and firms is confined to regions or takes place in the broader context of the national system. In other words, we are investigating the notion of a regional system of innovation (RSI).

Certain regions would appear to be self-sufficient in the sense that their firms often collaborate with local academic institutions. However, many regions appear to be "knowledge exporting", because their local scientific specialization is more aligned to industry in other regions, while some regions can be classed as "knowledge importing" because the firms within their region are forced to collaborate on scientific projects with institutions outside their territory, because they lack the relevant competence or it is not available in their immediate area.

Our study will illustrate the variety of regional innovation contexts that are involved. Only a small number of local regions encompass the array of actors and links that are involved in the innovation process: large and smaller enterprises in relationships with universities and public research institutes, "knowledge intensive business services" (KIBS), which capitalize on and diffuse technological knowledge and managerial skills among the other organizations, regional authorities capable of implementing (in coordination with national administrations) the relevant policies, etc. In short, few regions have a RSI, although many of them have important elements of such a system.

In this chapter we focus mainly on one aspect of the innovation system: university-industry collaboration, but our analysis casts light on the regional context in general and leads to a consideration of the specific role of KIBS.

In Section 1 we begin by defining a regional innovation system and describing the role of university-industry collaboration within such a learning environment. In the second section we describe the French doctoral training system – Cifre. Finally we construct regional indicators using statistical data on Cifre in order to analyze the differences between regions in terms of science-industry collaborations.

7.2 Regional Systems of Innovation

In a global economy, science and technology policies are designed and implemented at various geographical levels: national and European, but also sub-national (regional) levels. As a result of this multi-level governance structure, scientific production as well as technological and knowledge transfer must be analyzed using various levels of the innovation system. This requires a specific disciplinary approach, which encompasses the theory of innovation systems in a wider sense, the regional and geographic economy, and knowledge theory.

7.2.1 Different Systems of Innovation

Before addressing the idea of a RSI, we begin by defining in a very general way the concept of a *system of innovation*:
"*A* system of innovation *can be thought of as consisting of a set of actors or entities such as firms, other organisations and institutions that interact in the generation, use and diffusion of new-and economically useful-knowledge in the production process*" (Fischer 2000).

In other words, the different components of the system must interact. But, do they all interact simultaneously? Does their interaction follow a specific pattern? Are all these interactions within the system? The answers to such questions help to define the concept of a system, especially in the sub-national context. We want to stress that in addition to organizing simple "communication", the system must facilitate the creation and exchange of "knowledge". Sharing the same culture, the same languages, and the same routines is a positive factor for the exchange and creation of new knowledge. To take into account these characteristics of knowledge interaction leads to consideration of various notions of national, sectoral or regional innovation systems (Carlsson et al. 2002).

Applying the system approach at the national level, authors such as Nelson (1993), Lundvall and Borras (1997), and Lundvall et al. (2002) underline the fact that nations are typically the political and institutional framework that allows the different actors to produce knowledge based on a common language, culture, and political regulatory environment. Therefore, the national dimension seems to be most appropriate for analysis of the formation and development of an innovation system. Based on this same notion of innovation systems, other authors have developed the concepts of sectoral systems of innovation (Malerba 2002) and technological innovation system (Carlsson et al. 2002).

7.2.2 Systems of Innovation at Regional Level

One of the RSI models in the literature considers that the actors within the regional system share a history, language and culture, which promote relationships based on trust. In this model the actors are in close geographical proximity, enabling face-to-face contact and exchange of tacit knowledge. It is supposed that complex interaction involves a high degree of tacit knowledge exchange, which is typical of innovative networks and learning economies (Foray and Lundvall 1996; Lundvall and Borras 1997). The importance of this notion of RSI has increased in recent years due to the simultaneous processes of globalization and localization (i.e. the relative decrease in national regulation).

The term region in this chapter is not always used to identify a local administration. Techno-economic coherence can often be found at a sub-regional level (in urban areas for instance). In certain cases, trans-border regional systems of innovation exist in which cultural attitudes and sectors of specialization are the same. But, it is also true that political will plays an important role in the design of innovation systems. At the regional level in particular, the early stages of the construction of an innovation system sometimes depend on the specific actions of individuals in initiating such a movement (for example, the "regional developer").

From our point of view, a good definition of a region is: “A meso-level political unit set between the national or federal and local levels of government that might have some cultural or historical homogeneity but which at least has some statutory powers to intervene and support economic development, particularly innovation” (Cooke 2001, p. 953).

Our study will confirm that the existence of innovative structures does not automatically lead to a full-fledged regional system, or, if the concept of a RSI does apply, it will be shown to be a largely open system[1].

There are certain elements whose interaction is valuable for the generation of innovations, possibly leading to the creation of a RSI (Catin et al. 2001; Asheim and Isaksen 2002; Lung et al. 1999):

---

[1] We have tested the existence of such a regional system in previous works, especially in the case of the French region of Alsace (Héraud and Nanopoulos 1994; Nonn and Héraud 1995). It appears that innovation networks of firms were concentrated only to a limited extent within the region: regional partners accounted for less than 25% of innovative links. Furthermore, this degree of regional concentration varied greatly depending on the type of partner (another firm, public laboratory, technology centre, etc.). Therefore, the existence of a "regional" system of innovation is debatable, at least in the case of French regions.

- The industrial sector (possibly organized within a cluster[2]), composed of small and medium sized enterprises (SMEs) within the region and larger firms – often subsidiaries of national or international groups.
- The science-based sector, with public institutes and university laboratories forming the bulk of the institutions of technological infrastructure (ITI)[3].
- Regional government and other territorial institutions.
- Various institutions whose mission is to promote innovation, for example, technology centres, university technology transfer offices, etc.
- Private actors who act as "go-betweens" and play an important role in advanced regions: KIBS [4].
- The national scientific and institutional system (sometimes with local offices), and the European programmes that increasingly are focusing on regional capabilities.

---

[2] Industrial clusters are "geographic concentrations of interconnected companies and institutions in a particular field. Clusters encompass an array of linked industries and other entities important to competition. They include, for example, suppliers of specialized inputs such as components, machinery, and services, and providers of specialized infrastructure. Clusters also often extend downstream to channels and customers and laterally to manufacturers of complementary products and to companies in industries related by skills, technologies, or common inputs. Finally, many clusters include governmental and others institutions - such as universities, standards-setting agencies, think tanks, vocational training providers, and trade associations - that provide specialized training, education, information, research and technical support" (Porter 1998, p. 78).

[3] Academic institutions are important elements of the regional technological infrastructure, but other actors within the regional scene can fulfil their function, including certain large firms or high tech SMEs. For a presentation of ITIs and their role in the generation and diffusion of knowledge, see Bureth and Héraud (2001).

[4] This tertiary regional fabric composed of technological, legal, management, or marketing services, tends to build a non-institutional informal knowledge transfer structure in the regions. For understanding their increasing role for active regions in the process of globalization, see Strambach (2001).

7.2.3 The Role of University-Industry Collaboration and the Diversity of RSI

Innovation systems at regional level can involve a large variety of industrial structures and economic dynamics. A number of regions clearly exhibit a type of development based on knowledge and service activities. Braczyk et al. (1998) give the examples of California and Singapore, but they also cite Midi-Pyrénées as a potential innovation system based on knowledge and service industries. Varga (1997) includes regions such as Lombardy, Baden-Wurtemberg, Rhône-Alpes, and Catalonia ("the *Four Motors of Europe*") in a list of the same type, along with Silicon Valley, the Boston area, and Western Canada.

There are also regions that cannot be considered to be complete knowledge-based systems, but that nevertheless host important elements of the innovation system. For instance, the *Third Italy* districts described by Beccatini (1991) were presented in the literature as paradigms of innovative territories although they do not offer significant scientific facilities. Conversely, regional concentrations of S&T institutions are not necessarily linked to the local industrial fabric.

Besides firms, universities and scientific "competence centres" (Institutions of Technology Infrastructure: ITIs), as well as KIBS, play a crucial role in RSI. The functioning of the innovation system implies various transfers of knowledge. In particular, university-industry collaboration plays a major part in regional dynamics, by increasing the stock of knowledge and human capital, triggering technological or methodological spin-offs, and influencing the formation of networks (Gibbons and Johnston 1974; Salter and Martin 2001; Etzkowitz et al. 2000). Private business services are increasingly fulfilling the intermediary function of diffusion, adaptation and capitalization of cognitive assets between firms, particularly SMEs (see Muller 2001).

A number of case studies (Varga 1997; Atkins et al. 1999; Da Rosa Pires and Anselmo de Castro 1997; Fritsch and Schwirten 1999, Jones-Evans and Klofsten 1998; Lee 2000; Rip 2002) have shown the importance of regional cooperation for universities, and stressed the importance of bi-directional contact between them and other regional actors.

Various econometric studies have tried to evaluate the effects of geographic spillovers from academic institutions within a region (Jaffe 1989; Acs *et al.* 1992; Audretsch and Stephan 1996). The majority of these studies use patent citations analysis or large national or European innovation surveys. They focus on spillover effects on the firm side, but not on the bilateral effect of collaboration.

If universities are contributing to the generation of knowledge and are a critical component of the region's knowledge infrastructure (through collaborative and learning relationships with the other actors in the regional system, such as SMEs, big firms, regional administrations, etc.), they are networking to a large extent with actors in the national, sectoral, or technological innovation systems. The research performed by a university (and by a firm) can support the development of a region, but is never restricted only to the region. Public research institutions absorb knowledge from firms and research institutions in other regions and contribute to the innovation processes within their own regions (Fritsch 2001), but they also export the knowledge produced by local research institutions and firms, to other regions.

In this chapter, we conduct an exhaustive study of the French regions to examine the role of bilateral cooperation between academic and business organizations in order to establish whether universities and firms collaborate between or within regions. In so doing we consider two issues:

- the importance of regional collaboration *within* a RSI. The relevance of this issue is linked to the fact that regional excellence does not necessarily result in a closed innovation system. Quoting Landabaso et al. (2001, p. 252): "the regional dimension is important but not exclusive". What is important is to compare the role of intra-regional and inter-regional cooperation, and its impact on the development of the RSI;
- to what extent can the French regions claim to be real and consistent RSI? The importance of this question is stressed in Héraud and Isaksen (2001) and Héraud (2003). Different modes of regional development are possible, since not all regions are deemed to belong to the core group of "poles of excellence" in the new knowledge-based economy.

Statistical evidence from the Cifre doctoral funding system, will show the existence of relationships between universities and firms. We will use these statistics to test whether their interactive learning process operates within a purely regional system or nationally. ANRT[5] gave access to the complete set of Cifre agreements from the time that the system was set up in 1982. By comparing the location of firms and laboratories in the ANRT database we can answer some of the questions raised above.

[5] We wish here to express our gratitude to Philippe Gautier, who allowed us to use the ANRT database and whose expertise in managing it was invaluable.

7.3 The Cifre System

We describe below the Cifre system and evaluate it as an indicator of the science-industry collaboration.

7.3.1 Presentation of the Cifre System

The Cifre doctoral training agreement is a contract between a firm, a university research team (we will call it a "laboratory"), and a PhD student. The object is a research programme of common interest to all parties, leading to: innovative results for the firm[6]; scientific results, i.e. PhD dissertation, and a contribution to the research agenda of the laboratory; and professional training for the student (Quéré 1994). The rationale behind the policy was not only to provide an incentive for innovative work, but also to ease the transition between university and work[7]. The three types of actors and their roles are briefly described below.

- The firm hosts the student for three years, providing facilities for research and an annual salary of at least 20,215 euros. The firm receives a subsidy of 14,635 euros from the ANRT. Both large and small (almost half have less than 500 employees) firms have been involved in the scheme, mainly from industrial sectors such as electrical and electronic products, and chemistry. However, increasingly service sector firms (often consultants and other KIBS) are taking part in this kind of collaboration.
- The PhD student must be under 26, a recent graduate (5 years French university diploma or equivalent), with no previous professional experience. The student is required to work partly in the firm and partly in the laboratory, the proportion varying from case to case. The majority

[6] From the study of the Cifre system over 20 years ANRT (2001), it can be seen that 83% of the firms involved in a Cifre project have benefited from industrial spillovers such as know-how (39%), process (19%), product (17%), patent (14%), and prototype (11%).

[7] From the same study, we can see that 91% of the PhDs were successful (in the case of half of the remaining 9% the thesis could have been finished, but the student gave the preference to immediate employment). At the end of the doctoral project, 67% of the students found a job (40% in the same firm as their Cifre sponsorship), and 10% entered public research. 10% were initially unemployed, but after two years most had found a job. A small proportion (2%) set up their own firms.

students receiving Cifre sponsorship have come from engineering schools.

- The laboratory involved can be in a university (42%) or an engineering school (37%), public research institutes, or sector-specific technology centres. Foreign laboratories are eligible to take part in the scheme. The research fields have, in the past, been mainly confined to computer science, physics, and chemistry, but, more recently, Cifre sponsored students have been studying the human and social sciences, including economics.

The scheme is organized at the national level by ANRT, but applications are made to and scrutinized by the regional offices of the Ministry of Research and Technology (DRRT). This is an example of a national policy that is managed regionally, using the technological and economic expertise of the DRRT for evaluation of the firm in terms of financial capacity and ability to ensure good training conditions. National experts assess the feasibility of the research, and consider whether the scientific background of the student and the quality of the research team are appropriate for the project.

From its creation in 1982 to 2001, more than 10,000 Cifre agreements have been evaluated. Only 9% of applications were rejected. Each year, the number of applications increases and ANRT's target of 820 PhDs annually will soon be achieved.

Of the firms that benefit, 48% of them are independent SMEs or subsidiaries of large firms with less than 500 employees. This large percentage of small organizations involved in science-based projects reflects the promotion of policy to facilitate knowledge transfers to small organizations and underlines the "regional" focus of such policy. Other policies – aiming not only at technology and knowledge transfer to SMEs, but also at transforming attitudes towards and perceptions about innovation – are organized regionally: for example, the Cortechs agreements, involving the training of young technicians (see Héraud and Kern 1997). Experience from the Cortechs agreements, even more than the Cifre scheme, confirms Chabbal's (1995) observations about science policies and innovation policies that the first are mainly national policies, and the second are increasingly regional (focusing on SMEs). The Cifre scheme, however, involves both aspects – scientific impact and innovation networking, and the regional nature of the network has still to be assessed.

7.3.2 The Cifre System: a Good Indicator of Science-Industry Collaboration

We can use the Cifre contract statistics as indicators of the mediation between university and firm (Sander 2000). In the course of their PhD programmes, Cifre sponsored students act as a cognitive platform between the academic and industrial spheres. They stimulate the transfer and creation (by combination) of knowledge between these two worlds.

There are many types of links between universities and industries, and these interactions can be one-way (from science to industry), or two-way (collective learning). The role and the importance of any interaction are dependent on how the exchange is facilitated, i.e. by people, knowledge, technology and/or finance. In Table 7.1 we show the different types of interactions described in the literature (Schaeffer 1998; Schartinger et al. 2001; Scott et al. 2002; OECD 2002; Isabelle et al. 2003) in order to position the Cifre scheme in a more general framework of the various relationships between firms and universities.

Generally, the links represent "one way" transfers from universities to firms and not a real cooperation. But the Cifre system promotes complex relationships, with bilateral exchange based around the PhD student's activities. This young researcher is able to overcome many of the constraints that might hinder communication and allow knowledge and technology to be transferred across the two communities. The Cifre system demonstrates how PhD students can ideally act as a 'two-way bridge' (Meyer-Krahmer and Schmoch 1998) between the academic and industrial spheres. In some previous research, we tested the hypothesis that the students implement a bilateral knowledge exchange between firms and laboratories (Levy 2004).

7.4 Empirical Results

On the basis of the Cifre database, we can characterize the French regions in terms of their university-industry collaboration. The core of our analysis concerns the existence of regional innovation systems: the Cifre statistics are the basis for indicators of regional self-sufficiency in S&T to be constructed. Since a proportion of the firms becoming involved in Cifre agreements are business services, it is also possible to identify the growing role of KIBS in regional innovation networks.

Table 7.1. Different modalities of interaction between universities and firms

	Financial flows	Technological flows	Codified knowledge	Tacit knowledge	Personal flows
Research contract	++	(+)	++	(++)	
Technological co-development	++	++	(+)	(+)	
Co-publications			++	(++)	
Patents	++	++	++		
Prototype or technological artefact		++	++	(+)	
Biological and genetic material		+	+		
Cross-licensing	++	++			
Research project in partnership	++	(+)	++	(++)	(+)
Research consortium and network (including European Framework Programmes)	++	(++)	++	(++)	
Internship of graduate students		(++)	(++)	(++)	++
PhD in firm (typically: Cifre)	+	(+)	(++)	(++)	++
Training of industrial researchers by universities	++		++	(+)	
Recruitment of scientists by industry	+		(++)	(++)	++
Stay of academic researchers in industry			++	(++)	++
Seminars and conferences			++	+	++
Informal contacts				+	++

Sources : Schaeffer 1998; Schartinger et al. 2002; Scott et al. 2001; OECD 2002; Isabelle et al. 2003

+ and ++ indicate the degree of implication and importance of the different modes of interaction in the relationships between universities and firms. Bracketed symbols indicate that transfers are not systematic (the transfer could be made without the participation of people, knowledge, technology, and/or finance).

7.4.1 Towards a Typology of Regions

In an ideal RSI, universities and firms collaborate in a way that leads to relative closeness within the innovation system. This should be reflected in the Cifre database, with local laboratories being often associated with local firms. However, if the NSI does not consist of self-organized regions, but centrally manages the different functions across the whole country, then there will be no systematic geographic correlation between the location of the laboratories and the firms they collaborate with (at least no more than a bias towards proximity for practical reasons). Regions that do not exhibit well-balanced specific innovation systems, but, nevertheless, participate significantly in the NSI, may be strong in academic or industrial competencies. Regions where a large proportion of Cifre agreements are between laboratories in the region and firms from outside are classed as "knowledge-exporting". Firms contribute to knowledge creation; in using this term we focus only on the academic side. If the situation is reversed the region is classed as "knowledge-importing".

This empirical study examines the 10,002 Cifre agreements signed between 1982, when the system was first introduced, and 2001. We calculate two different indicators: one for the absolute balance of knowledge flows, and one for the self-sufficiency of the region. Figures 7.1 and 7.2 depict the number of Cifre contracts involving firms and laboratories in each of the 21 regions of France[8] (Figure 7.2 excludes Ile de France in order that the other regions are more fairly represented).

On average, each region has about 300 Cifre agreements in operation involving local firms and/or laboratories. But there are strong discrepancies between the regions in real terms, reflecting differences in region size, as well as academic and industrial endowments. The overwhelming weight of Ile de France is reflected in the Cifre statistics, as it accounts for about 30% of the laboratories and 40% of the firms. The French NSI is still very centralized, but the other regions also exhibit quite wide discrepancies, since Rhône-Alpes (mainly around Lyon and Grenoble), Midi-Pyrénées (Toulouse), and Provence Alpes Côte d'Azur (PACA), with Aix-Marseille and Nice-Sophia Antipolis account for about 25% of the firms and more than 30% of the laboratories. Not surprisingly, these regions are also among the largest and the richest.

---

[8] Two regions were excluded, Corse and the Overseas Territories. The reasons were twofold: their small size and the fact that localization indicators are not available. Also, for these two regions we did not have certain specific information (indicators of scientific and technological outcomes) that will be used later in our analyses.

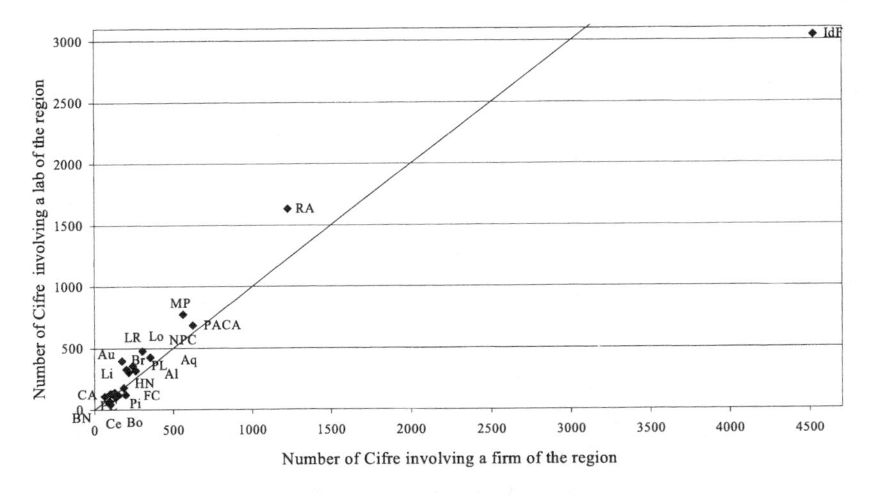

Fig. 7.1. Firms and laboratories collaborating in Cifre contracts in each region

It is also noticeable that some regions are mostly "knowledge exporting" (more regional laboratories are involved than regional firms) while others are "knowledge importing". To measure this differentiation we constructed several indicators, which are presented in Table 7.2 along with other basic information about regions, and plotted in Figure 7.3.

A very simple index (R1) is the number of regional firms involved in Cifre arrangements divided by the number of regional laboratories involved[9]. This index represents the balance of knowledge exchange. Languedoc-Roussillon is the typical knowledge exporter with R1=0,4286 and Champagne-Ardenne the typical knowledge importer with R1=2,2558.

The interpretation of the cases where R1 is close to 1 is ambiguous: are such regions “closed” innovation systems in which all the firms find academic partners locally, or is the number of laboratories and firms importing and exporting external competencies the same? In order to answer these questions, we consider the following ratio (R2): number of Cifre contracts linking partners within the region divided by number of Cifre

[9] $R1 = F / L$; where F is the number of Cifre arrangements involving a firm in the region and L is the number of Cifre arrangements involving a laboratory in the region.

contracts where only one partner is in the region[10]. This gives the self-sufficiency of the region.

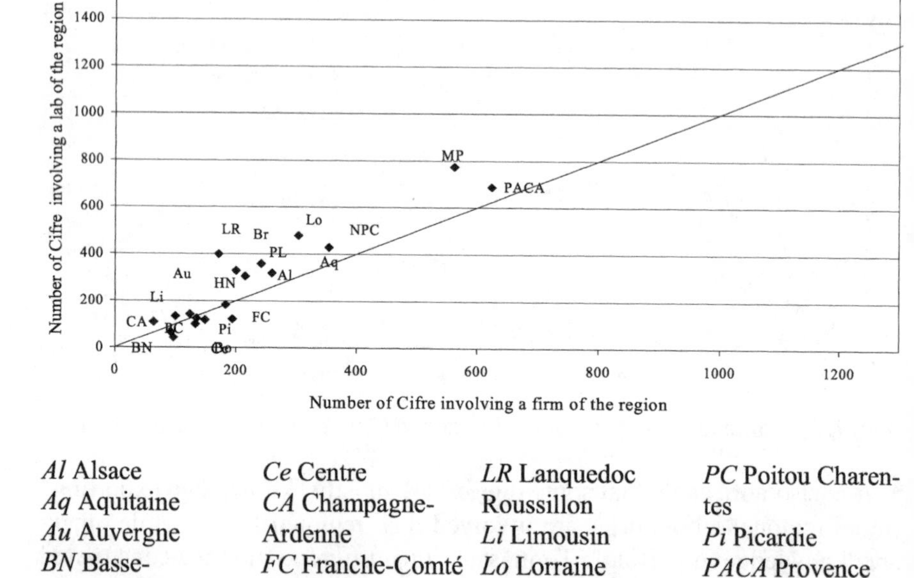

Al Alsace
Aq Aquitaine
Au Auvergne
BN Basse-Normandie
Bo Bourgogne
Br Bretagne
Ce Centre
CA Champagne-Ardenne
FC Franche-Comté
HN Haute Normandie
IdF Ile de France
LR Lanquedoc Roussillon
Li Limousin
Lo Lorraine
MP Midi Pyrénées
NPC Nord Pas de Calais
PL Pays de Loire
PC Poitou Charentes
Pi Picardie
PACA Provence Alpes Côte d'Azur
RA Rhône-Alpes

Fig. 7.2. Firms and laboratories collaborating in Cifre contracts in each region (excluding Ile de France)

To provide a more accurate test of the characteristic of self-sufficiency, we have considered another indicator (R2'), where the numbers of contracts within the region is weighted by dividing by the corresponding national figures. The values of R2' for the different regions are given in Table 7.2 along with R2. It can be seen that the introduction of this relative indicator does not produce any significant change in the ranking and classification of the regions.

Based on these two (or three) indicators, we can classify the 21 French regions (excluding Corse and Overseas territories) into four types:

---

[10] $R2 = \frac{(F \cap L)}{F + L - (F \cap L)} * 100$; where F∩L is the number of Cifre contracts involving both a firm and a laboratory in the same region

- Type 1 includes self-sufficient (or at least balanced) regions. There are eight regions in this category. Ile de France is a net importer of academic competencies, which is explained by the overwhelming concentration of firm headquarters in the Paris area[11]; the Basse-Normandie region is similar. Other important regional systems, such as Rhône-Alpes and Midi-Pyrénées, are net exporters of academic competencies. These two areas are model regions described in the literature as knowledge- and service-based regions (Braczyk et al. 1998; Varga 1997). They have developed their innovative clusters around the technological competencies of Lyon-Grenoble and Toulouse respectively. Whatever the relative importance of academia and industry, Type 1 regions are regional systems of innovation in the sense that they have apparently developed their internal networks. In this category are four other regions that show balanced flows of knowledge: PACA, Nord Pas de Calais, Aquitaine, and Bretagne. These four regions are not specialized in terms of either firms or laboratories; they tend to build university-industry links within their own territories, but, because of their size, cannot be considered real regional innovation systems.
- Type 2 regions are characterized by open territorial systems ($R2 \leq 20\%$) contributing to the NSI more through industrial demand than academic supply of knowledge ($R1 > 1.25$). Champagne-Ardennes is the best example of this type of region. Champagne-Ardennes has innovative industries, but in terms of academic competencies these are mainly to be found in the neighbouring region of Ile de France. The other regions in this category are also quite close to Paris (Centre, Haute Normandie, Bourgogne) as can be seen from the map in Figure 7.4.
- Type 3 encompasses regions with relatively open systems ($R2 < 33\%$) and net academic exports ($R1 < 0.75$). These regions contribute to the NSI by supplying academic competencies, but do not exploit them to any great extent within their own territories. The best example can be seen in Languedoc-Roussillon, which includes the Montpellier area, which is home to several important technological and scientific institutions grouped together in a large technopole (Voyer 1998), but where the industrial fabric is incomplete. Alsace is an example of a region where there is a highly developed basic science complex (mainly

[11] Indeed, the French national system remains largely centralized around its capital region. In 1998 this region accounted for 49.3% of employment of industrial researchers in France (OST 2002, p. 162) and 48% of total private expenditure on industrial R&D (OST 2002, p. 162). However, it can be seen that the introduction of a relative indicator R2' does not affect our typology.

around Strasbourg) and a significant industrial base composed of middle-tech SMEs and subsidiaries of multinational firms that are specialized in production rather than strategic functions. The other Type 3 regions are Lorraine, Pays de Loire, Poitou-Charentes, and Limousin.

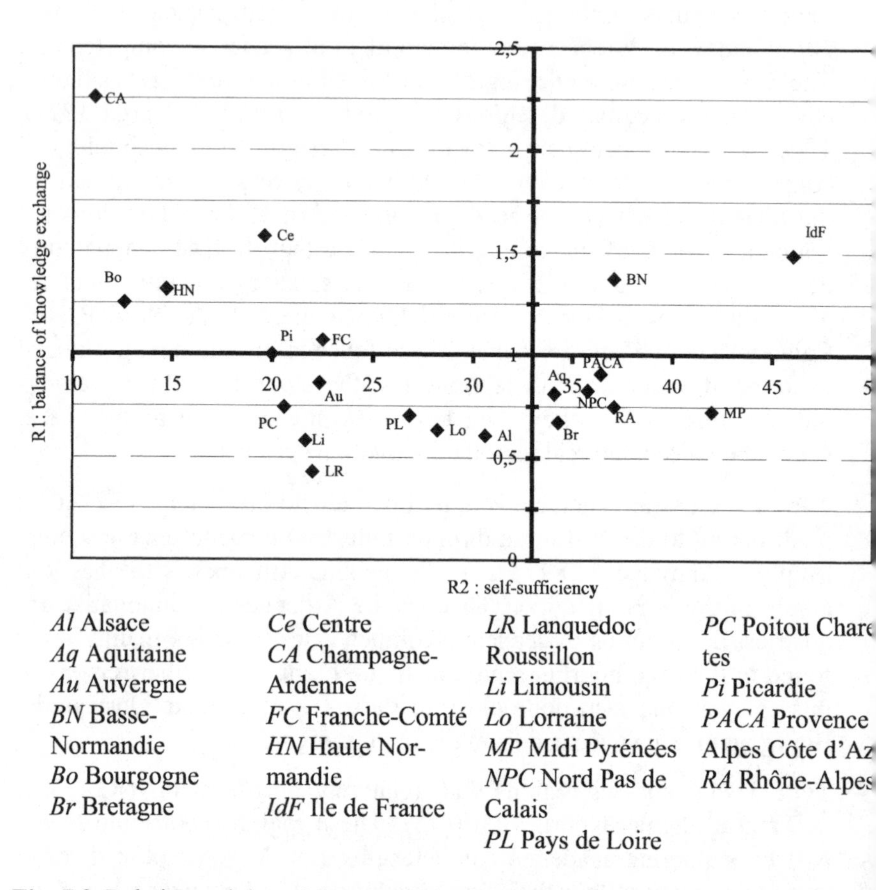

Fig. 7.3. Relative weights of firms and laboratories and self-sufficiency of regions

– Type 4 regions cover three areas where the knowledge flows are relatively balanced (R1 close to 1), but which are not very self-sufficient (R2 under 25%): Picardie, Franche-Comté, and the Auvergne. These regions do not fit into any standard "regional system" model[12]. This is not to say that these regions have no specific scientific assets or techno-

[12] Moreover, Picardie, Franche-Comté, and the Auvergne accounted for 0.5%, 0.5% and 1.1% respectively of the national expenditure by public institutions in France in 1998, and 1.8%, 2.2% and 2.2% respectively of industrial expenditure on research in France in 1998 (OST 2002, pp. 148, 163).

logical identity, but rather that the graph of the links between academia and industry is not restricted to the territory. These regions contribute to the national system in various ways, but without forming a sub-system.

7.4.2 The Role of the KIBS

As indicated in the literature review, a category of firms in the service sector plays an important role in the established systems of innovation: these are the KIBS. In the RSI, they contribute to knowledge flows in interactions between industrial firms and scientific institutions. They have a direct impact on the innovation processes in individual firms by performing their R&D (outsourcing of industrial research) and by improving firms' competencies to innovate (information diffusion, absorptive capacity building, organizational skills, legal advice, etc.). They also work as intelligent intermediaries based on their ability to learn and teach, constituting an indirect network of the actors in the innovation system, by capitalizing on and recycling knowledge (Muller 2001). Their presence and activity are an indicator of a well developed RSI.

Using the information in the Cifre database it is possible to test for the increasing role of KIBS in the past few decades and to characterize the various regions, in particular those supposedly organized as RSI. From such studies as Strambach (2001), we can see that Ile de France and Rhône-Alpes are the two regions of France with a relatively high density of KIBS: the former is comparable to Greater London, and the latter can be compared to the Stockholm or Madrid areas. In the case of Rhône-Alpes region, this confirms that the region is a knowledge- and service-based regional innovation system.

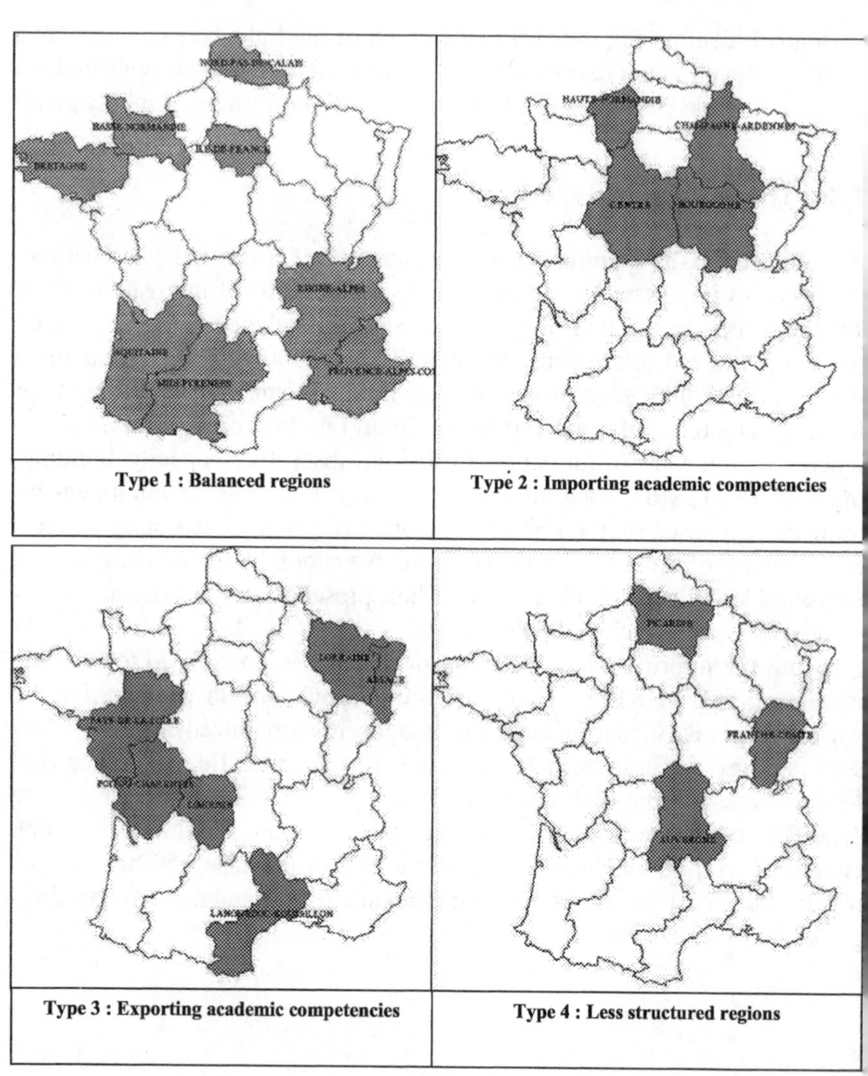

Fig. 7.4. Typology of regions

Group of regions	region	Legend	Firm	Lab	F∩L[a]	R1	R2 (%)	R2' (%)	KIBS (%)[b]	Technological density[c]	Scientific density[d]
Type 1: Balanced regions	Ile-de-France	IdF	4523	3040	2384	1,49	46	80	16	217	203
	Midi-Pyrénées	MP	562	774	395	0,73	42	73	13	71	118
	Rhône-Alpes	RA	1227	1634	774	0,75	37	65	14	175	122
	Basse-Normandie	BN	92	67	43	1,37	37	65	12	44	47
	PACA	PACA	624	687	350	0,91	36	64	23	69	95
	Nord-Pas de Calais	NPC	354	428	206	0,83	36	62	10	33	48
	Bretagne	Br	241	358	153	0,67	34	97	14	47	69
	Aquitaine	Aq	259	319	147	0,81	34	60	23	44	77
Type 2: Importing academic competencies	Champagne-Ardenne	CA	97	43	14	2,25	11	19	11	49	32
	Centre	Ce	194	123	52	1,58	20	34	6	85	50
	Haute-Normandie	HN	133	101	30	1,32	15	26	4	79	42
	Bourgogne	Bo	149	119	30	1,25	13	22	5	78	47
Type 3: Exporting academic competencies	Languedoc-Roussillon	LR	171	399	103	0,43	22	39	14	42	124
	Limousin	Li	64	110	31	0,58	22	38	8	34	56
	Alsace	Al	200	329	124	0,61	31	53	11	110	154
	Lorraine	Lo	303	478	172	0,63	28	49	7	62	74
	Pays de Loire	PL	215	305	110	0,70	27	47	15	43	48
	Poitou-Charentes	PC	100	134	40	0,75	21	36	15	43	43
Type 4: Less structured regions	Franche-Comté	FC	135	126	48	1,07	22	39	11	89	49
	Picardie	Pi	183	183	61	1	20	35	9	71	25
	Auvergne	Au	124	144	49	0,86	22	39	8	66	66
Other geographic areas[e]	/	/	52	101	42	/			/		
Total	/	10002	10002	5358	/	/	/	/	13	100	100

[a] Number of Cifre contracts where both the firm and laboratory involved are in the region. [b] Part of KIBS in the total number of Cifre contracts of the region.
[c] Number of patents per capita: index 100 = national average (OST, 2002, p. 167). [d] Number of publications per capita: index 100 = national average (OST, 2002, p. 149).
[e] This entry regroups Corse, overseas territories and the foreign areas (for laboratories only).

We will next examine the links between KIBS and academic institutions reflected in the database. Figure 7.5 shows the increase in the participation of KIBS in the Cifre system[13] since its creation. The trend shows an increase in relative terms, from around 6% of contracts during the first years that the system was in operation, to the present level of close to 20%. Therefore, it can be said that acquiring academic competencies is now a relatively common strategy for certain business services. In one-fifth of cases, industry-university collaboration will be indirectly developed through these links.

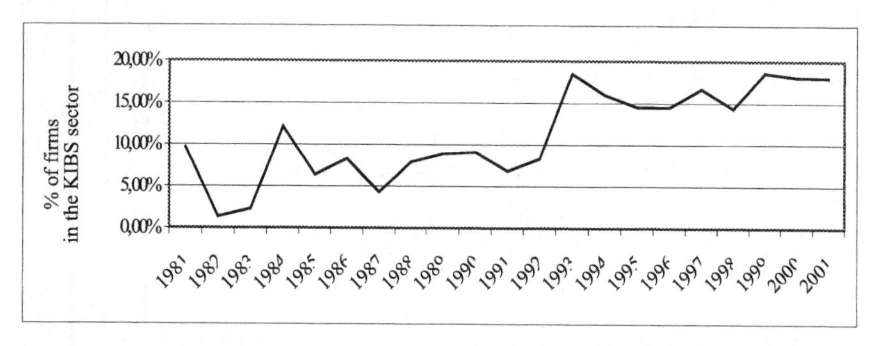

Fig. 7.5. Proportion of KIBS in the set of firms contracting Cifre agreements

Let us now look at where this catalytic role is strongest. The regions generally considered to be fully developed systems of innovation show relatively important proportions of KIBS throughout the period under examination – 16% for Ile de France and 14% for Rhône-Alpes, while Aquitaine and PACA have 24% of KIBS in the firms in their regions[14].

Our interpretation of these results is that while well-formed territorial systems (Type 1 regions) have necessarily developed an efficient fabric of knowledge-based business services, some regions with weaker innovation systems can also have a very high proportion of KIBS, which probably compensates for the lack of industrial partners. In the case of Aquitaine and PACA, which have significant scientific poles, but lack the industrial critical mass of Paris or Lyon-Grenoble, local political will and academic initiatives (science parks, start-up companies, etc.) may have had an influence. At the other extreme, Haute Normandie, with the lowest score of Cifre contracts with KIBS (4%), is a typical industrial region which imports

[13] In the database, we defined a subset of KIBS: R&D subcontractors, ICT services and various consultants.

[14] Cf. Table 7.2.

knowledge from universities outside its region (Type 2). It is probably too close to the Paris area to develop an independent innovation system.

7.4.3 Integrating Classical Indicators into the Analysis

We now compare our results based on the Cifre database, and, in particular, on the four regional types, with the classical indicators of scientific and technological production. For the French regions we use OST (2002) indicators of "scientific density", based on bibliometric data, and "technological density", based on European patent application statistics[15]. Technological density is particularly important as an indicator of success for a RSI; scientific density points to the nature of a regional system.

We start by observing that regional ranking by both scientific and technological density confirms our typology. As shown in the last two columns of Table 7.2 and in Figure 7.6[16], the four Type 2 regions (importing academic competencies) systematically display a scientific density that is lower than their technological density; the six Type 3 regions (exporting academic competencies) have a scientific density that is higher than (or equal to in the case of Poitou-Charente) their technological density. These regions then are clearly specialized either in firms' demand for, or in laboratories' supply of, academic competencies. The industry-university networks they form contribute to the NSI, but are not the basis for a regional system.

The Type 1 set of regions comprises different cases of scientific and technological development. If we compare our results with the two indicators of technological and scientific density, we can see that not all Type 1 regions are well developed RSI even though universities and firms within the region are collaborating.

Ile de France and Rhône-Alpes are the only regions with both scientific and technological indexes generally above 100. They are clear candidates for the title of "RSI"; it is interesting to note that they are also the only regions in this category where technological performance ranks higher than scientific performance.

Midi-Pyrénées has good scientific scores (118) but comparatively poor technological results (71): although the Cifre data indicate a balanced situation between firms and laboratories, the Toulouse area seems to be

---

[15] These regional indicators are normalized: the value100 corresponds to the national density (of the number of publications and the number of patents per capita respectively).

[16] The correspondence is also evident if Figures 7.2 and 7.4 are compared.

more of a sectoral cluster based around aero-space activities within the French NSI than a fully developed RSI.

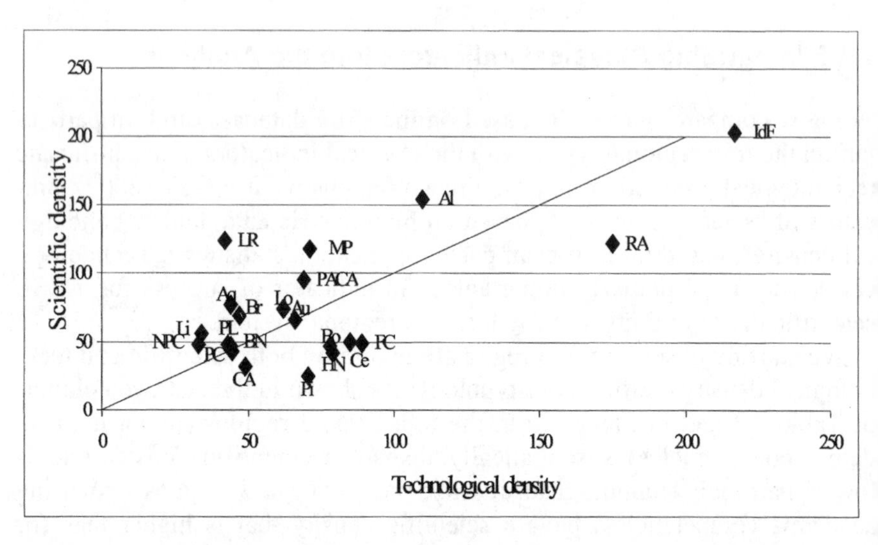

Fig. 7.6. Scientific and technological densities of French regions

Type 1	Type 2	Type 3	Type 4
IdF Ile-de-France	Bo Bourgogne	Al Alsace	Au Auvergne
RA Rhône-Alpes	HN Haute-Normandie	LR Languedoc-Roussillon	FC Franche-Comté
PACA	Ce Centre	Li Limousin	Pi Picardie
Aq Aquitaine	CA Champagne-Ardenne	Lo Lorraine	
NPC Nord-Pas de Calais		PL Pays de Loire	
BN Basse-Normandie		PC Poitou-Charentes	
MP Midi-Pyrénées			
Br Bretagne			

We now turn to the issue of global efficiency. As has been shown, the two regions at the top of the technological ranking are Ile de France (217) and Rhône-Alpes (175). We can definitely consider them to be well-formed and relatively autonomous systems of innovation. One result that is surprising is that Alsace is ranked in third position (110) while being a Type 3 region. The very high scientific score for Alsace (154), just below that of Ile de France, is explained by the academic concentration in the Strasbourg area, which has an international reputation for basic science

The industry in Alsace is active and efficient (leading to a good technological index, 110), but not very well connected to the local academic supply of knowledge and competencies, since most industry is "medium tech" SMEs and subsidiaries of multinational companies. This explains the Type 3 characteristics of Alsace, i.e. a net exporter of academic competencies. The region is very active in both science and innovation, but not as an integrated system. This territory is mainly the geographical location of a large number of actors of various innovation systems (national, international, trans-border, etc.) as several studies have shown (Nonn and Héraud 1995), and furthermore its industrial fabric and technological system are relatively split between the northern and the southern parts. Alsace has a long tradition of industry, and a large and diversified industrial fabric (often described as a big "production platform" interlinking large and small firms, subcontractors, etc.),

In contrast, although within the same category (Type 3, about the same number of Cifre contracts, high scientific density), Languedoc-Roussillon has a very low technological density (42). The main reason for this difference is the apparent lack of industrial critical mass. The existence of some high-tech firms around Montpellier is not enough to increase this.

Analysis of the empirical results allows us to examine the concept of RSI. Type 1 has been defined as a category of regions characterized by a relatively balanced involvement of local firms and laboratories (R1) and a significant proportion of Cifre contracts linking local firms with local laboratories (R2). However, this is not enough for these regions to qualify as RSI. For instance, Nord-Pas de Calais is in Type 1, but shows weak technological results overall (33, the weakest density of all the regions). Franche-Comté, a Type 4 region, is better technologically (89). In the case of Nord-Pas de Calais the strong participation in the Cifre system in our view is more an indication of a proactive policy than of a RSI; however, in the long run, such a policy could help to construct a RSI.

Most Type 2 regions have weak scientific density, but significant results for technology. The small number of Cifre laboratories is explained by the absence of important academic centres: the firms must find the research partners elsewhere. About one third of the laboratories associated with regional firms are located in the capital region of Ile-de-France. These regions are also characterized by a very small proportion of KIBS. We can conclude that such regions belong to larger systems of innovation: the French NSI or the Ile-de-France RSI. Their relatively high scores in terms of technological results probably reflect the performance of the larger systems, and the adequacy of the region to satisfy the needs and opportunities of the larger systems.

7.5 Conclusion

Our study of university-industry research collaborations, based on the Cifre database, has clearly confirmed some aspects of the French NSI. In this centralized system, there are few genuine subsystems. Outside the capital region Ile-de-France, Rhône-Alpes is the only region with a complete and balanced set of innovation actors. Other regions present interesting characteristics in terms of science and technology, but are generally either specialized in academic knowledge production, or have an efficient industrial network. Both types of regions contribute to the NSI, but without forming real subsystems. Some regions that show good performance in terms of innovation and knowledge creation are far from the model of an autonomous system. Conversely, we cannot support the hypothesis that closed regional systems are good examples of creative territories.

As intermediaries between industry and science, the advanced business services seem to play an important role. Their increasing involvement in the Cifre system is an indicator of this phenomenon and demonstrates a willingness to develop science-based activities. RSI rely strongly on such firms. Regional authorities should take cognisance of this in constructing their innovation policy.

Cifre PhD students are important in bridging academic and industrial communities. They create new knowledge by a recombination of qualitatively different sorts of knowledge and competencies. Whatever the geographic proximity of industrial firms and research laboratories, that sort of mobility of younger researchers between the two communities is a valuable contribution to collective learning. As a policy tool, the Cifre system has proved to be efficient and the French government recently decided to increase the grant to ANRT. Our study shows that geographic proximity is not a necessary condition for science-industry relationships. Therefore, in developing regional policies, science policy and innovation policy should be distinct from one another. There will certainly be links between them, but it would be a mistake to try to force the local science system to exactly match industrial demand.

Overall, we want to underline the importance of the link between academic science and industrial innovation. If this relationship is to be further reinforced within the knowledge-based economy regions should concentrate on a deliberate science policy alongside established innovation policies.

7.6 References

Acs ZJ, Audretsch DB, Feldman MP (1992) Real effects of academic research: comment. American Economic Review 82: 363-367.

ANRT (2001) 1981/2001, 20 ans de Cifre. Conference, ANRT, Paris.

Asheim BT, Isaksen A (2002) Regional innovation systems: the integration of local 'sticky' and global 'ubiquitous' knowledge. Journal of Technology Transfer 27: 77-86.

Atkins A, Dersley J, Tomlin R (1999) The engagement of universities in regional economic regeneration and development: a case study of perspectives. Higher Education Management 11(1): 97-115.

Audretsch DB, Stephan PE (1996) Company-scientist locational links: the case of biotechnology. American Economic Review 86: 641-652.

Becattini G (1991) Italian districts: problems and perspectives. International Studies of Management & Organization 21(1): 83-90.

Braczyk HJ, Cooke P, Heidenreich M (1998) Regional innovation systems. Routledge editions, London.

Bureth A, Héraud J-A (2001) Institutions of technological infrastructure (ITI) and the generation and diffusion of knowledge. In: Koschatzky K, Kulicke M, Zenker A (eds.), Innovation networks. Physica Verlag, Heidelberg, pp. 69-91.

Catin M, Lacour C, Lung, Y (2001) Innovation et développement régional. Introduction to special issue, Revue d'Economie Régionale et Urbaine 1 : 3-10.

Carlsson B, Jacobsson S, Holmen M, Rickne A (2002) Innovation systems: analytical and methodological issues. Research Policy 3: 233-245.

Chabbal R (1995) Characteristics of innovation policies, namely for SMEs. STI Review 16: 103-140.

Cooke P (2001) Regional innovation systems, clusters, and the knowledge economy. Industrial and Corporate Change 10(4): 945-974.

Da Rosa Pires A, Anselmo de Castro A (1997) Can a strategic project for a university be strategic to regional development? Science and Public Policy 24(1): 15-20.

Etzkowitz H, Webster A, Gebhardt C, Cantasino Terra BR (2000) The future of the university and the university of the future: evolution of the ivory tower to entrepreneurial paradigm. Research Policy 29: 313-330.

Fischer MM (2000) Innovation, knowledge creation and systems of innovations. The Annals of Regional Science 35: 199-216.

Foray D, Lundvall B-Å (1996) The knowledge-based economy: from the economics of knowledge to the learning economy. In: Employment and growth in the knowledge–based economy. OECD, Paris, pp. 11-32.

Fritsch M, Schwirten C (1999) Enterprise-university co-operation and the role of the public research institutions in regional innovation systems. Industry and organisation 6(1): 69-83.

Fritsch M (2001) Public research institutions in regional innovation systems: assessment and outline of a research agenda. In: Broecker J, Hermann H (eds.),

Spatial change and interregional flows in the integrating Europe – essays in honour of Karin Peschel. Physica, Heidelberg, pp. 89-100.

Gibbons M, Johnston R (1974) The economic benefits from science. Research Policy, 22: 220-242.

Héraud, JA (2003) Regional innovation systems and European research policy: convergence or misunderstanding? European Planning Studies 11(1): 41-56.

Héraud JA, Nanopoulos K (1994) Les réseaux de l'innovation dans les PMI: illustration sur le cas de l'Alsace. Revue Internationale PME 7(3-4) : 65-86.

Héraud JA, Kern F (1997) Les CORTECHS: innovations, apprentissage en coopération et dynamique organisationnelle. In : Guilhon B, Huard P, Orillard M, Zimmerman JB Économie de la connaissance et organisations. L'Harmattan, Paris: 383-399.

Héraud JA, Isaksen A (2001) Changing nature of knowledge, globalisation and European integration, relevance, effects and opportunities for European less favoured regions. Contribution to the CONVERGE programme, CISEP, Technical University, Lisbon.

Isabelle M, Guichard R, Fleurette V (2003) Analyse économique des modalités de transfert de savoir dans les grands organismes de recherche français. Working Paper de l'IMRI.

Jaffe AB (1989) Real effects of academic research. American Economic Review 79: 957-970.

Jones-Evans D, Klofsten M (1998) Role of the university in the technology transfer process: a European view. Science and Public Policy 25(6): 373-380.

Landabaso M, Oughton C, Morgan K (2001) Innovation networks and regional policy in Europe. In: Koschatzky K, Kulicke M, Zenker A (eds.) Innovation networks - concepts and challenges in the European perspective. Physica, Heidelberg, pp. 243-273.

Lee YS (2000) The sustainability of university-industry research collaboration: an empirical assessment. Journal of Technology Transfer 25: 111-133.

Levy R (2004) Les CIFRE: un outil de médiation entre les laboratoires de recherche universitaire et les entreprises. Document de travail du BETA.

Lundvall B-Å, Johnson B, Andersen ES, Dalum B (2002) National systems of production, innovation and competence building. Research Policy 23: 213-231.

Lundvall, B-Å, Borras S (1997) The globalising learning economy: implications for innovation policy. Final report for the TSER Programme, European Commission, Brussels.

Lung Y, Rallet A, Torre A (1999) Connaissances et proximité géographique dans les processus d'innovation. Géographie, Economie, Société 1 : 281-306.

Malerba F (2002) Sectoral systems of innovation and production. Research Policy 31: 247-264.

Meyer-Krahmer F, Schmoch U (1998) Science-based technologies: university-industry interactions in four fields. Research Policy 27: 835-851.

Muller E (2001) Innovation interactions between Knowledge-Intensive Business Services and Small and Medium-Sized Enterprises. Physica Verlag, Heidelberg.

Nelson R (1993) National innovation systems: a comparative analysis. Oxford University Press, Oxford.

Nonn H, Héraud JA (1995) Les économies industrielles en France de l'Est: Tissus et réseaux en évolution. Presses Universitaires de Strasbourg, Strasbourg.

OECD (2002) Benchmarking industry-science relationships. OECD, Paris.

OST (2002) Rapport de l'observatoire des sciences et technologies (Barré, Esterlé, eds.).Editions Economica, Paris.

Porter ME (1998) Clusters and the new economics of competition. Harvard Business Review November-December: 77-90.

Quéré M 1994, The 'convention CIFRE': a successful French incentive scheme for the management of human resources in research activity. International Journal of Technology Management 9: 430-439.

Rip A (2002) Regional innovation systems and the advent of strategic science. Journal of Technology Transfer, 27: 123-131.

Salter AJ, Martin BR (2001) The economic benefits of publicly funded basic research: a critical review. Research Policy 30: 509-532.

Sander A (2000) Les conventions CIFRE, un indicateur des relations entre acteurs de la création de connaissances dans les régions françaises 37ème colloque de l'association de Science Régionale de Langue française, Lausanne.

Schaeffer V (1998) Les stratégies de valorisation de la recherche universitaire, éléments d'analyse. PhD thesis, University Louis Pasteur, Strasbourg.

Schartinger D, Rammer C, Fischer MM, Frohlich J (2002) Knowledge interactions between universities and industry in Austria: sectoral patterns and determinants. Research Policy 31: 303-328.

Scott A, Steyn G, Geuna A, Brusoni S, Steinmueller WE (2001) The economic returns to basic research and the benefits of university-industry relationships: a literature review and update of findings. Report prepared for the Office of Science and Technology, Department of Trade and Industry. SPRU, University of Sussex, Brighton, UK.

Strambach S (2001) Innovation processes and the role of Knowledge-Intensive Business Services (KIBS). In: Koschatzky K, Kulicke M, Zenker A (eds.), Innovation networks. Physica Verlag, Heidelberg, pp. 53-68.

Varga A (1997) Regional economic effects of university research: a survey. Regional Research Institute Working Paper.

Voyer R (1998) Knowledge-based industrial clustering: international comparisons. In: De La Mothe J, Paquet G Local and regional systems of innovation. University of Ottawa Press, Ontario, Canada.

8 Research and Development Tax Incentives: a Comparative Analysis of Various National Mechanisms

Stéphane Lhuillery[1]

BETA, Strasbourg, E-mail: lhuillery@cournot.u-strasbg.fr

8.1 Introduction

It is now fairly well documented that private research and development (R&D) contribute to productivity growth (OECD 2001a). Therefore, it is usual to consider that governments have a role in stimulating R&D investments and activities in the face of market failures resulting from technology leakages. Subsidies and patent rights are popular incentive mechanisms to achieve these purposes.

A look at the recent evolution of science and technology (S&T) state policies in the OECD countries suggests two stylized facts: a reduction in public R&D support to firms (OECD 2003a); a shift in the structure of public R&D funding. On the first issue, the changes can be considered as an evolution toward a general enabling framework for innovation. This is evident in the recent attention being given to a more efficient venture capital system or a more efficient intellectual property rights (IPR) system, which needs few public resources. It can be interpreted as an evolution toward a market based S&T policy, which recognizes the importance of markets for technology.

On the second issue, the evolution towards a more decentralized system, in which direct R&D procurements are smaller and direct support for R&D scarcer, is significant. The process of decentralization from mission oriented R&D policies, can be considered through the increase in prize-based incentives and R&D tax incentives (RDTI thereafter). In the period 2001–2003, 16 OECD countries were offering RDTIs compared with only 12 in 1996 (OECD 2003b). Furthermore, seven countries (Australia, France, Ja-

[1] I assume sole responsibility for the contents of this chapter which does not reflect the views of the French ministry of education and research. I acknowledge the assistance received from office in charge of the French R&D Tax credit for access to the data.

pan, Norway, Portugal, Spain and the United Kingdom) have recently increased the attractiveness of their RDTI schemes.

The increase in RDTIs could be interpreted as a trade off between different S&T policy tools. However, this argument seems somewhat weak: traditional direct R&D funding has continued and is undisputed. Thus, the introduction of fiscal policies to promote innovation must be considered as a particular mechanism. Analysis of the provisions of R&D tax schemes is required to understand why, over the last 20 years, they have been introduced by several governments. Such an examination is necessary to provide a rationale for the interactions between S&T funding processes, and to provide better understanding of the overall efficiency of national S&T policies.

This chapter, therefore, provides an in-depth study of the specificities of national RDTIs. We intend to show the diversity of RDTIs that have been implemented, as well as their various targets, efficiency, problems, and solutions. The chapter is however not exhaustive. The limitations are the amount of information available and the impossibility of establishing how these complex fiscal systems and user practices have evolved over time.

The first section offers a brief survey of the different types of national R&D fiscal incentives, and their impacts. Section 8.3 deals with exogenous problems, which are the boundaries that are inevitably raised by the adoption and practice of a fiscal incentive. We go on to show that different provision is made by governments to shape private R&D or innovation investments (Section 8.4). Section 8.5 looks at the interaction between R&D tax schemes and other S&T policy tools and fiscal tools. Section 8.6 concludes.

8.2 R&D Tax Incentives: an Overview of National Schemes

8.2.1 A Spreading Mechanism

RDTI are the most recent tools within S&T policies[2]. With the exception of Canada (since 1944) and Japan (since 1967), it is only in the last 20 years that these tools have been introduced. The choice of RDTI is neither systematic, nor specific to R&D-intensive economies (e.g. Finland, Germany and Sweden have no RDTIs), nor are they restricted to the OECD countries (Cyprus, Hungary, Poland, the Czech Republic, as well as Brazil,

[2] Jean-Baptiste Colbert, the French Minister of Finance in the government of King Louis XIV, set up a system of tax exemption to establish and sustain more than 400 "manufactures royales" (from 1662 to 1666) that produced new and improved quality goods.

India, Malaysia, Mexico, Singapore and Taiwan have implemented similar R&D fiscal incentives).

The implementation of RDTIs is not systematically linked to the use of direct R&D funding, nor does it address a real incentive to invest in R&D. As shown in Table 8.1, there are major differences between countries with regard to the percentage of total government aid to R&D that tax incentives represent. Italy, New Zealand and Norway finance business R&D, but do not give preferential tax treatment. Japan and Mexico, even with RDTIs, do not as yet have a particular policy with regard to firms executing R&D. Direct funding dominates in France, the US and the United Kingdom, where quite favourable tax schemes are also in place. Canada and Australia are the countries that are the most reliant on RDTIs, although the incentives are highest in Portugal and Spain.

Table 8.1. R&D direct funding and fiscal incentives, by country

R&D Tax Incentives Direct R&D funding	Low	Medium	High
Low	Japan, Mexico	Canada, Australia	
Medium	Germany, Sweden, Belgium, Ireland, Finland	Denmark, Netherland, Austria, Korea	Portugal, Spain
High	Italy, New Zealand, Norway	France, USA, UK	

Based on OECD (2002).

Several econometric studies have estimates the price elasticity of R&D expenditure, and have suggested that the type of R&D fiscal framework in place has a positive impact on R&D spending (Hall and Van Reenen 2000). The results show that RDTI is not a sufficient condition to induce a significant surge in R&D investments. In Europe, in particular, these mechanisms will not be enough to allow the forecast overall target of 3% of GNP dedicated to R&D by the year 2010 to be achieved[3]. Several economic arguments can be put forward to explain the wide range of and differences in such R&D tools (EC 2003). RDTIs are decentralized tools – that leave investment decisions in the hands of firms. They can be ex-

[3] The tax price elasticity of total R&D spending is on the order of unity (see Hall and van Reenen (2000)). It thus would take a huge tax credit to reach the 3% EC target. Roughtly, taking the EC R&D expenditure, and keeping the share between private and public expenditure (about 56% and 44% in 2001, OECD, (2003)), the additional R&D tax concession required should thus be around 0.5% or 0.6% of EC GDP.

ploited at low administrative cost by firms with R&D investments. Fiscal tax incentives can be broadly defined and do not discriminate a priori between industry sectors, types of firms, locations, or fields of R&D. RDTI are therefore not the best way to achieve the highest social returns. It could be argued, therefore, that RDTI are more neutral than other technology policy tools. For example the EC explains that measures pursuing general economic policy objectives, through a reduction in the tax burden related to certain production costs (such as R&D), provided that they apply "without distinction to all firms" and to the production of all goods, do not constitute state aid (EC 1998).

8.2.2 The Different Types of R&D Tax Incentives

There are three major fiscal ways to sustain R&D investments by firms: accelerated depreciation; special allowances; and tax credit (see Table 8.2):

- Depreciation allowances for R&D expenses are considered to be RDTIs if they are allowed at a rate that is greater than the rate of economic depreciation. Several countries make special allowances for R&D expenditure in the form of machinery, equipment, or buildings (capital assets). Even though knowledge leakage and technology depreciation can be high in R&D activities, this kind of accelerated depreciation represents a subsidy for firms where 100% or more of annual costs are taken into account (as is the case in Ireland, Denmark, and the UK for buildings).
- Special allowances on R&D investments allow firms to deduct more from their taxable income than they actually spend on R&D. The allowance can be in proportion to the level of R&D or be incremental, or a combination of both. In this case, a firm is allowed to deduct from its taxable income its current R&D expenditures, and some fraction of the increase in its R&D expenditures over a base period. For example, since 2000, Austrian firms have been able to deduct 25% of their R&D investments from their profits. If R&D investment in a particular year is above the average of the preceding three years, then an additional allowance of 35% can be deducted from the taxable income.

Table 8.2. The different type of RDTIs, by countries in 2002 and before

country	Accelerated depreciation		Tax allowance			Tax credit			
	Machinery & Equipt	Buildings	Level	Increment	Combination	Level	Increment	Combination	Self selection
Australia					Yes				
Austria					Yes				
Belgium	3 years		Yes						
Canada[d]	100%					Yes			
Denmark	100%	100%	Yes						
Finland						Yes[c]			
France							Yes	Since 2003	Substitute[b]
Germany						Yes[a]			
Greece	3 years	12.5 years							
Ireland	100%	100%		Yes					
Italy			Yes						
Japan[e]									Complt and/or Subst
Korea									Substitute
Mexico	35%						Yes		
Netherlands	5 years					Yes			
Norway						Yes			
Portugal	4 years							Yes	
Spain	100%	10%						Yes	Complement
Taiwan									
England	100%	100%				Yes			
USA							Yes		

[a] Ended in 1991; [b] 1988-1990 only; [c] Ended in 1987; [d] On two occasions (1962–75, 1978–83), Canada introduced an incremental measure which it subsequently abandoned; [e]. Japan, in 1976, lowered the rate of its incremental mechanism and later brought in others that were based on volume.

- R&D tax credits are similar to tax allowances, but relate directly to the level of tax payable. Four different types mechanisms seem to be possible.

  - The first, the so-called "volume" mechanism, gives an incentive proportional to investment. A firm operating in a country that allows volume RDTIs at a rate of, say, 25% would therefore get a $25 tax credit for every $100 of R&D expenditure. Canada, Italy, Malaysia, the Netherlands, and Singapore each have a system based on this principle.

  - The second type of mechanism, incremental RDTIs, is more complicated in that it benefits only a firm's marginal expenditure. Incremental RDTIs lower the marginal cost of R&D, but only for outlays in excess of base-period expenditure. Under this system, and assuming an RDTI rate of 50%, a firm that spends $300 in base year *t* and $400 in year *t + 1* will get a tax credit of $50 [0.50 x ($400-$300)]. France, Taiwan and the US have opted for this type of mechanism, but apply different base periods.

  - The third mechanism is a combination within a single RDTI of the volume and incremental schemes. This scheme is an attempt to counter some of the criticisms of incremental tax credit, while also retaining its incentiveness. In Spain, annual R&D expenditure qualifies for a tax credit of 30%. In addition, there is an incremental tax credit of 30% if R&D expenditure for the period is greater than average R&D expenditure for the previous two years. Portugal has in place a similar scheme, but with greater emphasis on an increase in overall R&D expenditure (30%) rather than on the R&D level (only 8%). This 'combination' type scheme seems to be attractive to firms and policy makers (see EC 2002), but has been implemented only in Spain, Portugal and France (since 2003 in this last country).

  - The fourth mechanism is to allow firms to choose their preferred RDTI scheme. This however is rarely applied: France discontinued the use of this mechanism at the end of the eighties. It allowed a firm to choose between the "volume" and "incremental" RDTI schemes. Within this category is the Japanese system, which is extremely complex, and consists of combining several tax incentives for R&D, either incremental or volume. The Japanese allow 20% incremental tax credit for R&D expenses. In addition, in 2003, they introduced several proportional tax credits that are more generous, to cope with the decline in private R&D funding. Japan allows small and medium-size enterprises (SMEs) or firms to deduct up to 15% on all R&D expenses; a 15% tax credit is also offered for R&D expenses incurred in joint projects with academia (a 10% scheme exists in Spain regardless of firm size), business or government, or for R&D commissioned by government. Firms are able to choose among the different schemes: the 15% tax credit for collaborative working is additional to the basic 15% tax credit. More interesting, is the possible trade-off for SMEs, since a level tax credit (cumulated or not) can be applied instead of the 20% incremental R&D tax credit. A simpler scheme has been implemented in Korea where, in 2002, there was a 40% incremental tax credit and a 7% R&D Facility Investment Tax Credit. As in Japan, it is only SMEs that can choose between the two tax credit provisions. In the US, this type of auto-selection is possible since an alternate incremental tax credit is available to generate higher research credits for companies with significantly increased sales figures or otherwise stagnant research expenditures[4].

It is, of course, also possible to combine accelerated depreciation with R&D allowances or R&D tax credits. For example, R&D tax credits in Canada are related to current expenditure whereas accelerated depreciation applies to R&D capital expenditures. In Japan, in addition to the R&D tax credit, a special allowance (50%) is now available on R&D investments.

8.3 Defining the Base for RDTIs

A major issue is the definition of the tax base. Its definition can be a means to reach a target and to protect government from possible opportunistic behaviours from firms. There are two main aspects to this definition. The first is to distinguish between what a fiscal innovation policy does and does not encompass, and the second is to identify what is the frontier of the innovation process for an innovative firm embedded in scientific and technological networks.

8.3.1 Internal Dividing Lines

The initial problem for regulators is to define the R&D activities within a firm. How certain R&D outlays are categorized can create substantial dis-

[4] See http://www.irs.gov/businesses/small/industries/article/0,,id=97643,00.html#aic

tortions, since firms will tend to neglect investments that are not included in the base (Eisner et al. 1984). Governments are therefore compelled to draw a strict dividing line between expenditure that can and expenditure that cannot be regarded as relating to R&D. This line may be flexible (see the successive extensions in the USA in 1996 and 1999). Most countries define R&D on the basis of the *nature of the activities* carried out (e.g. France). For the purposes of RDTI, the definition of R&D is often in line with the OECD's "Frascati Manual" (OECD 1994). However, Canada, for example, has widened the definition of the uncertainty attaching to any research to include not only uncertainty as to results, but also the uncertainty of R&D costs, whereas the US acknowledges the absence of uncertainty in respect of some of the work that is done on technical systems (IFRI 1992). In some countries (e.g. Spain), the distinction is embodied in a *list* of activities considered to constitute R&D. The Netherlands officially lists both activities that qualify as R&D and activities that do not. The UK, France, the USA, and Japan do not have such lists, on the grounds that they might be incomplete. However, in practice, there are several examples, and lists of several exclusions available on web sites or help leaflets.

Another option is to define the *nature of R&D findings*. In general, R&D expenditures are industry-biased, in the sense that they are a better indicator of innovative activities in the manufacturing than in the service sectors. Consequently, R&D tax schemes also are industry-biased. Furthermore, R&D often does not produce results, but is explicitly covered by RDTI (e.g. in the USA).

Computation (and thus control) of qualified R&D also requires complementary provisions. At first is needed to calculate the amount of eligible R&D expenditure if staff is not assigned to R&D on a full-time basis. Personnel are considered to be doing research if they spend a percentage of their time on it. Likewise, in respect of equipment, it is necessary, as Canada has done, to set a threshold for the amount of investment considered eligible for a RDTI. However, such a threshold may be unsuited to (the increasingly frequent) project-based R&D, the horizontal structure of which makes it difficult to allocate equipment to specific tasks. The Canadian system became less restrictive in 1992: to qualify, equipment had to be in used for R&D for the equivalent of 50% of the working day, instead of 90% as previously, to be eligible for a credit. Another, more flexible, approach is to count only human resources. This limits allocation to R&D to the most visible and most easily controllable of corporate outlays. In the Netherlands, for example, only R&D-related wages and social security contributions are taken into account. In Belgium, the incremental number of researchers is taken into account. While this method is certainly simpler (Germany also bases its RDTI on researchers), it has two drawbacks: it does not compensate a firm for the entirety of its research and development effort; and it may induce labour/capital substitution with regard to R&D. The French system – a more elaborate arrangement, but one that retains this simplicity of calculation – pegs current expenditure at a flat 75% of wages and social security contributions, for research technicians and researchers only. Although it is a more accurate reflection of actual corporate expenditure, this method favours those companies – of which there are many in, say, the service sector – whose overheads constitute less than 75% of operating expenditure. In fact, wages and social security contributions represent only 41% of R&D expenditures declared by French firms. The Canadian system is similar, but seems fairer and more flexible: the rate is 65% of wages and social security contributions for all R&D personnel, and firms are free to opt for this flat-rate treatment, or not.

A second dividing line is needed, but is as difficult to apply as the first (IFRI 1992). Since R&D is just one facet of innovation, financial support for R&D is an incomplete way of encouraging this: R&D costs can never represent more than a fraction of the full cost of *technological innovation*, as shown by community innovation surveys.

The question therefore arises as to whether the base of the tax credit should be broadened to include, alongside research and experimental development activities, spending on industrial development, market development, and so on. Some of Pacific rim countries (Australia, Korea, Malaysia, and Singapore) have already opted for a wider vision of R&D, by extending the range of eligible innovative activities (to include, for instance, market surveys). The Spanish definition of R&D is very broad since it includes expenditures on quality certificates, know-how acquisition, industrial design (as in France), and production engineering. What is very interesting, however, is that, apart from Belgium, France, the UK and the USA, there has been no inclusion of IPR costs, although these expenditures are easy to monitor and their appropriation is an integral part of the innovative process. A second aspect is the cost of standardization and/or normalization. The French RDTI scheme is the only one that includes the costs related to normalization activities, even if they are restricted to meetings with official organizations (and have a ceiling).

Widening the base for RDTIs poses critical problems for practitioners. Apart from the problem of identification, that of indivisibilities is becoming increasingly acute, inasmuch as the process of innovation clearly cuts across corporate functions. Once these practical problems are out of the way, the feasibility and relevance of such a form of government assistance to innovation remain. Last, such tax incentives for innovation should be classified as competitive measures, which does not fit with EU regulations or the General Agreement on Tariffs and Trade (GATT)[5].

The introduction of RDTI schemes has a qualitative impact on firms: we can identify three consequences:

- Institution of *accounting systems*; for example, 87% of Australian firms say they have set up ad hoc accounting arrangements (BIE 1989).
- A *restructuring of R&D outlays* in favour of items included in the base (Eisner *et al.* 1984). Nevertheless, the distortion caused by the change in relative prices casts some doubts on its relevance.
- A *reclassification (relabelling)* of R&D-related expenses. It is estimated that in the US and Canada, some 14% of outlays have been affected (Mansfield and Switzer 1985). An Australian study confirms this order of magnitude with a rate of 19% (BIE 1989). Nevertheless, as we have suggested, this reclassification is a partially legitimate process stemming from the very organization of the process of innovation and from the irregular nature of R&D activity. Nevertheless, it seems that Germany abolished its R&D tax credit at the beginning of the 1990s due to problems of abuse (see OECD (2002)).

8.3.2 External Dividing Lines

Corporate R&D expenditure is not, however, limited to inhouse R&D. An external dividing line has to be drawn including, or not, external R&D services, R&D cooperative agreements, R&D within a group of firms or international financial flows.

Firms subcontract or cooperate in order to achieve research and development objectives, thereby generating R&D-related financial flows. Systems differ considerably from country to country to cope with the problem of double counting (on private and even public R&D funding). Several countries (e.g. France) have introduced licensing procedures in respect of R&D service providers. Even if such a procedure does not really limit the amount of double credit, it can favour a particular type of R&D partner. Some countries limit further double crediting by disqualifying R&D flows *between private firms*, in respect of the sources and/or uses of funds devoted to acquiring or producing R&D. Often, only R&D subcontracted to universities or specific research organizations is eligible (in Australia, Canada, the United Kingdom, and Spain). The problem is overcome sometimes by application of a specific RDTI (in Japan or Spain), or a special

[5] But does the European EUREKA initiative do so?

accounting procedure (USA[6], Australia) dedicated to R&D cooperation with a public organization. An even more radical solution to the possible double credit problem is to restrict eligibility to applied R&D tax credit only to subcontracted R&D, as has been the situation in Switzerland since 1995.

R&D is also often organized at the group level, which also introduces the problem of double credit when there are financial flows for R&D between affiliates. A simple rule is to consolidate the different R&D expenditures within a group to avoid paying twice for the same research. Such a procedure has been implemented in Canada, the US, the UK, and France. In contrast, Japan limits volume-based tax credits to independent small businesses. There are two intermediate systems: a tax credit against consolidated tax (e.g. Spain), and a groupwide ceiling which can be apportioned freely among subsidiaries (e.g. the Netherlands). Both systems are demanding, however, since they impose a very exacting exercise in consolidation for groups and tax inspectors alike.

In order to foster innovation, production and national competitiveness, work performed *abroad* is seldom included in the RDTI base. Canada's initial willingness to let firms include current R&D spending abroad was generous, but Canadian R&D tax credit has evolved towards a strictly national scheme, as is the case in France (except for normalization activities), the Netherlands and the US. Intermediate schemes can be found in Japan and Spain. In April 1995, Japan introduced a research tax credit applicable to cooperative R&D with foreign laboratories. Spain's RDTI base includes external R&D expenditures that are performed by foreign universities or government laboratories, but only those belonging to the European Community. Even when opened to private firms abroad, RDTIs are designed to limit the possible loss of fiscal income for government: in Australia, RDTIs were applicable to spending abroad up to a limit of 10% of total R&D, whereas the UK scheme gives tax allowances for overseas R&D expenditure for SMEs only.

8.4 Design of RDTI Mechanisms

Besides the definition of the tax base, there are three types of provisions that shape the range of RDTIs: those intended to limit revenue shortfalls and administrative costs primarily associated with new measures; those to ensure that firms with no tax liability can also benefit from tax incentive

[6] See http://www.irs.gov/pub/irs-regs/research_credit_basic_sec41.pdf

schemes; and those designed to include some selectivity – in terms of firm, technology and region – in tax incentive schemes.

8.4.1 Limiting the Risks Attached to RDTIs

Two main types of provision lessen the risks of budgetary overshoots and manipulation that accompany incremental tax incentives.

In addition to the rate of tax credits, crowding-out effects and budgetary overshoots can be limited by imposing a *ceiling* on tax benefits. Half the systems examined have such a ceiling. This can be achieved in two ways: either the ceiling can be a fixed deductible amount (Australia, Portugal, France, the Netherlands), or it can be a percentage. Spain has a rate of 35%. Switzerland opted for a flexible scheme whereby deductions are allowed for contracted future costs of R&D carried out by third parties, but limited to 10% of taxable profits, or CHF1 million, whichever is lower. The US also uses an original double ceiling mechanism, but firms cannot choose between them: not only must a firm spend no more than twice the previous year's base amount[3], but the average intensity of R&D expenditure (i.e. its ratio to turnover) must not exceed 16%.

By calculating the rate of increase over a base period of several years (the US, France, Korea, and Taiwan), the amount of tax benefits and, thus, governments' fiscal loss, can be *smoothed*. This kind of provision mainly concerns incremental RDTIs. In France, Korea, and Taiwan, base amounts are calculated based on the previous 2–4 years. The smoothing of tax credits, along with the fact that penalties are imposed if R&D spending is subsequently reduced, limits the scope for opportunistic behaviour on the part of firms. In the US, the base is the average ratio of R&D spending to turnover, which is more equitable in that RDTIs are indexed to the actual efforts a firm makes during the year. The computation is quite different in Japan, where the sustained base is the average of the three highest R&D expenditures in the five previous years.

The smoothing method, nonetheless, has two drawbacks. First, the price adjustment of past R&D expenditures raises problems (OECD 1993). The consumer price index or (in France's case) investment indices used by tax authorities, do not always accurately reflect cost trends in R&D (Mansfield et al. 1983). Second, as Hines (1994) emphasizes, the fact that the base on which a firm is to get a credit is calculated over several years will prompt it to postpone investment. Consequently, the calculated base will be lower and will yield a bigger credit in subsequent periods. It would appear, therefore, that budgetary planning considerations outweigh concerns over the cost of firms' opportunism, which may be underestimated.

In incremental tax schemes, there are two types of *negative credits:* a strict system whereby credits have to be paid back every year (as in France between 1983 and 1987, and in Belgium); and a system that allows negative credits to be carried forward (e.g. in the US and France today). While the first system is a disincentive, the second allows a firm to reap benefits on a one-off basis without incurring any real penalties; it thus facilitates a firm's initial efforts, but does not ensure that R&D will be continued. In the latter case, the inequity of incremental RDTIs also affects small innovative firms and firms in low-tech sectors.

Despite all the provisions, manipulation of R&D tax schemes may still be possible. For example, 15 firms in France are affected by the €6.1 million ceiling while only 13 are declaring in the fiscal file a similar amount of R&D to what is revealed by the annual R&D survey. What is the reason for this? Two firms out of the 15 realized that under an incremental tax scheme, they could under-value their annual R&D investments for the purposes of the RDTI administration and declare only an annual increase on their R&D spendings, that is, just twice the French ceiling (the tax credit is 50% in France). They thus receive the maximum tax credit regardless of actual R&D efforts.

8.4.2 Targeted Incentives

Disparities between firms are taken into account to varying degrees by the RDTI systems of each country, with some firms being treated better than others. RDTI mechanisms frequently feature three types of segmentation – by size, by technology, and by region. In this situation, RDTIs are more than a remedy for market failures; they constitute an instrument of industrial policy.

8.4.2.1 Small, Young and Poor Firms

RDTIs are more and more frequently being targeted toward SMEs and start ups, which are more likely to face credit problems. Firms are selected based on the number of employees or their level of tax liability.

Segmentation by *size* is to the benefit of smaller firms: either they receive credits at preferential rates (Australia, Canada, the Netherlands, Italy, Japan, Korea, and the UK) or credits are limited to SMEs. In Japan, SMEs with equity capital of less than ¥100 million and less than 1,000 employees receive tax credits equal to 6% of their R&D expenditure. The R&D policy toward SMEs is however not restricted to the rate of tax credit. The ceiling definition may be also important to define the scope of the RDTI. A level definition of a maximum (as in the Netherlands or France), helps SMEs in preference to large firms as long as the ceiling is not too high. In the Netherlands and France, however, few firms are affected by this threshold. The focus on SMEs, therefore, is more significant when implemented through a double relative ceiling, as in Japan: a SME may deduct up to 15% of its taxes, whereas for large firms the ceiling is lower (10%). UK and Italy are more focused on SMEs since the RDTIs in these countries are restricted to SMEs (in Italy there is enhanced deduction of 130% of all current and eligible R&D expenditure). The target on SMEs can be justified by the greater ability of large firms to spread fixed costs on large market sales, but also by the possibly better responsiveness of large firms to RDTIs (Koga 2003).

A firm must have *a tax liability* in order to benefit from RDTIs. As a result, SMEs and start-ups in particular, can be penalized. To solve this problem, most systems incorporate two main alternative arrangements. A system of tax credit *refunds* is often implemented (UK, Australia). In Austria, a cash refund is also possible, but at a flat 3% allowance rate when there is no tax liability. Here, the aim of the tax credit is wider than promoting technology. The US, given the method adopted for computing the base, offers start-ups a flat-rate base derived from turnover. Such firms receive a tax credit if they spend more than 3% of their turnover on R&D. Portugal, Spain, the US, and Australia allow tax benefits to be *carried forward*, so that a firm must improve its overall profitability before it can benefit. France and Canada implement both systems, but a cash refund is only available for specific firms: in France, the cash refund is restricted to young firms, whereas, in Canada, the federal tax credit is refundable only for small corporations. Lastly, in France, the tax credit can be used to underwrite a bank loan by a firm that cannot be refunded and does not want to wait for its credit. The Netherlands and Norway have adopted a different and interesting approach: they have tax incentive schemes in which tax credits are based on R&D labour costs instead of profits. Firms need not have incurred these R&D labour costs, but they may apply for a monthly tax credit on taxes on salaries. Under such a scheme, new firms and firms with no profit tax liability are not penalized.

8.4.2.2 Deserving and Specialist Firms

If governments try to limit the amount of their fiscal loss, they also often try to help firms making minimum R&D efforts, important efforts or effective efforts. The problem here is that provision is based on very different criteria.

A first advanced mechanism is a two tiered credit, giving a higher rate of super-deduction on spending up to some given amount of qualifying R&D expenditure, and a lower rate on spending above this level. Under this scheme, it is assumed that the first euros are more difficult to spend (and/or are more productive) than the marginal ones after a certain threshold. In the Netherlands, the rate of tax credit varies by "tranches" of R&D spending. The first tranche, under €90.756 qualifies for a credit of 40%, and the upper tranche for a credit of 13%. In Canada, firms earn credits at a rate of 35% on the first €1.322.489 of R&D. All other firms earn 20% credits for the rest.

Some countries establish a minimum threshold for eligibility for R&D tax credit. The justification here is almost the reverse, although compatible with the two-tiered mechanism. In this scheme the first euros are considered unproductive until a significant R&D base has been achieved, or, put more simply, the social rate of return on R&D investments must cover the government's administrative costs. In the UK (in 2002), unless a company had spent more than £25,000 on qualifying R&D expenditure (i.e. without capital assets) in a 12 months accounting period it did not qualify for tax relief. In Australia, the threshold is only AUS$20,000. In Japan, the 2003-2006 acquisition costs of R&D facilities, from a threshold of ¥2.8 million, open an accelerated depreciation of 50%.

Third, specialization can also be considered to be a guarantee of a better R&D outcome. In Spain, R&D expenditure on salaries for qualified researchers assigned full-time to R&D projects, and business R&D expenditure on projects that are contracted to universities or research organizations, are eligible for an extra tax credit incentive of 10% and are thus considered as more effective than part-time R&D jobs. Specialization is also rewarded by the current (2002–2003) Australian RDTI in two ways. First, R&D activities are required to adhere to an approved R&D plan. The idea behind this is that it will "encourage Australian companies to think strategically about their research and development as a critical and ongoing part of their business" (ATO 2003, p 31). Second, specific consideration is given to plant dedicated to R&D: all plant will indeed be eligible for life depreciation deduction at 125% for the period that the plant is used for R&D activities.

8.4.2.3 High Tech Firms, High Tech Sectors and Beyond

Tax measures are frequently designed to achieve more neutrality. Whether this holds true hinges on how much credence is given to empirical studies that point to substantial disparities between sectors. A large body of econometric research reveals major differences between intra- and inter-sectoral externality rates and, thus, social rates of return for R&D (Mohnen and Mairesse 1996). As direct R&D support, RDTIs can thus be used to fund specific technologies. Since 1985, Japan, for example, gives a 7% credit – which can be combined with other credits – to firms that carry out R&D into energy-saving measures, and certain drugs (AIST/MITI 1994), and to firms that do R&D in basic industrial technologies, such as robotics, electronics, advanced engineering, biotechnology, and new materials (Warda 1994). Korea also used tax measures at the beginning of the 1990s as a means of promoting certain industries, by raising the credit rate to 30% for industries categorized as high-tech (Kim 1993). The UK system specifically rewards certain types of research, such as vaccine research (EC 2003). Vaccine research relief is an associated measure targeted specifically at R&D into vaccines and medicines for the prevention and treatment of the so-called "killer diseases" of the developing world (TB, malaria, HIV, and certain forms of AIDS), which mainly occur in less developed countries. The initiative thus goes far beyond the usual emphasis on national competitiveness.

Another measure is to foster high level or fundamental research. As already mentioned, special provisions cover R&D that is carried out jointly with universities (e.g. Japan, the US). Basic research, or high level research, can be considered even if it is not conducted in public R&D organizations: For high tech Japanese corporations, i.e. those with a higher proportion of R&D expenses, up to 2% of additional tax credit is available. Currently, France is introducing a provision to boost employment of postdoctoral students: operating expenses are considered to be higher for PhDs. During the first 12 months, operating costs are computed at a rate of 100% instead of 75% of salaries.

High level research and high tech industries are not the only targeted sectors. Singapore has designed a special R&D tax credit to encourage financial institutions to develop new and innovative financial products. RDTI can even be opened to non-technological innovations, as in France, where the costs for clothing designs are included in the RD tax credit base.

8.4.2.4 Local and National Firms

Another fairly common practice is to vary tax credits by *region*. This is practised by federal states, such as Canada and the US, but also by countries such as Italy, Spain, and Korea. Canada has a two-tier system of R&D tax credits. Several provinces (British Columbia, Manitoba, New Brunswick, Newfoundland, Nova Scotia, Ontario, Québec, and Saskatchewan) have their own credits, which can be combined with federal ones. Assistance of the kind the Canadian federal government provides to certain regions can also be found in Spain (the Basque country, Navarre, and the Canary Islands) and in Italy. In the US, several states currently offer special tax credits, on top of the federal tax credit, for biotechnology research. The local RDTI scheme can also be proposed as an alternative computation to the federal RDTI provisions. A firm can therefore choose between a regular federal tax credit and the national computation method (Paff 2003). However, regional exclusions can also apply: up to 2003, RDTI did not apply to Korean start-up companies in the Seoul metropolitan area that were established after January 1, 1990. In 1995, France tried to move towards a regional structure of tax credits. However, such unequal treatment of individual firms induced constitutional problems. A new procedure defining high tech clusters qualifying for high tax relief, is expected to be implemented in 2005.

Finally, one of the risks for any country is that allowances will be given for R&D that will be exploited in other countries. As already mentioned, the national dimension is often defined by the dividing line between national and international R&D sub-contractors. However, a further discrimination toward national interests occurs in Canada where Canadian-controlled private corporations (minimum 50% Canadian ownership) are given preferential tax treatment. It is clear that foreign multinationals' R&D activities are seen as being relatively less profitable for the country (i.e. giving a lower social rate of return).

8.5 RDTIs and Their Environment

A RDTI system cannot be analysed independently of its institutional environment. RDTIs interact not only with other R&D subsidies, but also with other types of tax incentives and corporate taxation in general. Therefore, while the focus in this chapter so far has been on the "internal coherence" of RDTIs, in this part we will look primarily at their "external coherence".

8.5.1 Overlapping R&D Policy Tools

As mentioned in the introduction, most countries use budget provisions to promote R&D. In a nutshell, four conventional types of assistance are used to varying extents, depending on the country: the usual R&D *subsidies*, assistance with the *financing* of R&D (e.g. France, Spain), *government contracts* for R&D (especially defence R&D) (e.g. France, the UK, the US) and *awards*[7] (e.g. the US, the UK, France), which increase the expectations of gain for firms investing in R&D. The coherence between RDTI mechanisms and the other four measures designed to promote R&D has been little studied in the academic literature (Maurer and Scotchmer 2003). Furthermore, this relationship between RDTI and other innovation programmes is not highlighted by governments, especially if the introduction of the RDTI induces some rationalization, or grant programmes collapse (Hawkins and Lattimore 1994), explicitly in relation to Australia.

The interaction between RDTI and other innovation programmes is managed through the definition of the R&D tax base. A core principle of coherence is the *non-redundancy* of assistance measures and other incentives. An RDTI is thus presumably combinable with other forms of assistance. As in the case of financial flows between private firms, the risk is that the same research will be paid for several times over. Defensible as this principle may be in the context of promoting cooperative research, governments seem to be limiting the pluralities by restricting tax incentives. R&D subsidies, therefore, in most cases, are deducted from the R&D budgets of firms. In France, this principle sets the real ceiling for the RDTI mechanism, since large firms get so much government money that fewer than 15 companies were affected in 2001 by the €6.1 million cap. Similarly, R&D funds received from government agencies for R&D contracts are generally deducted. In France, however, a firm with a surge of €1 million due to Department of Defence R&D procurement was allowed an additional €500,000 tax credit. The French government, in this case, may thus be paying at a level 1.5 times the value of the R&D results. The requirements have become much stricter with the recent introduction of a RDTI computation at the group level[8].

The usual separation of the bodies responsible for granting subsidies or low-interest loans from those responsible for awarding RDTIs, their varying degrees of centralization, and their lack of coordination, makes overlaps unavoidable. Ensuring coherence entails a proper balance between RDTIs and other R&D funding mechanisms. If the proportions of both are roughly the same, there is bound to be more duplication and conflicting objectives.

Coordination among types of R&D funding is a matter for governments. On the firm side, beyond opportunism due to a lack of coordination by government, or to bad provisions, the main burden on firms is to trade off

---

[7] See Maurer and Scotchmer (2003) on the modern renaissance of awards.

[8] In 2004, a new ceiling will be implemented at €8 millions. This ceiling will affect very few firms. So this higher ceiling seems rather a consequence of the recent consolidation at the group level, reducing the induced loss of tax credit, than a general signal to firms.

between the different available sources of R&D funding in order to maximize their expected gains. The chase may be difficult when several complex measures are taken at the same time, at national[9], regional, or city levels, dealing, for example, with innovation or R&D, targeting R&D execution, cooperation, or even networks.

In order to deal with this issue, we investigate, through a simple multivariate probit model, the possible complementarities between the R&D tools available in France in 2001. Controlling for size of firms and sector activities, the model is used to evaluate the link between the four main different sources of R&D supports: at the national level, R&D procurements, R&D direct funding, RDTI, and at the international level, EC funding (see Appendix A for definition of variables and econometric specification).

Table 8.3 shows that non-redundancy of assistance implemented in the R&D tax credit scheme is a sufficient condition for substitutability when direct funding is considered. Accounting for size and activity, firms involved in the R&D tax credit scheme are thus less likely to receive finance from other institutions in charge of direct R&D funding. The substitutability is found to be quite low, but significant, at the French and EC levels. The substitutability between RDTI and R&D procurements is not significant. Firms with R&D contracts, mainly in the aeronautic and electronic industries around defence programmes, are thus not likely to find complementary or substitutive support through the French R&D tax scheme. However, these firms are very likely to benefit from direct R&D funding coming from French or even EC sources. Complementarities can be found between EC and French direct funding. Such positive correlations among the different sources suggest that a firm able to get money from one sponsoring body will be able to tap other resources even if it is not for the same R&D project. Table 8.3 thus shows that there are two main ways of getting support for R&D in France:

- to be involved in the chase for R&D direct funding and/or procurements. Firms are here likely to get money from different sources of support for R&D.
- to be eligible to receive the R&D tax credit.

[9] And at the international level for multi national enterprises (MNEs).

Table 8.3. Correlation among access to different French 2001 R&D funding (N=6276 firms)

	French R&D procurements	French Direct funding	EC direct funding
French Direct funding	0,506*** (0,063)		
EC direct funding	0.570*** (0.054)	0,605*** (0,045)	
RDTI	-0.591 (0.058)	-0.106*** (0.025)	-0.113*** (0.041)

Reported values are estimated correlations among residuals (See Appendix A)
*, **, *** significant respectively at the 10%, 5% and 1% level
Standard errors in parentheses.

The issue becomes more complicated as new R&D policy tools are implemented over time or as some are removed. The main problem is that there are no reliable data on which to assess the impact on firms' behaviour since annual R&D surveys only gradually include new targeted firms and do not furnish a fine measure of the different evolving procedures managed by a single public organization. From the French experience over the two last decades, what can be said is that there is a learning curve when the R&D support system first evolves. Firms learn how to get money from institutions as well as learning to trade off between the different available R&D funding sources. Since the introduction of incremental tax credit, many high tech firms with increasing R&D budgets have rejected traditional R&D direct funding procedures with their higher administrative costs and often low expected returns. The recent strengthening of the French capital venture system, or even the various EC initiatives, have increased this crowding out effect. Many high potential projects involve important R&D investments (especially in biotechnology and new information and communication technologies (NICTS)). Several traditional procedures are thus now restricted to low tech or not very innovative projects that are also chased by local bodies (regions, departments, and even cities).

8.5.2 From R&D to Technology Fiscal Incentives

Tax incentives are not confined to corporate R&D outlays, or even to the defined R&D tax base. Industrial development, and the problems encountered in producing and marketing new processes and products, represent knowledge that a firm can accumulate, and development usually takes precedence over research. As a result, different tax incentive mechanisms are implemented in order to deal with the acquisition, building and maintenance of competences that are complementary to R&D investment.

Technology diffusion is a first target for governments interested in growth. There are two kinds of tax incentive for knowledge diffusion: those that promote acquisition of technological knowledge and those that promote its transfer.

The first type is much more common than the second, and the targets are firms undertaking little or no R&D. Spain has tax incentives for companies that do not carry out R&D, through a volume based tax credit (10%) with a ceiling. The Korean tax scheme is broader, allowing companies that acquire patents, or even utility model rights, technical know-how, or technology, a tax credit of 3% for large firms and 10% for SMEs, of the acquisition cost.

Technology acquisition or a technology production tax scheme is often technology oriented. Tax credits act as incentives for more traditional fields, such as energy conservation and environmental protection (Germany, Australia, the US, Italy, Japan), which can also be regarded as technology diffusion. NICTSA often attract dedicated tax provisions (Spain, the UK, Korea). For example, Korean corporations investing in "Software Configuration Management" or "Customer Relationship Management" are eligible for a 3% tax credit rate (7% for SMEs).

Often, the kind of investment tax incentives are biased towards foreign products or investments (FDI). In Australia, for example, technology acquisition policy is restricted to duty-free entry of goods imported for use in "space projects" or on chemicals, plastic, and paper raw materials, and intermediate goods, and certain food packaging, which has "a substantial and demonstrable performance advantage" over those produced in Australia (AusIndustry 2001, p.2). This kind of direct support is quite rare. There are no or very few general-purpose inward investment incentives at national level. However, there is a intense competition for inward investment at the infra-level (states in the US, regions in France, the länder in Germany, etc.).

Technology transfer can be also induced by incentives to knowledge suppliers. Such mechanisms though are rare, although Ireland and France encourage the licensing-out of technologies by a preferential tax rate on patent royalties.

The fiscal incentive also applies to *human capital*. Denmark has a special 32% tax credit (including contribution for social insurance) for the hiring of foreign high skilled labour, instead of the 63% rate for domestic labour. The tax incentive is general, temporary, and not confined to R&D employment. Danish employees working abroad may also qualify provided they did not go abroad specifically to apply for this scheme. The foreign employee's gross salary must not be less than US$7,000 per month after deduction of social insurance contributions. By way of exception this threshold of US$7,000 does not apply if the employee is engaged in an approved R&D project. A similar scheme has been implemented in Sweden. A foreign researcher who is expatriated in Korea is granted a non-taxable income allowance of up to 20% of his normal monthly income. The introduction of such a preferential income tax to attract foreign researchers (only) has been a recurring issue in France since 2000.

Vocational training can be seen as a complement to successful innovation even though it is also considered to be a way of improving productivity more generally. The UK, Sweden, and Spain have different tax credit schemes for educational, vocational, and training costs, often taking the form of co-financing. In France, such schemes are restricted to SMEs and represent only about €14 million (compared to €482 millions for the 2001 French R&D tax credit). Currently, in 2003 the US has no federal training tax credit, although some individual states operate such a scheme (e.g. Ohio, California, Rhode Island). Since 2001 in Arizona, there is a training tax credit dedicated to NICT to encourage people to enter technology careers even if such careers are difficult to delimitate).

Fiscal provisions on *stock options* are also considered as important for innovation success and sustainability; in this case, stock options are considered as a tool of R&D management especially for young SMEs with high growth potential, or even for MNEs keen to reward their researchers. With the exception of Italy, the national incentive in the countries examined seems quite marginal compared to other fiscal mechanisms in place (EVCA 2003).

Finally, many countries are operating specific tax incentive schemes to boost investments in new innovative firms through *venture capital funds* (VCF) or selected stocks (France, UK, Italy, Denmark, Spain, Ireland, the Netherlands).Generally, these fiscal schemes concern dividends and capital gains for investors in venture capital funds. The first aspect concerns dividends paid on VCF shareholders. In Spain and France and the UK, no corporate income tax is payable by a venture capital firm on any dividend income remitted by a target company in which the VCF has a participating interest. Second, very low taxes are imposed on capital gains made by the VCF on any profitable sales of its shareholding in a target company (0% in France and the UK, 1% in Spain). Third, the tax rates on dividends paid to shareholders of VCF or investors in the different kinds of innovation mutual funds are low. In France, dividends remitted by the VCF to individual shareholders are subject to a flat rate tax of 16% instead of a progressive rate. The fiscal scheme is more interesting in Spain since outgoing dividends remitted by the VCF to its shareholders are exempt from tax. In many countries, however, (e.g. Germany), direct support to venture capital companies is still preferred (see EVCA 2003).

8.5.3 RDTIs in an Overall and Global Tax Policy

In an even more wider perspective, RDTI are embedded in corporate national tax policy. Tax coherence may be gauged from two perspectives: how the RDTIs in with other tax incentives dedicated to knowledge creation and diffusion, and how RDTIs fit in with the overall business tax system[10].

The principle of *non-redundancy* applies again here: if governments do not want to reward businesses more than once for the same R&D, they must ensure that tax credits cannot be combined with any of the other types of tax breaks mentioned above. Analysis of RDTIs as they relate to corporate taxation provides insight into possible *interdependence* and, to a lesser degree, the possible effects of such a mechanism on corporate behaviour.

Another aspect is the interdependence between RDTIs and corporate tax rate. The higher the rate of corporate income tax, the greater the incentive for firms to make use of tax incentive systems to reduce their liability. In contrast, in countries with low corporate tax rates (e.g. Korea, Ireland, Finland, Sweden) the incentives are small[11]. The interdependence between RDTIs and the tax system would appear to involve more than the incentive or disincentive effects of corporate tax rates. In their general reform of corporate taxation, Finland, Germany, and Sweden simply did away with RDTIs. This denotes a certain *de facto* primacy of the rationality of tax policies over that of S&T policies in countries where the ratio of private R&D to gross domestic product (GDP) is high. The assumption is that the tax reform will ultimately benefit the firms best able to pursue R&D. The complexity of these interactions makes it harder to delve any deeper into the coherence of the tax system as a whole, or even into the real impact of RDTIs.

---

[10] See Warda (1994, 2001, 2002) for a comparison of the national level of incentives induced by an R&D tax scheme taking account of certain other national tax devices.

[11] Korea's corporate tax rate stands at 15% for companies with an annual taxable income below 100 million won ($82.325) and 27% for those with income at or above 100 million won. Hong Kong has the lowest corporate tax rate in Asia at just 16%, compared to Singapore's 22%, Taiwan's 25%, Malaysia's 28%, Japan's 30%, China's 30% and Thailand's 30% rates. These last rates are close to the EC average rate reported in Devereux et al. (2002).

National R&D tax policy is also dependent on foreign tax schemes. The more open and competitive environment of recent decades has encouraged several countries to make their tax systems more attractive to investors. Tax policy towards FDI, in combination with targeted tax incentives, is often regarded to be the main source of the Japanese economic miracle in the 1970s and Irish economic recovery in the 1990s (Walsh 2000). RDTI mechanisms are thus part of the lowering of overall tax rates and are designed to settle or attract foreign R&D investors. If multinational firms, and the more profitable firms, are more likely to be interested in the gains that tax competition generates (Fuke 1997; Devereux et al. 2002), then this evolution will have two disadvantages: national R&D tax schemes, embedded in the new international fiscal Nash equilibrium, do not induce any significant and long term competitive effect even if a global effect occurs (OECD 2001b) on subsidies). Second, the general lowering of tax rates can induce a general loss of fiscal receipts that will curb public R&D activities, especially in periods of economic slowdown. An extreme mechanism, used by Malaysia, Singapore, and Taiwan, is to exempt a firm from tax under certain conditions. This *exemption* is often long-term. For example, if the firm undertakes to perform R&D for ten years, it may be granted this benefit. The purpose of such exemptions is undoubtedly to obtain a regularity of R&D efforts over a long period and to avoid investment cycles. From an international standpoint, such a policy may be considered to be harmful when combined with other devices (OECD 1998). However, the tax issue for countries is far from simple. Thus, the numerous criteria that enter into R&D investment decisions lessen the likelihood of significant tax-induced international transfers of R&D activities. In particular, when multinationals relocate their research facilities the aim is seldom to maximize tax benefits, but more likely to be closer to markets, and skilled labour, and to provide technical assistance to production units (Pearce and Singh 1992; Hines 1994).

8.6 Conclusion

Public support for S&T is often seen as the provision of discrete subsidies for R&D. This chapter shows that even RDTI schemes are shaped to fit the goals of programme designers. RDTIs are complex and cannot be applied "without distinction to all firms" with clarity and simplicity (EC 2002). Furthermore, we would suggest that the introduction of RDTIs allows governments to support firms that do not qualify for public funding in the shape of subsidies, procurements, prizes, etc. We would suggest also that RDTIs induce shifts within these traditional public funding procedures.

Besides the issue of the rate of return of RDTI schemes compared to other policy tools, one question that needs to be addressed is why RDTIs are becoming so popular in the OECD countries. The success of RDTIs in the eyes of policy makers relies mainly on their flexibility. RDTI amounts are positively correlated with fiscal incomes. Even if this pro-cycle aspect may be regarded by governments as a problem, such a self regulating mechanism is much easier to manage than direct subsidies during "bearish" periods. Second, as has been shown, the provisions are easy to change or adapt on a yearly basis. Policy makers can thus monitor the scale (amount) and the scope of incentives. Last, RDTIs can be managed by a very small staff and therefore, neither the introduction nor the withdrawal of RDTIs requires additional human capital or introduces social problems.

RDTIs are popular with governments because they touch a lot of firms, are easy to make visible, and their effect is probably seen more quickly than that of other types of R&D funding (Guellec and van Pottelsberg 2000), which may fit more easily into political cycles.

We suggest in this chapter that the success of RDTIs relies on the lack of institutional changes needed to already implemented policy tools. The permanence of direct instruments is a condition of the success of RDTIs. First, S&T policy makers do not consider that an indirect tool, such as RDTI, is sufficient for the creation of start-ups in new strategic technology fields, such as biotech, nanotechnologies, or NICTS. Second, although many RDTIs are oriented toward SMEs, the standard elitist S&T policy of "picking the winners" among national large firms through grants, subsidies, and procurements may still be carried on.

Even if the addition of RDTI does not suppress the availability of traditional R&D funding, we suggest that it usually reduces the field of intervention of direct R&D subsidies. However, RDTIs provoke weak opposition from traditional S&T policy makers, which can be explained by three factors. First, when annual budgets are considered, the different administrations see the allocation of resources among players as a zero sum game. The direct effect on the weight that is given to S&T policy among the state policies, and the indirect effect in that public S&T traditional actors (other S&T policy makers or even public research organizations), makes the rise of the RDTI scheme a threat to their future resources. Also RDTIs do not induce a financial flow toward the administration responsible for this new allocation of money dedicated to R&D. Thus, the introduction of RDTIs demands a new political actor, not hitherto involved in S&T policy (the fiscal administration or the budget). The person with responsibility for RDTIs may not be seen as being powerful since they have no means to influence the allocation of the amounts of money that they are responsible for. Third, RDTIs are easy to introduce because, in general, R&D policy makers are unfamiliar with this new kind of tool and only later become aware of its influence on their activities.

From an economic point of view, few existing works have looked at the interaction and coordination issues involved in the different S&T policy tools, although many authors have investigated the balance between IPRs and public knowledge. A critical issue for S&T policy makers are the interactions and balance between the different R&D funding tools.

8.7 References

AIST/MITI (1994) Untitled mimeo. Agency of Industrial Science and Technology/Ministry of International Trade and Industry, Tokyo.

ATO (2003) Guide to the R&D Tax Concession.Version 2, November, Australian Taxation Office and AusIndustry.

AusIndustry (2001) Certain Inputs to Manufacture (Items 57 & 60), Policy & Administrative Guidelines. The Australian Customs Service, the Department of Industry Science & Resources and AusIndustry, 26 February.

BIE (1989) The 150% tax concession for research and development expenditure - interim report. Program Evaluation Report 7. AGPS, Canberra.

Devereux, MP, Griffith R, Klemm A (2002) Can international tax competition explain corporate income tax reforms? Economic Policy October: 451-495.

EC (1998) Commission notice on the application of the State aid rules to measures relating to direct business taxation, 10.12.1998, Official Journal of the Commission 384, Luxembourg, p 3.

EC (2002) Corporation tax and innovation: issues at stake and review of European Union experiences in the nineties, by Asesoría Industrial Zabala, S.A., European Economic Development Services, Ltd., Michel Goyhenetche Consultant and MCON Consulting, Innovation papers No. 19, EC , Bruxelles.

EC (2003) Raising E.U. R&D productivity improving the effectiveness of public support mechanisms for private sector research and development: report to the European Commission by an Independent Expert Group. Directorate-General for Research, Bruxelles.

Eisner R, Albert SH, Sullivan M (1984) The new incremental tax credit for R&D incentive or disincentive? National Tax Journal 37: 243-255.

EVCA (2003) Benchmarking European tax & legal environments: indicators of tax & legal environments favouring the development of private equity and venture capital in European Union member states, European Private Equity and Venture Capital Association, Zaventem, Belgium, March.

Fuke T (1997) The restructuring phase of tax law in Japan. An issue of legitimacy over a more equitable and fairer system towards the 21st century. In: Zhang

Y, Fuke T (eds.) Changing tax law in East and South East Asia: towards the 21st century, Kluwer Law International, The Hague Hardbound, , pp. 320.

Gourieroux C (2000) Econometrics of qualitative dependent variables. Cambridge University Press, Cambridge.

Guellec D, van Pottelsberg B (2000) The impact of public R&D expenditure on business R&D. Economics of innovation and new technologies 12(3): 225-244.

Hall B, van Reenen J (2000) How effective are fiscal incentives for R&D? A new review of the evidence. Research Policy 29: 449-469.

Hawkins B, Lattimore R (1994) The Australian experience in evaluating the R&D tax concession. Presented at Conference on government recipes for industrial innovation, Vancouver, 20-21 October.

Hines JR (1994) No place like home: tax incentives and the location of R&D by American multinationals, tax policy and the economy. MIT Press Review 8: 65-104.

IFRI (1992) A comparative study on the tax treatment of research and development expenditures. International Fiscal Research Institute.

Kim L (1993) National systems of innovation: dynamics of capability building in Korea. In: R Nelson (ed.) National Innovation Systems: A Comparative Analysis, Oxford University Press: New York, pp. 357-383.

Koga T (2003) Firm size and R&D tax incentives. *Technovation* 23(7): 643-648.

Mansfield E, Switzer L (1985) The effects of R&D tax credits and allowances in Canada. Research Policy 14: 97-107.

Mansfield E, Romeo A, Switzer L (1983) R&D price indexes and real R&D expenditures in the United States. Research Policy 12: 105-112.

Maurer SM, Scotchmer S (2003) Procuring knowledge. NBER Working Paper No. 9903, National Bureau of Economic Research, Inc, New York.

Mohnen P, Mairesse J (1996) Externalités de la R&D et croissance de la productivité. OECD STI Review, 18 : 39-66.

OECD (1993), Taxation in OECD Countries, OECD, Paris.

OECD (1994) The measurement of scientific and technical activities, "Frascati Manual". Revised Edition. OECD, Paris.

OECD (1998) Harmful tax competition: an emerging global issue. OECD, Paris.

OECD (2001a) Science, technology and industry outlook - drivers of growth: information technology, innovation and entrepreneurship. OECD, Paris.

OECD (2001b) Competition policy in subsidies and state aid, DAFFE / CLP (2001) 24, November. OECD, Paris.

OECD (2002) Tax incentives for research and development: trends and issues, science and technology industry. directorate for STI, OECD ed., Paris.

OECD (2003a) Main science and technology indicators, Vol. 2003-2 (Complete Edition). OECD, Paris.

OECD (2003b) Tax incentives for research and development: trends and issues. OECD, Paris.

Paff L (2003) Does the alternate incremental credit affect firm R&D?, Technovation 24(1): 41-52.

Pearce R, Singh S (1992) Internationalisation of research and development among the world's leading enterprises: survey analysis of organisation and motivation, in Granstrand O, Håkanson L, Sjölander S (eds.) Technology management and international business. Wiley, London, pp. 137-179.

Walsh B (2000) The role of tax policy in Ireland's economic renaissance. Canadian Tax Journal 48(3): 658-673.

Warda J (1994) Canadian R&D tax treatment: an international comparison. Report by the Conference Board of Canada, No. 125-94. The Conference Board of Canada, Ottawa.

Warda J (2001) Measuring the value of R&D tax treatment in OECD countries. OECD STI Review 27: 185-211.

Warda J (2002) A 2001-2002 update of R&D tax treatment in OECD countries. OECD Directorate for Science, Technology and Industry. OECD, Paris.

Appendix A: Complementarity or substitutability among R&D funding sources: a multivariate probit model

In order to test the complementarity hypothesis, we introduce simultaneously equations explaining the probability to benefit from at least one euro from public organizations.

The data set is the result of merging the R&D survey data and the RDTI data set for 2001. All firms with a positive R&D tax credit in 2001 are not included in the R&D tax survey. 5,908 firms are part of the French R&D tax credit scheme among them, 3652 with a positive credit in 2001. Only 944 firms with positive R&D tax credit are thus available in the 2001 R&D tax survey. Furthermore, firms with declared subventions included only in the tax credit data set were removed since we were unable to identify the sources of funding. The final data set includes four R&D public funding sources for 6276 firms. A firm could use any or all of these sources. The multivariate probit model estimates the association between size and sector activities (NACE 2 level) while controlling for the effects of the covariates and correlation among the R&D funding bodies variables.

More precisely, considering the observed level of received financial flows as indicators of latent variables measuring the funding strategy of firms, we introduce a multivariate probit model with 4 equations : $RDPUBR^*_{ik} = \gamma_k z_{ik} + \xi_{ik}$, k=1,..,4 where $RDPUBR_{ik} = 1$ if $RDPUBR^*_{ik} > 0$, and $RDPUBR_{ik} = 0$ if $RDPUBR^*_{ik} \leq 0$, for k=1,..,4 and z_{ik} the set of explanatory variables for the 4 equations. RDPUBR includes:

- financial funds on R&D procurements (mainly from DoD, CNES, ONERA or CEA). PROCUREMENT is a dichotomous variable that is one (0 otherwise) when a firm earns money in 2001 from R&D contracts with French organizations.

- R&D direct incentives (Mainly, Ministry of Research, Ministry of industry, ANVAR, Ministry of Transportation, Ministry of Environment, Regions). DIRECT is where a firm gets direct R&D funding from French organizations (18% are in this category).

- EC funds (without financial funds coming from ESA…). EC variable is one (0 otherwise) when funding was received from the EC in 2001. 4% of firms are involved.

- R&D tax credit (Ministry of Finance). TAX CREDIT is 1 (0 otherwise) when the firm has a positive tax credit in 2001 (38% of the sample). This last variable thus does not represent a financial flow.

51% of firms in our final sample obtained public funding for their R&D or for their sub-contracted R&D.

The z set of explanatory variables is similar for the different equations. We assume that the probability to benefit from public funds depends roughly on the size of firm (SIZE = number employees taken in log) and its belonging to a sector (A set of 21 sector dummies is introduced as SECTOR). A second order term is introduced when necessary to fit non linearity with size. Thus the multivariate probit model can be summarized as a system of four equations:

$$\begin{cases} DIRECT_i^* = \beta_{11}\, SIZE_i + \sum_{s=2}^{s=23} \beta_{s1} SECTOR_{is1} + \xi_{i1} \\ TAX\ CREDIT_i^* = \beta_{12}\, SIZE_i + \beta_{22}\, SIZE^2{}_i + \sum_{s=3}^{s=24} \beta_{s2} SECTOR_{is2} + \xi_{i2} \\ PROCUREMENT_i^* = \beta_{13}\, SIZE_i + \sum_{s=2}^{s=23} \beta_{s3} SECTOR_{is3} + \xi_{i3} \\ EC_i^* = \beta_{14}\, SIZE_i + \beta_{24}\, SIZE^2{}_i + \sum_{s=3}^{s=24} \beta_{s4} SECTOR_{is4} + \xi_{i4} \end{cases}$$

($SIZE^2$ is SIZE squared that is introduced as mentioned)

One way to test the interdependent hypothesis is to assume that the error terms of each decision to ask for public funds (direct or indirect) are correlated in the form $Cov[\xi_{ik}, \xi_{ik'}] = \rho_{kk'}$ with k=1,..,4 and k'=1,..,4 with k≠k' (See Gourieroux 2000). If $\rho_{kk'}>0$, the complementarity between the two considered endogenous variables is obtained. Interdependence of R&D funding is rejected if the $\rho_{kk'}$ variable is significantly different from 0. A third case may be that the different R&D tools are substitutes ($\rho_{kk'}<0$).

The analysis of the expected sign of explanatory variables as well as the comments on the estimated coefficients is beyond the scope of this paper. However, focusing on complementarity or substitutability among R&D funding we can estimate of the disturbance covariance matrix (see Table 8.3).

9 Twenty Years of Evaluation with the BETA Method: Some Insights on Current Collaborative ST&I Policy Issues

Laurent Bach and Mireille Matt

BETA, Strasbourg, E-mail: bach@cournot.u-strasbg.fr
BETA, Strasbourg, E-mail: matt@cournot.u-strasbg.fr

9.1 Introduction

As the number and variety of forms of public actions supporting Science, Technology and Innovation (ST&I hereafter) has increased, despite their reliance on ever-scarcer public financial resources since the 1980s, evaluation studies have multiplied. At the same time, this period has seen an increasing sophistication and refinement in analyses of the research and innovation processes, and the accompanying rationale for and tools of state intervention.

This has led to a somewhat paradoxical situation[1] where: i) although an impressive number of evaluation reports have been published, they address several different, generally quite context-specific, but sometimes overlapping, aspects of these public initiatives, with no coherent framework to explain their relevance (in terms of level of evaluation or rationale); ii) very simplistic, but complex and theoretically grounded evaluation tools are being applied; and iii) apparently new approaches are being taken to addressing the same problems and questions that have already been addressed and analysed in other works, while at the same time new and crucial issues related to ST&I processes are being neglected. There is, thus, some discrepancy between the fields of evaluation studies and ST&I analysis. Broadly speaking, a lot of these new issues are emerging because ST&I processes are increasingly resulting from the collaborative activities of various actors.

---

[1] Despite the many rich and promising attempts to synthesize and summarize the different evaluation approaches and the related scope of relevance (see for instance OECD 1997; Georghiou et al. 2002; Fahrenkrog et al. 2002).

Against this background, the BETA approach is rather specific in that it has been used to evaluate a large variety of public programmes over the last twenty years. However, these assessments have addressed specific aspects, in particular the so-called "indirect effects" of public research and development (R&D) programmes. These effects are at the heart of the innovative processes fostered by these programmes and the fact that the evaluations have covered different types of programmes and related institutional set-ups has highlighted the results and mechanisms linked to some of the issues currently preoccupying policy analysts (often from different disciplines) and policy-makers dealing with public support for collaborative ST&I activities.

The objective of this present contribution is to provide some insights on three issues, central to cooperative R&D programmes: university-industry relationships, the role and performance of small and medium sized enterprises (SMEs) in collaborative innovation activities, and the design of partnerships between actors involved in such activities. We propose, therefore, to select for examination only those public programmes supporting R&D cooperation between the participants. Lessons and results from these studies will be linked to the results from some other recent academic and empirical studies in corresponding fields. However, each BETA evaluation relates to a specific public programme that is often highly context-dependent (sector, economic conditions, etc.). Therefore, translating these results and conclusions to other contexts must be done with caution. It should also be noted that each BETA evaluation involves a vast number of quantitative results, which, for reasons of space and simplicity of presentation, are not included here.

Parts 9.3, 9.4 and 9.5 of this chapter are devoted respectively to the three issues mentioned above. Following on from the comments made at the beginning of this introduction, Part 9.2 describes the BETA method, its relevance and its main methodological features, as well as briefly introducing the different studies performed using this approach.

9.2 Positioning, Methodology and Overview of Empirical Studies

9.2.1 Positioning

The relevance of the BETA approach is illustrated by four main perspectives, derived from some of the major distinctions with reference to the field of "Evaluation of ST&I policies".

First, there is often an analytical distinction between the evaluation of: a) researchers (at individual level); b) teams or organizations; c) projects or programmes; d) specific mechanisms or institutional settings; and e) overall sets of policies or system of innovation. Clearly, the BETA method deals with the evaluation of programmes and projects.

Second, different types of programmes are designed and implemented by policy-makers, with what falls under the term "programme" being very frequently rather fuzzy and vague. The BETA approach is concerned with the evaluation of a limited range of programmes, namely the public Research, Training and Development (RTD) programmes, with the following characteristics:

- they are partially or wholly funded by public money;
- firms are involved possibly in cooperation, with a university and/or other institutions (thereafter called participants or partners);
- R&D is performed;
- there is agreement between the parties involved on R&D topics and on operational goals;
- they have a limited time horizon.

Therefore, pure basic research exclusively conducted by academics and Public Research Organizations (PRO) is beyond the scope of the BETA approach. However, both basic research as a part of an R&D project, and the involvement of PROs in collaborative projects are considered.

Third, the term evaluation encompasses different exercises, from different perspectives, that mobilize different stakeholders and use a different range of methods. Following the traditional definition (OECD 1997; Georghiou and Meyer-Krahmer 1992), one usually talks about the analysis and assessment of:

- the goals of the programme and the decision process leading to them;
- the institutional set-up of the programmes (instruments, rules, etc.), their design and their implementation;
- the results, outputs, outcomes and other impacts of the programme.

Coupled with this distinction is the time perspective that is adopted by the evaluators, leading to another useful breakdown usually proposed between *ex ante*/strategic evaluation, and monitoring, and *ex post* retrospective (recapitulative) and/or proactive (or endo-formative) evaluation.

The Beta approach focuses on the latter aspects of each of these two distinctions. Aiming at evaluating specific types of impacts, it provides a minimal estimation (in monetary terms) of their importance, and helps to address such question as the programme's efficiency, effectiveness, impact and sustainability. But it also enables a better understanding of how these effects are generated and, more generally, how innovation processes are generated by large RTD programmes and are creating economic value. In this sense, it fulfils one of the basic requirements of any evaluation work, that is, to enable the incorporation of lessons learnt into the decision-making processes of policy-makers, programme managers, and participants.

Finally, from a more theoretical standpoint, in many respects the BETA approach is a mix of standard market-oriented approaches and systemic/evolutionary approaches. The indicators adopted by BETA to quantify the effects (changes in market shares, added value, cost reductions, time savings, etc.) are related to the neoclassical framework (cf. 9.2.2.3). However, the type of impacts evaluated, which correspond mainly to learning effects, refer to the second type of approach. Basically, knowledge creation, processing, and diffusion, but also externalities within and between actors and network creation, are among the phenomena under scrutiny. The first three chapters of this book analyze the policy rationales taking into account evolutionary (see Chapter 1 by Bach and Matt), systemic (see Chapter 2 by Metcalfe) and knowledge-based (see Chapter 3 by Cohendet and Meyer-Krahmer) approaches to the process of innovation. They provide the theoretical context for, and challenge of, the economic evaluation of public R&D programmes.

In addition, some further classification of evaluation approaches and studies relates to the scope or identity of the actors involved (basically participants vs. non-participants) and to the level of evaluation (micro, meso, or macro, with the possible additional investigation of dual and articulated perspectives). Regarding these last categories, the BETA evaluation is limited to the participants in the programme: what is evaluated is the economic impact (or the economic effects) generated by, and affecting, these participants. It is, therefore, based on a microeconomic approach, but the results obtained at this level of investigation are aggregated in order to provide an evaluation at the level of the set of participants as a distinctive population.

9.2.2 Methodology

The first specific feature of the BETA approach reside in the distinction it makes between what are termed the direct and the indirect effects of programmes, a distinction that is based on their objectives. The second feature is the way that these indirect effects are defined and evaluated, and is at the core of the BETA approach.

9.2.2.1 Direct Effects

Direct effects are those effects directly related to the objectives of the programmes and projects, as defined at the start of each project. The nature of direct effects may change depending on the public policy aims and the institutional arrangements adopted for its implementation.

For instance, technology procurement policy (such as a new space satellite, a new transport infrastructure, or a means of producing energy) supposes that technology or technological artefacts will be delivered by the body that is awarded the public contract, and then put at the disposal of the whole of society under specific rules. This form of ST&I policy leads to direct effects that are related both to the contract between the public authorities and the delivering entity (that is, the very activity of producing the technology or the technological artefact), and to the use of the technology or technological artefact by society.

In the case of financial subsidies provided to a single firm (or to a consortium of firms and PROs) for its RTD activities, if the objective is to develop a new product (or a new process), the sales of such products (or the economic effects of the use of this new process) can be considered as direct effects. This holds for more fundamental research-oriented projects: direct effects are related to the application of new scientific knowledge or new technologies in the field envisaged at the beginning of the project; however, the range of possible direct effects may be enlarged, since the fields of application may have been broadly defined.

It is clear then that the direct effects for different types of programmes cannot easily be compared, and often require the use of different methodological approaches and metrics. The BETA evaluation method is not very innovative in terms of how these direct effects are identified and evaluated, and is not always best suited to making such identification and evaluation. In addition, evaluation studies based on the BETA approach, which accounts for indirect effects, do not always take direct effects into account.

9.2.2.2 Indirect Effects

Indirect effects are those that are beyond the scope of the objectives of the project. Generally speaking, indirect effects are derived from application of what has been learned during the execution of the project, through the activities of the project participants, but which is not directly related to the objectives of the project. All types of learning leading to the creation of all types of knowledge are taken into account: technological, organizational, networking, management, industrial, and so on. This is probably the main feature of the BETA approach since it provides a very detailed view of how an RTD activity performed within the framework of a public programme affects the learning processes of participants. Indirect effects can be broken down into four sub-categories:

- Technological effects concern the "transfer" of technology from the project to other activities of the participant (the transfer is thus "internal" across the scope of the participant's activities, possibly involving different teams or divisions). Following the definitions proposed by "evolutionary economics" and "knowledge economics", the term "technology" here encompasses artefacts (products, systems, materials, processes) as well as codified, tacit, scientific, and technological knowledge (apart from methods, see Organization and Method Effects). What is transferred can therefore be of a very diverse nature, from scientific expertise to a worker's know-how, including technology laid down as a blueprint, or new theories or "tricks of the trade". This broad approach represents part of the originality of the BETA methodology. The transfers lead to the design of new or improved products, processes or services, allowing participants to carry out further research projects in the same field, or to contribute to research activities in more or less related domains.
- Commercial effects fall into two categories, not directly or necessarily linked to a technological learning process. First, network effects refer to the impact of projects on the cooperation among economic actors (establishment of business links between participants in a consortium, which lead to the continuation of commercial or technical collaboration after completion of the project; cooperation between participants and organizations or firms not involved in the project, for instance, with the supplier of another participant, or as the result of a conference or workshop organized by the public organization managing the programme being evaluated). Second, there is the label effect, which results from the reputation acquired by participants working on behalf of a particular public programme: this good image or reputation is often used afterwards as a marketing tool.

- Organization and Method (O&M) effects occur when experience gained through the project allows a participant to modify its internal organization and/or to apply new methods in project management, quality management, industrial accounting, and so on.
- The last type of indirect effect, termed Competence and Training (or Work Factor) effect, differs from the first three. Competence and Training effects describe the impact of the project on the "human capital" of the participant. Each of the participating organizations masters a certain range of competences related to more or less diversified scientific and technological fields, which form what has been called the "critical mass" or the "knowledge base" of their organizations. The impact of the project on this "critical mass" constitutes the work factor effect. Thus, the BETA approach aims to differentiate between routine work and innovative work, which increases or diversifies the technological abilities of the participant.

9.2.2.3 Quantification of the Economic Effects[2]

It is assumed that most types of indirect effects (as well as the direct effects of some RTD policies or programmes) can be expressed in terms of added value generated by sales and cost reductions, which, in turn, have been achieved as a result of the knowledge gained by the participants during the programme being evaluated. Whenever an indirect effect is observed, an attempt is made to perform such quantification. But this is not possible in all cases, and a significant number of identified effects generally remain unquantified.

Accordingly, technology transfer basically leads to the design of new or improved products, processes, or services, which allow the participant to increase sales, to protect market share, to benefit from a reduction in costs, or to win new research contracts. Commercial effects lead to new sales or research contracts, and O&M effects lead to cost reductions.

When the transfer of technology or method, or the commercial effects are only partly responsible for increased sales or cost reductions, the value of the corresponding effect only amounts to a share of those sales or cost reductions. This share is in proportion to the influence on those sales or cost reductions of work performed within the framework of the evaluated project ("fatherhood coefficients" are thus used). When such an evaluation is complex, a two-step process is used: first, the influence of one parameter

[2] A detailed description of the quantification is provided in (Georghiou et al. 2002) Part 4. Case studies. 4.2 Maket-oriented case study: http://les.man.ac.uk/PREST/

is evaluated (influence of technological aspect, commercial aspect, and so on, following the logic of the indirect effects classification), then evaluation of the specific influence of the evaluated project on this parameter is made.

There are two exceptions to this quantification method, and in both cases proxy values are used. First, in the case of patents that are not protecting existing products or processes, the indirect effect is estimated by the amount expended by the participant to register and keep the patents "alive". The second exception relates to quantification of the work factor. The method consists of: 1) isolating those people (engineers and technicians) who contribute to the technological capacity of the participating firm as the result of an increase in their competences (which may enlarge or diversify the 'critical mass' of the participant firm); 2) evaluating the time spent by these people on truly innovative activities; 3) for reasons of homogeneity, quantifying the effect in monetary terms by taking into account the average cost (including overheads) of these engineers or technicians over the time period estimated in the previous step.

9.2.2.4 Gathering Information

Information about the effects (as well as the quantification process) is gathered through interviews with a group of participants forming a sample of the whole population of participants in the evaluated programme. The individual data are kept strictly confidential by BETA, and only aggregated results are released.

Besides the information related to the effects themselves and their quantification, a large number of other data are collected to account for the specificities of the programmes under scrutiny. Typically, these data help to characterize the programme (for example, specific domains of research, duration, funding, etc.), the participants (such as firms, laboratories, etc.; type of production, sector, size, nationality, and so on), and the involvement of the programme participants (such as type of research, level of responsibility or position in consortia, type of IPR adopted, own funds invested, etc.). The scope of these data varies from one programme to another, and, therefore, can be finely adjusted to the particular characteristics of each programme and according to the focus that the evaluators/policy-makers want. A large part of the value of the conclusions drawn from this type of evaluation comes from the coupling of the effects observed and possibly quantified, and these characteristics.

9.2.2.5 The Principle of Minimum Estimate

The measurement of economic effects performed by the BETA group must be considered as a minimum estimate for the following reasons. All estimations of figures provided by participants are systematically minimized. That is to say, these estimations are expressed mostly as a range from which only the lower boundary is used for computation. Some effects cannot be measured, for instance, because the influence of the project exists, but is not separable from other factors. In spite of the time spent in interviewing people, some cases may also escape the interviewers, for instance when the technical aspects are very complex or when the firm has "forgotten". Finally, in spite of the guarantee of confidentiality provided to all interviewees, they might still be reluctant to disclose strategic information.

More generally, it should be underlined that the BETA methodology aims to assess only the economic effects for the participant and not the benefits to the rest of the economy. These latter are mid- or long-term effects, which may consist in diffusing technology (through imitation, technology transfers, staff mobility), or increasing consumer satisfaction. Thus, the measured effects are only a subset of the global economic effects of the projects.

9.2.2.6 The Sampling: Projects, Participants and Time Coverage

As mentioned above, the methodology is based on a sample of participants, which should be representative of the whole population of programme participants (i.e. all firms and research organisations involved in the R&D activities supported by the programme being evaluated). To achieve this, it is necessary to choose a set of objective parameters, which can be obtained independently of the results and which also correspond to certain criteria operating in the definition or in the objectives of the R&D programme. In the case of programmes subdivided into numerous projects, the projects might be the basis of the sample (selecting a representative sample of projects).

Interviews usually involve two members of the BETA team. The managers interviewed include those with overall responsibility for the project; in many cases, especially in SMEs, this person will be the General Director of the firm or the Technical, Engineering or R&D Manager. Other managers (on the Marketing, Accounting or Sales sides) are also interviewed. Prior to each interview, the BETA team sends some information about the study and the evaluation methodology to the interviewees. Information gleaned from the interviews is frequently augmented through postal and telephone exchanges.

For each project studied, the time period covered by the evaluation obviously starts from the beginning of the project, since some effects may appear right at the outset of the research work. As for the end of the time period, two years of forecast are generally considered.

9.2.3 Overview of Empirical Studies Using the BETA Method

Since the end of the 1970s, many programmes have been evaluated using the BETA approach. They are listed in Table 9.1. As mentioned above, these programmes all belong to the category of public RTD programmes, and many are of a collaborative nature. But they differ in terms of their design, some important aspects of the institutional arrangements that public authorities have put in place to implement them, and their interactions with society as a whole. Without addressing this in too much detail, one can distinguish between different forms of public RTD programmes using a simple description of the roles of the actors involved in the programme. These are:

- to provide funding;
- to design the operational (technical, temporal, etc.) objectives of the programme and/or projects;
- to design the modalities, rules, etc.;
- to manage the programme and/or projects;
- to perform R&D activity;
- to control the diffusion of the direct results of the programme and/or projects;
- to use the direct results of the programme and/or projects.

A detailed analysis of the roles of the various actors shows that R&D activity, and the control of access to the direct results and their use vary across the programmes evaluated using the BETA approach. However, the BETA approach focuses on the learning processes triggered by the R&D performed in each of the programmes and, consequently, on the initial beneficiaries of these processes, evidenced by the emergence of indirect effects. It therefore allows the innovation involved in these programmes to be measured, while also taking into account the various contexts in which the R&D activity is designed and managed, and its results used. The set of complementary data mentioned above enables this.

However, this means that comparisons of the overall results of projects, expressed in terms of effects (or ratio), must be treated with some caution. These results are not sufficiently comprehensive to completely grasp the variety of the effects generated, their meaning, and the determinants of the processes that have produced them or that have hindered or limited them. The results based on cross-analysis of effects and the various characteristics and the qualitative/interpretative analysis are undoubtedly far richer and more informative. As mentioned in the introduction to this chapter the following section highlights results related to three topics of importance in cooperative R&D programmes: university-industry relationships, the role and performance of SMEs in collaborative innovative activities, and the design of partnerships. Given their greater scope in terms of the numbers of projects evaluated and partners covered, and given their collaborative nature, we decided to focus on the studies within the European Commission Framework Programmes (ECFP), the European Space Agency (ESA) programmes and, to a lesser extent, projects that were part of the Material Ireland programme.

Table 9.1. Evaluations based at least in part on the BETA evaluation approach

Name of programme or organization in charge	Type of programme	Main features year of study release; sector; number of participants covered (by interviews unless otherwise stated)	Bibliographic source
EUROPEAN SPACE AGENCY			
All activities	Procurement policy through public agency	1980; space; 128	(BETA, 1980) (BACH et al., 1992)
Danish participation	Procurement policy through public agency	1987; space; 7	(BETA, 1987)
All activities	Procurement policy through public agency	1988; space; 67	(BETA, 1988) (BACH et al. 1992)
Canadian participation	Procurement policy through public agency	1989; space; 10	(BETA/HEC Montréal, 1989)
Technology transfers	Procurement policy through public agency	1989; space; 67	(BETA, 1989)(BACH et al., 1992)
Canadian participation	Procurement policy through public agency	1994; space; 8	(HEC Montréal/BETA, 1994)
Tech. transfers / Life science	Procurement policy through public agency	1996; space; 14	(BETA, 1996a)
Tech. transfers / Microgravity experimentations	Procurement policy through public agency	2000; space; 21	(BETA/NOVESPACE, 2000) (BACH et al., 2002)
SMEs in space activities	Procurement policy through public agency	2003; space; 196 (postal survey)	unpublished report from Bramshill Consultancy Ltd
CERN	Procurement policy through public agency	1975; suppliers of CERN infrastructure; 127 1985 (1); suppliers of CERN infrastructure; 160	(SCHMIED, 1975) (BIANCHI-STREIT et al., 1985)
EUROPEAN COMMISSION			
BRITE-EURAM	Subsidize for RD activity	1993-1995; all; 310	(BETA, 1993, 1997a, 1996b) (BACH et al., 1995, 2003)
ESPRIT	Subsidize for RD activity	1997; NTIC; 45	(BETA, 1997b)
AFME - France	Technical centers dedicated to sectors	1990; all; 12	(BETA, 1990)
Private Company - France	Private programme	1992; construction; 1	unpublished report from BETA
ANVAR (on region Alsace) - France	Subsidize for RD activity	1995; all; 22	(BETA, 1995a)
MATERIAL IRELAND - Ireland	Technical centers dedicated to sectors	1995; all; 31	(BETA, 1995b)
ECOPETROL - Colombia (1)	RD projects of public company	1996; software; 18	(GARCIA et al., 1996)
PETROBRAS - Brazil	RD projects of public company	1999; oil industry; 9	(FURTADO et al., 1999)
CBERS - Brazil (1)	Procurement policy through public agency	2001; space; 8	(FURTADO et al., 2001)
National R&D program for Medical and Welfare Apparatus - Japan	Subsidize for RD activity	2003; medical instrumentation; 34 (interviews) + 115 (postal survey)	unpublished report from PREST Univ. Manchester et al.

(1) without BETA laboratory participation

9.3 University-Industry Collaboration in R&D Activities

During the 1980s several national governments implemented new technology policies designed to promote university-industry partnerships[3]. Universities were put under pressure from government to change their function in society and to play an active role in the process of innovation. Thus, besides their classic mission of teaching and researching, universities were expected to diffuse new knowledge and to deliver new technologies to industry, in order to enhance the competitiveness of the economy. These policy initiatives have had a variety of consequences and impacts for both sets of actors. In this section, we show how the research results of universities or PROs have influenced the innovation capabilities and/or processes of firms in several programmes. We also illustrate how the research activities of the academic actors benefited from participation in collaborative innovation projects. We compare our results with the conclusions from existing empirical studies related to the topic.

9.3.1 The Impact of Scientific Research Results on Industrial Partners

During the 1990s, a large number of studies, based on surveys, patents, and publications, underlined that universities and PROs produce substantial R&D spillovers. For instance, Mansfield (1991) and Beise and Stahl (1999) show that in the absence of academic research a percentage of new products and new processes would not have been developed or would have come later. These positive impacts of university research are confirmed by studies based on the geographical dimension of spillovers[4] (cf. Jaffe 1989). These studies generally highlight that geographical proximity matters and that the benefits from academic research decrease in line with the distance between industry and university (cf. Mansfield and Lee 1996, Narin et al. 1997) underline that the knowledge flows between the two worlds increased threefold in the US between 1987 and 1994. If spillovers seem to be significant in the innovation process, firms do not consider PROs as important contributors to the creation of new ideas or to innovation completion (Cohen et al. 2002; Fontana et al. 2003). However, firms exploit research results in order to innovate believe that cooperative agreements or

[3] To understand the antecedents of policies facilitating university-industry relationships see, for instance, Poyago-Theotoky et al. (2002).

[4] Feldman (1999), Salter and Martin (2001), and Breschi and Lissoni (2001) contain valuable surveys of this literature.

formal contracts are important in accessing fundamental research (Meyer-Krahmer and Schmoch 1998). Some studies highlight the characteristics of firms with the highest propensity to benefit from public research results via cooperation (Mohen and Hoareau 2002; Fontana et al 2004), but very few analyses underline the potential benefits generated by firms participating in these kinds of agreements. In this section, we analyse the impact of scientific research results on the innovation abilities of firms participating in publicly financed R&D cooperative agreements.

In the course of evaluating the Euram, Brite and Brite-Euram I[5] (1993), we identified that the presence of a PRO (universities, institutions such as CNRS, Max Planck, CNR, etc) in a consortium had a positive influence on firms' generation of economic effects. The companies allied to PROs generated more direct effects than those not allied with PROs. The presence of a PRO in a consortium allows the process of generating indirect economic effects to be accelerated: in the short-run firms with an academic partner generate greater indirect effects than others; over the long-term the results are approximately the same for both types of firms. Association with PROs also influences the type of indirect effects generated by firms. Membership of a university in a consortium favours product rather than process transfers, network effects rather than reputation effects, organization rather than method effects, and competence building rather than training. These results are somehow consistent with those obtained by Mansfield (1991): 11% of new products and 9% of new processes could not have been developed or would have come much later in the absence of academic research.

In the course of the same study, and in HPCN (1997), we analysed how the nature of the research conducted by each partner impacted on their economic results. During the interviews partners were asked to define their particular research contributions: more fundamental (i.e. upstream), or more applied research or development oriented (i.e. downstream)? The group of organizations involved in upstream research (not just PROs) produced twice as many indirect effects as those conducting downstream research, but they generated far fewer direct effects. Moreover, downstream research providers generated a very small proportion of technological effects. These results may be explained by the fact that downstream research is more specialized and focused and, thus, less generic than upstream research. The more generic the knowledge, the more it is likely to be diffused within the organization. At the consortium level, it is interesting to analyse the differences between collaborations where fundamental re-

[5] These programmes focus mainly on new materials and industrial technologies. High-tech sectors, such as the pharmaceutical, biotech and NTIC sectors, are not well represented in these programmes.

search is associated with the other stages of innovation and those involving only downstream activities. The participants involved in consortia engaged in fundamental research are far more successful in terms of both direct and indirect effects. They also generate a large proportion of technological transfers while those working in more downstream consortia generate mostly human factor effects. These results underline the crucial role played by fundamental research in the innovation process (Kline and Rosenberg 1986).

As a result of deeper analysis of the direct and indirect effects, we can highlight which disciplines or scientific areas support economic effects. In Euram, Brite and Brite-Euram I, the direct effects were mainly generated in relation to three disciplines: mathematics and informatics, applied physics, and material science, which accounted for 93% of the cases of direct effects and 98% of the total value of direct effects. Effects in the area of mathematics and informatics were a surprise: Brite-Euram was merely intended to support the development of new materials and industrial technologies. This scientific area consisted mainly in developing mathematical tools for modelling and simulation; it became mature at the time of the programme and was able to be transferred successfully from universities to industries. Concerning the indirect effects, the main difference is in the relatively poor performance of mathematics and informatics. One of the reasons is probably that the main objective of projects supported by mathematics and informatics was to directly transfer the knowledge created by these disciplines to the intended process or product; less concern was given to the generation of indirect technological transfer. Material sciences and applied physics mainly supported the generation of indirect technological effects that were particularly successful in economic terms.

9.3.2 The Impact of Collaborative Innovation Projects on PROs

Universities and PROs may benefit from sources such as licensing, equity revenue, sponsored research, donations, and also from technologies developed by firms (Brooks 1994; Siegel et al. 2002). Links with industry may help universities to update curricula, to initiate new programmes and to create job opportunities for students (Stephan 2001). However, negative effects on education (content and quality of teaching), students and on the culture of "open science" (Nelson 2001; Dasgupta and David 1994) may arise. Increased commercialization of university research can delay the publication of scientific findings and reduce the willingness of faculties to disclose their research agenda and results (Blumenthal et al. 1997). Technology transfer could encourage a shift from basic to more applied research. Some empirical evidence refutes this hypothesis and shows that entrepreneurial scientists achieve higher productivity scores than their non-entrepreneurial colleagues (Louis et al. 2001). Our aim is to show that universities participating in publicly funded R&D programmes do generate economic effects and, especially, indirect effects.

In Euram, Brite and Brite-Euram I, we found that universities did not generate direct effects: the role of universities is not to directly sell a product on the market. However, they generated indirect effects mainly through new research contracts from industry, national governments, and the EU, and through increased competences. This latter aspect is considered by participants to be very important: it enables the discovery of new fields of research directly connected to the industry and thus increases the potential for future contracts. The researchers who acquire these new competences work in the laboratory for long periods and hold senior academic positions. The new research contracts will mainly find application in sectors that are close to the mainstream disciplines (physics, chemistry, material science) of universities: electricity / electronics, materials processing, chemistry / petro-chemistry / pharmacy, power generation / energy, precision equipment / instrument / sensors, etc. These sectors are not the most prolific as technology transfer receptors (cf. automobile, civil engineering) and thus it cannot be expected that universities generate large amounts of money from these kinds of activities.

9.4 Role and Performance of SMEs in Collaborative Projects

The key role of SMEs in all aspects related to innovation, and, correspondingly, the support the State can or should provide to these actors, are recurrent questions in the field of STI policies (Audretsch 2002; Comité Richelieu 2003). SMEs are often assumed to exhibit specific features regarding their capacity to innovate and to contribute to the link between innovation and economic development, however “positive” (favouring innovation) or “negative” (hindering innovation) these features may be (Dodgson and Rothwell 1994; Munier 1999; Hoffman et al. 1998; Julien 1994). In contrast to the Schumpeter, Galbraith and Schumacher contributions, the most common view of SMEs, which goes beyond the view of SMEs as "small big firms" (Welsh and White 1981), is that, on the one hand SMEs benefit from their inherent characteristics, which favour their capacity to adapt to the evolution of the environment (flexible organization, proximity to clients, entrepreneurial attitude), but on the other hand suffer from lack of resources (notably in terms of funding and competences), and from various forms of dependence leading to fragility (dependence vis-à-vis key personal, clients, assets, etc).

In consequence, SMEs have long been a favourite target for ST&I policies[6]. It was therefore, not surprising, that in most of the programmes studied using the BETA approach, policy makers and, more generally, all stakeholders, paid great attention to SMEs' performance, behaviour, constraints, opportunities, etc. Some interesting results shed some light on these aspects.

9.4.1 Do SMEs get More Benefits than Large Firms?

The related questions here are: do SMEs exhibit their supposed advantages and shortages as regards innovation in the context of publicly supported collaborative R&D activities? Do SMEs generate more innovations than big firms in such activities? If so, why? Contrasting results were observed in the programmes evaluated by BETA[7].

Generally speaking, big firms obtained better results than SMEs in terms of effects generated. This was especially true in relation to direct effects from large cooperatively funded types of policy. For instance, in the cases of EC Brite-Euram as well as Esprit, it was obvious that, at this level, SMEs could not compete with big firms. However, the difference between big firms and SMEs was not so great in terms of indirect effects. In the case of Brite-Euram, big firms achieved more indirect effects, whereas in Esprit, SMEs displayed more of these effects. Moreover, it is interesting to note that in programmes that were "closer" to firms, such as technical centre activities (cf. the Material Ireland Programme), the direct effects were at least as large for SMEs as for bigger firms. This could indicate that in terms of directly generating market application from sponsored research, SMEs perform better when they are involved in programmes whose design and implementation are more specifically oriented towards them and/or towards day-to-day needs. But, again, the conclusion seems different when

---

[6] We will neither detail here the variety of policies adopted across countries, nor insist on the debate about the adequate levels (sector vs. general, local vs. national etc) at which policies towards SME are/should be conducted (OECD, 1998) (North et al., 2001).

[7] See the different reports for the precise definition of SME adopted. The studies having been performed at different points in time, the exact definition varies from one study to the other, which means that i) results are not strictly comparable in quantitative terms ii) comments provided here are intended to show some tendencies contextual to the environment (type of programme, sector, etc).

indirect effects are considered. In the case of Material Ireland, indirect effects were important for SMEs. In terms of the research being conducted by the Technical Centre, this result highlights the existence within SMEs of an absorptive capability that was used to exploit the outcome of the R&D sub-contracted to the centre.

In most instances, the programmes evaluated helped firms to access new markets and new networks, which demonstrated some kind of additionality of policy for SMEs. In contrast, critical mass effects were comparatively more limited, which could indicate that long term support is the most suitable (or is a necessary complement), rather than "spot support" to help SMEs to build such capabilities.

Another particularly interesting result concerns the correlation between direct and indirect effects. Whenever it was possible to build statistically significant sub-samples (particularly in the case of EC programmes), it appeared that big firms were the only ones able to generate both direct and indirect effects from the same project; SMEs were only able to generate one type of effect. Big firms, therefore, are able to conduct different research projects in parallel, to directly exploit results, and to make internal knowledge transfers from one project to another (provided that the internal strategy and organization allow this). SMEs are generally limited to working on a single project or a small set of related projects. If the project is commercially exploitable they concentrate on directly exploiting it (ignoring indirect ways of developing it); if it is not, they either forget this project and pass on to another or try by whatever means to derive indirect effects from it. But they are not able to do both.

The role of the SME in the project must also be considered in terms of technical success: it has been shown that even when a project is a technical failure (and then ipso facto a commercial failure) big firms are able to derive indirect effects. This capability denotes an "R&D culture" that enables learning from successes and failures. SMEs seem to be much more limited in their ability to learn from technical failure in any systematic and organized way, that is to "pick up" the pieces of knowledge that could be useful for other projects.

In Brite-Euram, SMEs that were "Prime" contractors (coordinators of the project) were likely to generate fewer, but greater direct effects, while they generated few indirect effects. In other words, focusing on the prime activity reflects either a strong commitment to research, which could be a risky but winning strategy with high gains, but which limits the ability to value the activity other than through its direct consequences.

The reason for this apparent dominance of big firms in exploiting collaborative programmes is in part to be found in the classical explanations in the literature:

- Critical lack of financial resources: in the case of EC projects, for a big firm, in most instances the extra investment necessary to put the product of the research on the market was very small when compared to total turnover; for the SME, even for extra investment equal to the amount of money previously invested in the research project, the barrier was too high and the technologically successful project was abandoned.
- Lack of R&D capacity: some SMEs have participated in EC programmes without having budget or staff dedicated to R&D activities. These firms generated neither direct effects nor technology-related indirect effects.
- Lack of critical mass of competences and the lack of diversity of competences: even when projects were quite successful from a purely technical point of view, complementary assets (technical, commercial, productive, marketing, etc.) were often insufficient to go the step further needed for commercial exploitation.
- Lack of resources combined with a lack of flexibility: do not allow resources and attention to be allocated over the long term to different projects in parallel, but only to switch over time from one activity to the other.

Other factors have also been identified as barriers to the creation of effects. For instance, the problem of gaining the confidence and trust of potential partners and clients seems to be crucial. In a significant number of cases, big organizations were reluctant to become the first clients for the innovative products or services developed by SMEs because they did not trust their mid-term or long-term ability in terms of quality, cost, and delays. This limited the possibility for SMEs to economically value the developments they made during the projects. SMEs experienced difficulty in setting up and securing adequate intellectual property protection (IPR), which limited commercial exploitation in many cases.

9.4.2 Do all SMEs Perform Equally?

Finally, there is the question of heterogeneity of SMEs. Are the previous results relevant to all SMEs? Are SMEs similar in terms of their supposed specificities as regards innovation? In relation to the latter point, very schematically again, two related lines of arguments have been pursued in the literature in the last twenty years. The first tries to account for the heterogeneity of the population of SMEs. The argument was initially based on differences according to the age of the SME (following different growth models), but has conceded that "all SMEs are not future big firms". The research has led to numerous and often ad-hoc classifications based on various innovation-related variables[8].

A second way of tackling this question directly deals with the existence of specificities in SMEs, obviously beyond the simple observation of their size, whose relevant measurement is always subject to debate (GREPME 1997; Torres 1998). From a strict opposition between SMEs and big firms in terms of certain variables used to characterize any organization (such as centralization, formalism, hierarchy, information system, etc.), the research has moved on to the idea that these variables can take different values along a continuum, and, consequently, that firms can exhibit different configurations of characteristics, which more or less depart from the SME and big firms "reference" configuration. More generally, recently some researchers have argued that the delineation between many SMEs and big firms will become more and more blurred, as will the respective advantages/disadvantages of SMEs. On the one hand, different forms of convergence (often related to different organizational aspects (Lanoux 2001)) will be observed more and more frequently, although still remaining highly context-dependant[9]. On the other hand, the complex game of externalization of activities and multiplication of alliances and partnerships of all kinds has led to a growing decentralization of activities and a reinforcement of the links between the firms, with an increasing dominance of big firms (Sakai 2002). More SMEs are joining groups and are thus no longer

[8] For instance distinguishing technology developers, users, and followers (OECD 2001); "fast-expanding" firms including mostly young SMEs, investing more in R&D and having a bigger and more diversified portfolio of alliances, especially for innovative activities (see OECD (2000)); firms as regards their strategic orientation such as innovation, niche, etc. (OECD 2001); young SMEs, mature SMEs, very innovative SMEs, growth SMEs (EC 2002a,b); passive or traditional SMEs, dominated by big, imitative, technology based, high tech (Rizzoni 1991); etc.

[9] The generalization about quality management, which makes SMEs able to formalize and standardize many aspects of the organization that were previously rather informal; the pervasive use of ITT (Information and Telecommunications Technologies) and Integrated Information System which help them to rationalize and optimize all information processing aspects and allow big firms to avoid a too centralized management of information; operations management methods (Just-In-Time and Toyotism concepts, etc.), which help them to rationalize and optimize the production processes; internal strategies of big firms consisting, for instance, in organizational change towards less bureaucracy; external strategies of big firms consisting of the externalization of a large range of activities related to research, conception, production, etc.

fully independent, while others are joining networks to get access to resources that otherwise would be limited for independent SMEs.

Rather than being driven by the intrinsic features of SMEs, the characteristics of SMEs (especially their advantages/shortages as regards innovation) would be largely determined by contextual factors, among which, beside sectoral or technical factors, interactions with big firms play a prominent role. To what extent then is this reflected in collaborative publicly supported activities?

In the empirical studies based on the BETA approach, these debates were reflected by the observation of varying results among the population of SMEs, according to some of their characteristics.

For instance, in the Brite-Euram study (and at that time in absence of a strict definition of SME except in terms of small number of staff), it appeared that at least three classes of SMEs could be distinguished, all exhibiting quite different results. SMEs that were part of large groups dominated by big firms, generated far more effects (especially direct effects). SMEs that were linked together within small groups ("holdings" of SMEs sharing some resources and participations) could be ranked in second position, and largely dominated the third category composed of purely independent single SMEs. Clearly, funding was one of the key factors, but also marketing and strategic capabilities provided by the big parent-companies (while holding structures could allow scale and scope economies) were also important. More or less captive or preferential markets (for instance, related to sub-contracting activities) and credibility vis-à-vis partners and clients, played a major role.

In the case of EC programmes, the SMEs that acted as "technology producers" produced the most effects (direct as well as indirect), ahead of the pure researchers or those that combined production, research, and use of technology (big firms "integrating" those different roles were the most successful of all the firms in the study). Again, what is probably behind this phenomenon is the position that some specialized SME-technology producers hold as key players in established networks, or as sub-contractors of big firms.

In the case of the ESA programme, another type of relationship between big firms and SMEs also surfaced. Apart from "pure" SMEs, a lot of small space departments or teams from non-space oriented big firms were present. They often acted as quite independent companies forming "Space SMEs within big companies". For reasons similar to those outlined above, they were quite successful in generating indirect effects. But, to some extent, since they were technologically more "isolated" from the rest of the company, the volume of their technology-related effects was greatly dependent on the organizational set-up favouring internal cross-fertilization.

9.5 The Design of Partnerships and the Performances of Actors

Access to complementary assets and knowledge is very often the motivation for entering into a cooperative agreement, or joining an R&D consortium. For instance, Richardson (1972) suggested that complementarity was the essence of inter-firm cooperation. According to Teece (1986) the assets of suppliers, clients, and competitors are important for both the consortium and the participating firm. He also stresses the important role played by complementary external assets in the creation of successful innovation and, therefore, that complementary assets could justify the formation of strategic alliances. Along the same lines, Sakakibara (1997) shows that Japanese firms enter public R&D consortia more for "skill-sharing motives" and less for "cost-sharing ones". The exploitation of complementary assets inside a partnership gives access to new competencies and generates the creation of new knowledge. Mothe and Quelin (2001) show, for instance, that "the access to complementary assets (which are further downstream in the value chain) gives the partner firm a greater chance of creating new products and of accumulating knowledge" (p. 131). We want to emphasise similar results from our study.

However, the design of partnerships may be directly or indirectly influenced by policy. For instance, in some procurement or mission-oriented policies, where the technological objectives correspond to a complex artefact, such as a satellite or a high-speed train, the structure of the group of participants is often imposed by the technological artefact. In other words, the technology requires a hierarchy between the partners, typically with a prime contractor responsible for the project and integration of the artefact (including all the interfacing and management problems involved), partners that are responsible for systems and sub-systems, equipment providers, and also a series of service/support providers. This hierarchy exists, but in a less well-defined way, in more diffusion-oriented programmes such as those implemented by the EU. Very often in EU projects one partner plays the role of the general coordinator of the project (the prime partner), while the others are simply contributors to the final objective. What we first aim to show is that the different positions/responsibilities of partners in a consortium may induce different types of learning. This was especially clear in the ESA projects. Next, we focus on the link between the complementarity between partners, regardless of any hierarchy.

9.5.1 The Design Imposed by the Public Programme

The ESA related activities are good examples of R&D programmes that are structured in a very hierarchical way, based on the technical content of the programme. Three levels can be distinguished: prime contractor (typically responsible for a satellite), system suppliers (in charge of providing systems such as propulsion or telecommunication systems), equipment manufacturers (chips, electronic boxes and devices, mechanical parts, etc.), and service providers supporting the whole network. In the 1988 BETA study, the correlation between these four levels of responsibility of ESA contractors and the indirect effects they generated was studied, and the results clearly indicated the influence of the actors' positions in the network on their learning patterns. The prime contractors, and to a lesser extent the system developers, tended to concentrate their efforts on the space market and used the ESA activities to maintain develop and diversify a high-skilled workforce (expressed in the competence and training effects). They also gained experience in managing complex international projects (O&M effects) that could subsequently be exploited in other programmes. Prime contractors tended to diversify more (creation of new activities or new divisions), no doubt because their size and financial position allowed this. The firms generating the greatest indirect effects (and this was true of firms outside the space sector), were equipment developers, generally innovative, medium-sized firms or large firms with small space departments, using generic technologies to manufacture components, and the capability to move to mass-production. They were in "direct contact" with product and process technologies and thus generated technological effects. Few indirect effects were observed in relation to service providers, because they apparently make use of previously accumulated knowledge; however, almost all their effects were linked to administration and management innovation related to methods, quality control procedures, software standardization, and the like, confirming the influence of their role in ESA work (studies, consultancy, assistance, and maintenance) upon their learning profile. These analyses (presented in more detail in Zuscovitch and Shachar (1990) and extended in Zuscovitch and Cohen (1994)) also clearly showed that there was some form of irreversibility in the choice to act at a given level of responsibility. In particular, it seemed very difficult, in terms of resources management and know-how specificity, to resume a lower level of responsibility once having experienced a higher one.

As mentioned earlier, in the European programmes and particularly in Brite-Euram and Esprit, the participants agree upon a prime partner, that will be responsible for the coordination of the technological work and will be the main channel of communication with the European Commission project officer. Our main concern in this regard was: do prime-partners generate more economic benefits than others? In both Esprit and in Brite-Euram, the partners generated more direct and indirect effects than the prime partners. In Brite-Euram, this negative effect for prime partners was much worse for SMEs. The huge amount of administrative work required by the EC is generally seen as the main handicap. In interviews, most prime partners admitted that they would avoid this responsibility in subsequent contracts. At the start of the project some prime partners had the expectation that they would be able to give a particular technological orientation to the project, but the reality was completely different for most of them. Moreover, in the Esprit programme, some of the prime partners complained about the lack of authority used by the EC project officers to manage free-riding behaviour: it is difficult for the partners to an agreement to retaliate against bad behaviour by a partner.

9.5.2 The Exploitation of Complementarities

The positive impact of universities within a collaborative arrangement (cf. 3.1) can be considered as a first illustration of the importance in the innovation process of exploiting complementarities. We also mentioned that the presence of partners conducting both fundamental and applied research was beneficial for the participants.

The importance of user-producer relations (Lundvall 1992) in the process of innovation has been extensively recognized in the literature. In Brite-Euram we divided the partners into four groups according to the nature of their contribution (production). The first group included vertically integrated firms, which are simultaneously Producer, User, and Researcher/Tester. The second group included firms that are only Users or Users and Researchers/Testers. The third group involved firms that are only Producers or Producers and Researchers/Testers. The last group comprised the partners (mainly universities and research centres) that were Researchers/Testers. In Brite-Euram, the vertically integrated firms were four times more successful in terms of direct effects and three times more in terms of indirect effects, than the non-integrated firms. Users and producers (non-integrated) showed similar results for direct and indirect effects. These figures show that it is easier to market research results directly and indirectly, if all the competences are combined within the same company. In other words, complementarities inside a firm enable better performance than do non-integrated activities. But what about the association of non-integrated users and producers in an agreement, compared to those that are not associated? We found that the associated users and producers generated twice as many direct and indirect effects as users and producers that were not associated. These results show the existence of a complementarity or consortium effect, although of less importance than the integration effect.

In the study conducted in the mid-1990s examining technology transfers generated by Brite-Euram partners, we analysed more precisely the origin of the direct and indirect effects measured in the previous evaluation studies (BETA 1993, 1996b). We gathered information concerning the industrial sector, the scientific disciplines, and the technologies that originated the transfers. We were able to show that the association of some scientific disciplines, or of some industrial sectors, allowed the creation of new knowledge and innovation, which otherwise would not have occurred or would have occurred later. Mathematics and informatics, applied physics and material sciences, constitute the scientific disciplines at the base of most of the direct and indirect effects. The positive relationship between applied mathematics and a large variety of industrial sectors was one of our most important findings.

The exploitation of complementarities was also analysed to some extent through the interaction between specific sectors in the process of generation of technology transfer (cf. BETA 1997a). We focus on the indirect effects for the three sectors that generated the highest amount of technology transfer: i.e. the automobile, civil engineering and electric/electronic equipment sectors. The automobile sector is an important receptor of technologies generated by other sectors, such as electronic components, aeronautics, and informatics, but this sector does not transfer many technologies to other sectors. Only three sectors benefit from its technological developments (electronic equipment, aeronautic, and material processing). Civil engineering is the biggest receptor in our sample, and the transfers are mainly from the electronic/electric equipment sectors), but does not transfer technologies to other sectors. The electronic/electric sector is the biggest source of technologies, which are transferred mainly to engineering, automobiles and within its own sector. These examples underline that the association of some industrial sectors is an important factor in technological development and economic opportunities.

9.6 Conclusion

Among the richness and the variety of results obtained from the studies conducted using the BETA method, this chapter focuses only briefly on the outcomes related to public cooperative R&D programmes, but tackles more specifically the importance of university-industry interactions, the role and performance of SMEs, and the impact of the partnership design on the economic performances of actors.

Our findings and their policy implications can be summarized as follows. Scientific research results are complementary to technological related knowledge in the sense that associating PROs and firms, or upstream and downstream research, within an agreement fosters innovation and increases the economic impact of the outcomes. Moreover, PROs participating in cooperative innovation oriented programmes receive benefits (new research contracts with industry, new fields of research). More generally, the way different types of partners are associated (combining users and producers, some scientific disciplines with relevant industrial sectors or several industrial sectors) has an impact on innovation performance: "complementary skill-sharing objectives" should be seen from a political point of view as more important than "cost-sharing motives". Cooperative programmes often impose a hierarchy among the partners: the positions of partners in this hierarchy induce different types of learning. Finally, SMEs receive benefits from participating in such programmes, but do not perform as well as large companies. We underlined the main barriers explaining this weaker ability to generate effects. We also highlighted types of SMEs (those that are part of a large group, "technology producers") that perform better than others. Although policies seem to be important for SMEs to access new markets, or networks, they do not help to overcome the main disadvantages that these kinds of firms experience. Other public instruments should be introduced to complement these cooperative programmes.

More generally, related domains could be approached on the basis of the interpretation of the quantitative and qualitative information gathered in each implementation of the approach. The overall role and impact of public action in the emergence and structuring of industry and of connected technologies would be an interesting area warranting further study. Evaluation of the Esprit programme (on a new generation of computer architectures), of the Petrobras Procap programme (on creation/enforcement of equipment suppliers for oil industry), and, more generally, of successive generations of ESA programmes (on the creation of the European space industry and related competencies), could have been presented. Matters related to public policy design to support/orient technology transfers might be examined.

Nevertheless, the results presented here contribute to the understanding in three crucial areas where quantitative data to support qualitative observations and monographs are often lacking, and for which analyses are not always developed in a context of evaluation of public action. It must also be recalled that the results from these BETA studies, partially confirming previous statements or invalidating others, or possibly opening some new research perspectives, are some sort of secondary (but obviously important) utilization of data that were primarily designed to account for the creation of economic effects.

Certainly, there is a case for more interaction between the academics conducting research in related fields, and policy-makers. More involvement of academics/researchers (especially greater involvement of researchers dealing with ST&I or ST&I policies in evaluation work) would help to produce richer evaluation frameworks, metrics, and interpretation grids, better related to recent development in the ST&I fields, and that go beyond a simple cost-benefit analysis, which remains the basis of most studies. More interaction with policy-makers, including at the stage of the elaboration of the evaluation methods and approaches, would greatly help in the development of a "learning policy-making", which is more often than not only wishful thinking.

9.7 References

Audretsch DB (2002) The dynamic role of small firms : evidence from the US. Small Business Economics 18: 13-40.

Bach L, Cohendet P, Lambert G, Ledoux MJ (1992) Measuring and managing spinoffs: the case of the spinoffs generated by ESA Programs. In: Greenberg JS, Hertzfeld HR (eds) Space economics. Progress in Astronautics & Aeronautics, New York, US, p. 144.

Bach L, Condé-Molist N, Ledoux MJ, Matt M, Schaeffer V (1995) Evaluation of the economic effects of Brite-Euram Programmes on the European industry. Scientometrics. 34: 325-349.

Bach L, Cohendet P, Schenk E (2002) Technological transfers from the European space programmes : a dynamic view and a comparison with other R&D projects. Journal of Technology Transfer 27(4): 321-338.

Bach L, Ledoux MJ, Matt M (2003) Evaluation of the Brite–Euram Programme.In: Learning from science and technology policy evaluation: experiences from the United States and Europe. Shapira P and Kuhlmann S (eds.) Edward Elgar, Cheltenham (UK) and Northampton, Massachusetts (USA), pp. 154-173.

BETA (1980) The economic effects from the ESA contracts. Final report for the European Space Agency (4 volumes), ESA, Paris.

BETA (1987) The indirect economic impact of ESAs contracts on the Danish Economy. Final report for the Danish Research Administration, Copenhagen

BETA (1988) Study on the economic effects of European space expenditures, Results (Vol. 1) and Report on investigation theory and methodology (Vol. 2). Prepared for the European Space Agency, Paris, October.

BETA/HEC Montréal (1989) Indirect economic effects of ESA contracts on the Canadian economy. Final Report for the Canadian Space Agency, Montréal. BETA (1989) Analyse des mécanismes de transfert de technologies spatiales : le rôle de l'Agence Spatiale Européenne. Final report for ESA, ESA, Paris.

BETA (1990) Mesure des impacts des centres techniques : un essai méthodologique. Report for the AFME, Agence Française de la Maîtrise de l'Energie, Paris.

BETA (1993) Economic evaluation of the effects of the Brite-Euram programmes on the European Industry. Final Report, EUR 15171. CEC, Luxembourg.

BETA (1995a) Évaluation de limpact de laide à linnovation dans les PME soutenues par l'ANVAR. Étude de la région Alsace. Final report for ANVAR, Strasbourg.

BETA (1995b) Evaluation of the economic effects generated by R&D services provided to industry by Materials Ireland. Final report for Materials Ireland–Forbairt, Dublin (Ireland).

BETA (1996a) Transfert de technologie entre lespace et les sciences de la vie en Europe. Final report for the European Space Agency, Paris.

BETA (1996b) Evaluation of the economic effects of the programmes EURAM, BRITE, and BRITE-EURAM I -Portugal, Ireland, Greece, Spain and SMEs. EUR 16877 EN, EC - Directorate-General Science, Research and Development, Brussels.

BETA (1997a) Evaluation and analysis of the techological transfers generated by the programmes EURAM, BRITE and BRITE-EURAM, EUR 16878 EN. Final Report for the DG XII-C EC, Brussels.

BETA (1997b, Pilot economic evaluation of Esprit HPCN results. Final report for the DG III-Industrie F1, Brussels.

BETA/NOVESPACE (2000) Study on transfers of technology and spin-offs realised in framework of Space Station and Microgravity Programmes. Final report for the European Space Agency, Paris.

Beise M, Stahl H (1999) Public research and industrial innovation in Germany. Research Policy 28: 397-422.

Blumenthal D, Campbell EG, Anderson MS, Causino N, Louis KS (1997) Withholding research results in academic life science: evidence from a national survey of faculty. Journal of the American Medical Association 277: 1224-1228.

Bianchi-Streit M, Blackburne N, Budde R, Reitz H, Sagnell B, Schmied H, Schorr B, (1985) Utilité économique des contrats du CERN (deuxième étude). Rapport Final pour le CERN, Geneve.

Breschi S, Lissoni F (2001) Knowledge spillovers and local innovation systems: a critical survey. Industrial and Corporate Change 10: 975-1005.

Brooks H (1994) The relationship between science and technology policy. Research Policy 25: 477-486.

Cohen WM, Nelson RR, Walsh J (2002) Links and impacts: the influence of public research on industrial R&D. Management Science 48: 1-23.

Comité Richelieu (2003) Livre blanc des PME innovantes - Vers un small business act Européen? Paris.

Dasgupta P, David, PA (1994) Towards a new economics of science. Research Policy 23: 487-521.

Dodgson M, Rothwell R, (1994) Innovation in small firms. In: The handbook of industrial innovation. Dodgson M, Rothwell R (eds.) Edward Elgar, Cheltenham, pp. 310-324.

European Commission (2002a) High-tech SMEs in Europe, Observatory of European SMEs, 2002/N°6. Enterprise publications, Luxembourg.

European Commission (2002b) Sixth Report - Observatory of European SMEs 2002/N°2 - data 2000. European Commission, Luxembourg.

Fahrenkrog G, Polt W, Rojo J, Tubke A, Zinöcker K (eds.) (2002) RTD evaluation toolbox assessing the socio-economic impact of RTD policies. Report of EC Strata Project HPV 1 CT 1999-00005, EUR 20382 EN. IPTS and Joanneum Research. Manchester.

Feldman MP (1999) The new economics of innovation, spillovers and agglomeration: a review of empirical studies. Economics of Innovation and New Technology 8: 5-25.

Fontana R, Geuna A, Matt M (2004) Firm size and openness: the driving forces of university-industry collaboration. In Caloghirou Y, Constantelou A, Vonortas N (eds.) Knowledge flows in European industry: mechanisms and policy implications, Routledge, London, forthcoming.

Furtado A, Suslick S, Pereira N, De Freitas A, and Bach L (1999) Economic evaluation of large technological programmes: The case of Petrobras Deepwater Programme in Brazil – Procap 1000. Research Evaluation: 8(3): 155-163.

Furtado A, Costa Filho E J (2001) Avaliação dos impactos econômicos do programa CBERS: Um estudo dos fornecedores do INPE. Final Report, DPCT/IG/UNICAMP, Campinas, November.

Garcia A, Amesse F, Silva M (1996) The indirect economic effects of Ecopetrols contracting strategy for informatics development. Technovation 16: 469-485.

Georghiou L, Meyer-Krahmer F (1992) Evaluation of socio-economic effects of European Community R&D programmes in the SPEAR network. Research Evaluation 2(1): 5-15.

Georghiou L, Rigby J, Cameron H (eds.) (2002) Assessing the socio–economic effects of the EC RTD Framework Programme –ASIF project. Final Report for CE DG Research PREST in collaboration with BETA, ISI, AUEB, Joanneum Research, IE HAS and Wise Guys.

GREPME (1997) Les PME - bilan et perspectives, 2e ed., sous la direction de Pierre-André Jullien. Les Presses Inter-Universitaires Québec/Economica, Paris.

HEC Montréal (1994) Les effets économiques indirects des contrats de lASE sur léconomie canadienne - deuxième étude, 1988-1997. Rapport préliminaire final pour l'Agence Spatiale Canadienne (in collaboration with BETA).

Hoffman K, Parejo M, Bessant J, Perren. (1998) Small firms, R&D, technology and innovation in the UK : a literature review. Technovation 18(1): 39-55.

Jaffe A (1989) Real effects of academic research. American Economic Review 79: 957-970.

Julien P-A (1994) PME, Bilan et perspectives, GREPME ed. Economica, Paris.

Kline SJ, Rosenberg N (1986) Innovation: an overview. In: Landau R, Rosenberg N (eds), The positive sum strategy. Academy of Engineering Press, Washington, pp. 275-305.

Lanoux B (2001) L'adoption par les PME/PMI des systèmes de gestion par processus. Thèse de Doctorat ès Sciences de gestion, BETA, Strasbourg.

Louis KS, Jones LM, Anderson MS, Blumenthal D, Campbell EG (2001) Entrepreneurship, secrecy, and productivity: a comparison of clinical and nonclinical faculty. Journal of Technology Transfer 26: 233-45.

Lundvall B-Å (ed.) (1992) National systems of innovation: towards a theory of innovation and interactive learning. Pinter Publishers, London.

Mansfield E (1991) Academic research and industrial innovation. Research Policy 26: 1-12.

Mansfield E, Lee JY (1996) The modern university: contributor to industrial innovation and recipient of industrial R&D support. Research Policy 25(7): 1047-1058.

Meyer-Krahmer F, Schmoch U (1998) Science-based technologies: university-industry interactions in four fields. Research Policy 27: 835-852.

Mohnen P, Hoareau C (2002) What type of enterprise forges close linkswith universities and government labs? Evidence from CIS 2. Managerial and Decision Economics 24: 133-145.

Mothe C, Quelin BV (2001) Resource creation and partnership in R&D consortia. Journal of High Technology Management Research 112: 113-138.

Munier F (1999) Taille des firmes et innovation, thèse de Doctorat ès Sciences économiques. BETA, Strasbourg.

Narin F, Hamilton K, Olivastro D (1997) The increasing linkage between US technology and public science. Research Policy 26: 317-330.

Nelson RR (2001) Observations on the Post-Bayh-Dole rise of patenting at American universities. Journal of Technology Transfer 26: 13-19.

North D, Smallbone D, Vickers I (2001) Public sector support for innovating SMEs. Small Business Economics 16(4): 303-317.

OECD (1997) Policy evaluation in innovation and technology - towards best practices. Proceedings of the Conference organized by OECD's Diectorate of Science, Technology and Industry. OECD, Paris.

OECD (1998) Best practice policies for SME - 1997 edition, OECD, Paris.

OECD (2000) The OECD SME Outlook. OECD, Paris.

OECD (2001) Enhancing SME competitiveness. The OECD Bologna Ministerial Conference. OECD, Paris.

Poyago-Theotoky J, Beath J, Siegel DS (2002) Universities and fundamental research: reflections on the growth of university-industry partnerships. Oxford Review of Economic Policy 18(1): 10-21.

PREST (2003) The evaluation of the national research development program for Medical and Welfare Apparatus. Final report for the Ministry of Industry and Trade (METI) Mitsubishi Research Institute, Manchester.

Richardson GB (1972) The organization of industry. Economic Journal 82: 883-896.

Rizzoni A (1991) Technological innovation and small firms : a taxonomy, International Small Business Journal 9(3): 31-42.
Sakai K (2002) Industry issues - Global industrial restructuring: implications for small firms, STI Working Papers 2002/4. OECD, Paris.
Sakakibara M (1997) Heterogeneity of firm capabilities and cooperative research and development: an empirical examination of motives. Strategic Management Journal 18: 143-164.
Salter AJ, Martin BR (2001) The economic benefits of publicly funded basic research: a critical review. Research Policy 30: 509-532.
Schmied H (1975) Etude de l'utilité économique des contrats du CERN, Rapport Final pour le CERN, Geneve.
Siegel DS, Waldman D, Link A (2002) Assessing the impact of organizational practices on the productivity of university technology transfer offices: an exploratory study. Research Policy 31: 1-22.
Stephan PE (2001) Educational implications of university-industry technology transfer. Journal of Technology Transfer 26: 199-205.
Teece DJ (1986) Profiting from technological innovation: implications for integration, collaboration, licensing and public policy. Research Policy 15: 285-305.
Torres O (ed.) (1998) PME - De Nouvelles Approches. Economica - série Recherche en Gestion, Paris.
Welsh JA, White JF (1981) A small business is not a little big business. Harvard Business Review 59(4): 18-32.
Zuscovitch E, Shachar Y (1990) Learning patterns within a technological network. In: Dankbaar Groenewegen B, Schenk H (eds.) Advances in industrial organization, Kluwer, Amsterdam.
Zuscovitch E, Cohen G (1994) Network characteristics of technological learning: the case of the European space program. Economic Innovation and New technology 3: 139-160.

Part IV The Relevance of R&D Strategic Management in Policy Design

10 The Organizational Specificities of Brite-Euram Collaborative Projects: Micro-Analysis and Policy Implications

Mireille Matt and Sandrine Wolff

BETA, Strasbourg, E-mail: matt@cournot.u-strasbg.fr
BETA, Strasbourg, E-mail: wolff@cournot.u-strasbg.fr

10.1 Introduction

The aim of our chapter is to examine and characterize EU-sponsored R&D collaborations at the micro-analytical level, and to derive some policy implications. The existing literature on European framework research programmes relies, in the main, on quantitative, statistical information. We think that the relevance and the efficiency of such programmes may well require a deeper understanding of the cooperative practices adopted by the companies that participate in government-sponsored collaborations. Our contribution can be considered as an attempt at opening this specific "inter-organizational black box" at the micro level of the firm[1]. More precisely, we will explore the internal mechanisms of government-sponsored collaborations[2] by comparing them to those of spontaneous, privately funded research collaborations[3]. Our work is based on the existing literature con-

[1] Apart from Hagedoorn and Schakenraad's (1993) article comparing private versus subsidized R&D partnerships between big information technology firms, we could not find explicit mention of similar issues in the literature. Although Hagedoorn and Schakenraad do not identify substantial differences between the two kinds of agreements in terms of the general shape of the networks, we contend that sharp differences are observable if organizational mechanisms are examined.

[2] The subsequent observations and theoretical propositions of our chapter are relevant for the "diffusion-oriented" EU research programmes. Direct application to the so-called "mission-oriented" programmes or to other international/national programmes could be misleading.

[3] See Matt and Wolff (forthcoming 2004) for a comparative case study between a government-sponsored and a spontaneous R&D collaboration in the field of fuel cell powered electric cars.

cerning inter-firm alliances and on qualitative, empirical information obtained through numerous interviews relating to some of the research and development (R&D) projects within the EU's Brite-Euram[4] programme.

As a conceptual framework, we propose to consider inter-firm technological collaborations as particular forms of organizations that aim at creating new knowledge via the association of the resources of two or more independent firms[5]. We adopt the approach of March and Simon (1993) that organizations are "systems of co-ordinated actions among individuals and groups whose preferences, information, interests, or knowledge differ". Thus, an organization coordinates the actions of agents characterized by different knowledge bases and different interests. As a consequence, we suggest that any organizational mode (but particularly inter-firm collaboration) must fulfil the following three functions: coordination, incentive, and learning[6]. Our chapter applies this three-dimensional grid in order to understand the micro-mechanisms of inter-firm technological collaborations.

The chapter is organized as follows. In the first part we synthesize the relevant results from the literature about strategic management of technological alliances (focusing mainly, although not exclusively, on spontaneous research collaborations). Important issues generally concern: the motivations of the partners in connection with potentially opportunistic behaviour; the mechanisms of inter-organizational learning and value creation; the influence of contractual terms and other internal coordination devices of alliances. This allows us to apply our previously mentioned, three-dimensional, analytical grid (in terms of incentives, learning, and coordination) to explain some important aspects of the micro-rationale of inter-firm collaboration emerging from the literature.

In the second part of the chapter we again apply this grid in order to identify the specificities of "our" Brite-Euram research projects, and then try to compare them with ideal spontaneous collaborations. We contend that the two types of collaborations show rather contrasting rationales: government-sponsored collaborations most often concern peripheral activities, submit to pre-defined rules, and favour exploratory, unilateral learning; by contrast spontaneous collaborations concern activities closer to core competences, create their own operating rules, and may trigger an interactive learning process, which generates valuable collective specific assets. Evolution pathways also differ: government-sponsored collaborations

---

[4] Basic Research in Industrial Technologies for Europe – EUropean Research on Advanced Materials

[5] Independence means here that the partners are legal entities with separate identities, enjoying autonomy in economic and strategic terms.

[6] See Avadikyan et al. (2001) for a specific application of this grid.

seem to be more stable in the short term (no premature end), but less persistent in the long term. Finally, we derive some policy implications resulting from the identification of two different collaborative patterns.

10.2 A Review of the Literature: Incentives, Coordination, and Learning in R&D Collaborations, and Some Dynamic Implications

In this section we briefly present some important issues concerning the micro-mechanisms of inter-firm technological alliances, that can be found in the literature on strategic management, sociology of organizations, the competence-based view of the firm, transaction cost economics, and, to a lesser degree, on industrial organization. We do this using a three-dimensional analytical grid incorporating incentives, learning, and coordination.

The first of these, incentives, concerns the motivations and the rationale behind inter-organizational strategies (cf. 10.2.1). Learning and combining competences with external, complementary resources seem to be key motivations for collaborating with other non-affiliated companies. Consequently and unsurprisingly, learning mechanisms, and especially interactive learning mechanisms, are a major topic in this literature (cf. the cognitive issue discussed in section 10.2.2). The way the work of the partners is defined and effectively organized, via contract conditions and/or informal coordination rules, is also a recurrent focus of scholarly attention. Coordination devices are supposed to have a major influence on the efficiency, the outcome (success or failure) and, more generally, the evolution (stability, durability) of the collaboration. This "coordination" issue and the subsequent dynamic considerations will be developed respectively in Section 10.2.3 and 10.2.4. Although for the sake of clarity they are presented separately here, the three building blocks of our microanalysis are generally interdependent.

10.2.1 The Incentive Issue: the Motivations of Inter-Firm Technological Cooperation

Intrigued by the dramatic increase in technological inter-firm collaborations from the end of the 1970s[7], numerous scholars in economics and

[7] Cf. Hagedoorn (2002) for a recent and up-dated statistical analysis of the phenomenon.

management have attempted to explain the motivations of firms entering into collaborative strategies. Why should independent companies cooperate in a competitive world?

A first possible answer, presenting some analogy with the traditional explanation of collusions and cartels, is based on a "power" argument. Some authors consider technological alliances as just another way to compete, to eliminate rivals, and to obtain market power (e.g., creating entry barriers, specifying technical standards shared by few companies, free-riding while absorbing the core competences of the partner, etc.). These opportunistic behaviours are well analyzed in game theoretical models. While we recognize the existence of short-term, opportunistic actions, we do not place them at the heart of the alliance rationale: systematic null-sum games do not seem consistent with the longevity of the observed phenomenon at the macro- and sometimes micro- levels. Rather, we will consider "opportunism and power" as a second order motive in the case of technological collaboration, located far behind a first order motive consisting of quick and reciprocal access to external resources and/or competences.

Some preliminary empirical work revealed that the phenomenon was much more pronounced in industries characterized by rapid technological change, rising R&D costs, increasing complexity, and demand variety: computer and telecommunication, biotechnology, new materials, aeronautics. Thus technological innovation is probably at the root of most cooperative strategies. More precisely, the turbulence of the environment leads high tech companies to seek rapid access to external competences and/or to pool certain critical resources. "Access to external competences" is a rather broad motivation that can be split into – at least – three categories:

- to achieve critical mass via the pooling of similar resources;
- to combine complementary, dissimilar resources in order to create value;
- to acquire reputation and other "network" assets.

To achieve critical mass via the pooling of similar resources

R&D cost sharing and other cost minimizing strategies resulting in economies of scale in R&D, agreeing on a common technical norm to impose a de facto standard, enlarging commercial outlets via access to partners' markets – all of these "scale-based" motivations are central in the traditional economic analysis of cooperative R&D, as proposed by scholars in industrial organization (Katz and Ordover 1990; De Bondt 1997; Salant and Shaffer 1998). It should be noted that within this stream of the literature the partner firms have homogeneous capabilities, are competitors, and cooperate for cost minimization motives.

To combine complementary resources to create value

Early empirical analyses of alliance motives, such as the pioneering works of Mariti and Smiley (1983) or Hagedoorn and Schakenraad (1992) in the information technology industry, emphasized that one of the main motivations for alliances was technological complementarity. Access to complementary assets and knowledge is often a necessary – but not sufficient – condition for exploiting and benefiting from a technological innovation (Teece 1986). In his seminal contribution, Richardson (1972) also suggested that complementarity was the essence of inter-firm cooperation. Observing that cooperation (and not market) was the dominant mode of organization of economic activities, he proposed that dissimilar, but closely complementary, activities had to be articulated via explicit, *ex ante* cooperative mechanisms between firms.

Richardson's vision, according to which the frontier of the firm depends partly on the capabilities and know how of its human resources, can be considered as a precursor to the "competence-based view of the firm". The firm is seen as a portfolio of strategic, distinctive core competences (Prahalad and Hamel 1990; Teece 1992), i.e., pieces of collective knowledge that are the main source of its competitive advantage, but which are built through a time-consuming, cumulative process. Current competences constrain the scope of the future activities of the firm. In this perspective, the distance from the core determines the type of external growth chosen: a core activity is highly strategic and critical; it is not tradable on the market and has to be quasi-internalized, whereas a peripheral activity can be out-sourced. According to Amesse and Cohendet (2001), between the zone of core competences and peripheral knowledge lies an intermediary zone "where the firm holds significant pieces of knowledge but needs access to complementary forms of knowledge held by other firms to be able to develop and use the knowledge efficiently. This zone is characterized by networks" (Amesse and Cohendet 2001, p. 147) However, the notion of complementarity is even more interesting in an explicit dynamic perspective. Alliances may also be used as coordination devices, not only for exploiting existing complementary activities, but also for exploring new technological options (future core competences). If we go beyond the idea of a static exploitation of well-defined complementary assets and consider the opportunity to actually associate competences, learning mechanisms appear to play a central role in alliance rationale. Sharing the skills of heterogeneous firms can provoke a new combination that creates new valuable knowledge for one or for all the partners[8]. Exploration appears essential in a turbulent environment. But learning is a rather blurred notion, which has important implications for alliance dynamics. A deeper analysis of inter-organizational learning through alliances is conducted in section 10.2.2.

To acquire reputation and other "network" assets

By network we mean a microcosm of several interacting organizations[9]. In some cases R&D collaboration is an admission ticket to a broader collaborative, information network. Entering into such a network might be desirable *per se* because, in addition to a quick access to numerous and scattered resources, it provides one or a combination, of the following advantages:

- specific relational abilities vis-à-vis the other members of the network: the sharing of a common language/codebook, communication channels, coordination procedures, and other intangible assets result directly in decreasing transaction costs between the participating companies;
- reputation and visibility vis-à-vis unknown agents and/or potential partners. This motivation also deserves specific consideration since it is another kind of intangible collective asset, playing a crucial role in the case of a signalling and/or "networking" strategy[10].

Probably, these three categories of motives – critical mass, complementarity, and reputation – are not mutually exclusive and the distinction might not be so clear-cut in practice. But we contend that access to complementary resources in order to create knowledge is the main motivation of firms entering into technological strategic alliances. Power and cost minimizing motives, though not marginal, should not be seen as a priority for firms. In circumstances where "learning motives" prevail, power and cost minimization dimensions would probably even exert a counter productive effect. But learning does not only refer to an incentive dimension. It also exerts a strong influence on alliance dynamics. This is the reason

[8] Numerous authors insist on this "learning rationale". See for instance Koza and Lewin (1998) and Doz (1996).

[9] A network concerns a group of more than two companies, each of which is not systematically directly connected to <u>all</u> the others. An alliance concerns a small group of companies that are all directly interconnected via formal, contractual links. In this chapter we focus on individual alliances of the Brite-Euram network, i.e., a network of alliances.

[10] The "networking" strategy aims at expanding the number of business contacts and commercial links. The signalling strategy aims at advertizing to other companies and organizations the specific abilities of the firm as well as a willingness to cooperate in a given field.

why we develop this cognitive and knowledge dimension further in the next sub-section.

10.2.2 The Cognitive Dimension: Different Types of Learning Processes

It is worth noting first that the importance and impact of learning on collaboration dynamics are strongly emphasized in the literature and that this applies well beyond the case of pure research collaborations. Learning also matters for any type of alliance, even those with no technological dimension. At least four types of knowledge appear relevant in this respect: (i) obtaining information about the firm's environment; (ii) creating new technological and/or commercial competences; (iii) acquiring knowledge about the partner, possibly provoking the emergence of trust; (iv) accumulating agreement management experience.

Even when the creation of knowledge is not intentional (i.e., it does not explicitly belong to the set of initial agreement objectives), the activation – or not – of learning processes can exert a strong influence on how the collaboration evolves and its eventual success. This is probably related to the distinctive properties of learning. On the one side, learning may bring novelty, which may change the initial conditions of an alliance considerably. On the other side, it is a costly, time-consuming process with strong features of irreversibility (it generates high sunk costs). Although a kind of consensus seems to have emerged in the literature about the relationship between the dynamics of learning and agreement, if we consider the impact of learning upon the durability of the relationship then ideas differ.

In addition to a more precise specification of the meaning of "stability" (cf. 10.2.4), it seems to us that a clear recognition of the variety of types of learning that are involved must be considered. Learning is not a homogeneous process. It may refer to the maintenance of, as well as to incremental improvements to, existing know-how, or it may refer to the creation of totally new knowledge. Some qualification is needed in order to derive dynamic considerations. Economic and organizational theories distinguish between different types of learning[11] depending for instance on the nature of the knowledge being considered (tacit/codified), the configurations of the learning agents (individual/collective), the degree of novelty of the process (level of learning) and the origin of the knowledge (external / inter-

[11] For the best known definitions of different types of knowledge and/or learning see: Malerba (1992), Arrow (1962), Rosenberg (1982), Dosi (1988), Argyris and Schön (1978) and Ancori et al.(2000).

nal). In the case of inter-firm agreements, we find it relevant to make a distinction between unilateral and interactive learning that combines several of the previously mentioned criteria.

Unilateral learning

Unilateral learning occurs when one individual party acquires and uses knowledge that has resulted from the cooperative process. Ciborra (1991) mentions a type of unilateral learning particularly relevant in the case of technological alliances: radical and exploratory learning (i.e. "learning to learn" abilities). This high level learning may be related to the best way to manage agreements, or may provide guidelines for how to manage a turbulent, highly competitive environment. In other situations the agreement may lead to the genesis of new, valuable, technological knowledge that is largely redeployable by a particular partner. This supposes that this partner absorbs[12] some kind of – fairly – generic knowledge and is able to adapt it to a field of application other than that initially specified in the collaborative agreement.

Unilateral learning by one party may occur to the detriment of the others, operating in the frame of a "learning race" (Hamel 1991). The type of learning emphasized in this chapter does not for the most part concern the creation of entirely new knowledge, but is more concerned with the transfer of existing competences from one organization to the other. More precisely, asymmetric learning can modify the relative bargaining power of the partners, thus transforming a situation of bilateral interdependence into a situation of unilateral, non-viable dependence. Cooperative agreements, therefore, make it possible to internalize opportunistically the technological skills and even the core competences of the partner. The effectiveness of such technological "hold ups" depends on the transparency and receptivity of the partners, as well as on the appropriability regime (Teece 1986) of the technological competences considered. If the critical core knowledge of a company needing access to complementary assets has a weak appropriability regime (i.e., it is easily imitable and/or cannot be efficiently protected by a patent), then this company would be well advised to avoid cooperation and to integrate the complementary partner.

The impact of unilateral learning on alliance duration can be negative as well as positive. On the one hand, asymmetric learning linked to opportunistic behaviour and the search for power may lead to conflicts and a premature ending of the collaboration. On the other hand, learning is a time-consuming process that requires minimal durability.

[12] Cf. the notion of absorptive capacity developed by Cohen and Levinthal (1989).

Interactive collective learning
Interactive learning *à la* Lundvall (1988) refers *stricto sensu* to real reciprocal learning, agreed by the actors. We talk then about learning "with" partner(s), i.e., learning together in the course of the tasks to be carried out within the partnership and within the cooperative process. The knowledge thus created is often largely tacit and favours a strong appropriability regime in the sense of Teece (1986, cf. Section 10.2.1). Moreover, it may entail a strong collective dimension. In fact, one of the possible outcomes of interactive learning - if it is effective – is the emergence of new, collective, and indivisible competences, and other specific assets endogenous to the agreement itself. According to Williamson's (1989) definition, asset specificity refers to the degree to which an asset cannot be redeployed to alternative uses and by alternative users without its productive value being sacrificed. In our view, asset specificity should not be considered as a once for all and well specified, exogenous factor. Rather, it should be seen as an endogenous, dynamic factor, which evolves and grows as the interaction proceeds. More precisely, it is the outcome of a cumulative process of collective knowledge creation, of the progressive specification of a common language and also of the emergence of trust between the parties. Here the word "trust" requires some qualification. Trust acts as a cumulative process of investment in "transactional capital" (Palay 1984) based on a "reciprocity of favour" principle. It thus leads to the creation of a particular kind of valuable, intangible asset resulting from behaviour in the past. Unlike reputation, transactional capital is highly specific to the partners, and cannot be easily transferred to other agents (Ouchi 1980; Butler and Carney 1983).

The positive influence of asset specificity on agreement durability is twofold. First, the newly created assets may generate a highly valuable, "relational quasi-rent" (Aoki 1988; Dyer and Singh 1998), such that if the relationship broke down, most of the benefits from learning would definitely be lost. Second, if the newly created asset consists mainly of collective, indivisible, tacit knowledge, most of it is incorporated in human resources via progressive encoding in the organizational memory of the alliance. Such a process of routine creation, because it is associated with a kind of lock-in phenomenon, usually favours continuity of the relationship between the partners (Wolff 1992; Ring and Van de Ven 1994).

More generally, we contend that the articulation and the respective weight of different types of learning (unilateral versus interactive) may have a decisive influence on each partner's willingness to pursue the cooperation or not. Many authors stress the influence of different types of learn-

ing and/or their combination on the dynamics of agreements[13]. Our mai
assertion in this respect is that the effective implementation of interactiv
learning is a factor favouring the continuity of the inter-firm relation
whereas unilateral learning may, in some cases, lead to the cooperatio
breaking down (Bureth et al. 1997).

10.2.3 The Coordination Dimension: Flexibility, Formal and Informal Mechanisms

External knowledge, assets and competences may be obtained throug
means other than technological agreements, for instance, mergers or com
pany acquisitions. A full understanding of the micro-rationale underlyin
inter-firm cooperation is assumed to explain why and when such coordina
tion devices are preferred to full integration. It also involves examinatio
of the possible implications of coordination upon the evolution of alli
ances.

Transaction costs versus flexibility

We present below two alternative views:

- the first, the transaction cost approach, emphasizes static efficiency;
- the second emphasizes the flexibility of these agreements compared t full acquisition, and focuses on dynamic efficiency.

According to Williamson's comparative institutional analysis (1979)
there are three broad modes of coordination of economic activities: market
hybrid modes (including inter-firm alliances), and hierarchies, each corre
sponding to a given law doctrine (Macneil 1974). The first doctrine is tha
of classical contract law, based on complete presentation in formal docu
ments, that fits well with anonymous market transactions and contingen
claims contracting. It minimizes transaction costs for a stable environment
weak asset specificity, and low uncertainty. The second law doctrine i
neo-classical contract law, which corresponds to incomplete long-tern
contracts designed to preserve flexibility. Third party assistance (arbitra
tion) is often used to resolve disputes among the parties. This trilatera
governance is supposed to be efficient for occasional transactions and me
dium to high levels of asset specificity. An important advantage of arbitra
tion over litigation is that it preserves the continuity of the relationship. Fi
nally, the third doctrine is that of relational contract law, which relies o

[13] See, for instance, Parkhe (1991), Doz (1996), Child (1997) and Larsson et al (1998).

norms and past behaviours to provide efficient adaptation mechanisms in the case of very complex, uncertain, and recurrent transactions. Two subcategories of governance mode may be distinguished here. Hierarchy offers distinctive adaptive properties when assets are idiosyncratic; relational bilateral governance between autonomous parties is most efficient in the case of intermediate asset specificity.

Because it is in the main a static approach, transaction costs economics might not provide an ideal conceptual framework for analyzing technological alliances in fast changing environments[14]. Within a dynamic efficiency perspective, we suggest that alliances are better considered as flexible and exploratory tools, which help to create strategic and technological options (Vonortas 2000; Kogut 1991). They favour proactive R&D exploration without the high sunk costs of a totally internal development. In this way, they partly preserve the company from radical, potentially destructive, uncertainty. However, alliances are in no way perfectly reversible holding positions. Learning cannot occur without a minimum level of commitment (irreversible, tangible as well as intangible investments; cf. Bureth et al. 1997). Conversely, too much rigidity can impede and block exploratory learning. This leads to the idea that the more or less detailed way that the coordination of the collaborative work is specified may exert a strong influence upon learning effectiveness and hence upon alliance dynamics. Here, again, it seems to us that transaction cost economics provides interesting insights, which are discussed in the next paragraph.

Contractual safeguards versus informal mechanisms

While we might question the theoretical proposition that technological alliances are selected because they minimize transaction costs, Williamson's distinction between classical contracting and relational contracting is especially interesting as far as the – internal – flexibility of inter-firm agreements is concerned. To put it briefly, there seems to be a dilemma between the necessity to formalize written contractual terms and to promote the creation of informal, tacit, rules and routines.

Definition of coordination and division of labour can be achieved through formal, detailed contractual terms and safeguards specified via a costly negotiation process. But this formal specification of the allocation of the tasks, obligations, and outcomes for each party is often very inflexible

[14] The presumed superior adaptive properties of hierarchies should lead to a multiplicity of mergers and acquisitions (M&A) in the turbulent context of high technology industries. Moreover, hybrid form efficiency implies that asset specificity is located at an intermediate level. But we can express some serious doubts about the non-idiosyncratic nature of core competences and close complementary assets in the case of technological alliances.

and, thus, inefficient in a quickly changing context. On the other hand, re lational or psycho-sociological contracts based on routines, trust, and in formal coordination processes make any "contractual gaps" less importan In this sense they preserve flexibility within the tasks to be accomplishe The problem is that time is needed for routines, trust and informal coord nation processes to emerge, since they are built on past behaviour. There fore, initially a reciprocal commitment, formal contractual safeguard and/or exchange of hostages (Williamson 1985) may be necessary in orde to provide a stable basis for the actions (Bureth et al. 1997). Reference t this formal documented agreement will become less and less necessary a and when the collaboration evolves and grows. Informal rules may ofte substitute for formal explicit contractual terms.

The emergence of relational capital and tacit collective routines, in add tion to learning in general, also requires relatively frequent and direct in teractions between the participants, such as the implementation of specifi communication channels. The implementation of an effective interactio process is an especially important prerequisite if collective, indivisibl competences and specific assets have to be created. The creation of a joi facility may have to be programmed in this perspective (as opposed to a *ex ante* division of the tasks between the partners and their subseque separate execution by each member firm). Finally, as mentioned in th previous section, relational capital introduces the possibility of inertia an lock-in in the long run. To summarize, a fine balance between formal an informal mechanisms needs to be maintained throughout the life of the co laborative relationship (Ring and Van de Ven 1994).

10.2.4 Implication on the Evolution of Alliances

Before going further, it is necessary to define our concept(s) of stability. I fact we will define two types of stability: short-term stability and long term stability. By short-term stability, we mean that its short term is no due to the premature ending[15] of a cooperative agreement. Generally con nected also to the notion of success, this type of stability applies both t contracts with an *ex ante* precise time horizon and to agreements of unlim ited duration. Thus, it can be clearly distinguished from the notion of dura tion. Short duration does not signal instability or failure, it may simply b the direct outcome of an *ex ante* contractual specification. Beyond th short-term stability of a single collaborative agreement, it is relevant t take into account the continuity of the global relationship, that is, th

[15] Premature end usually corresponds to contractual breakdown.

whole set or sequence of formal and informal agreements between the parties. This we consider to be "long-term stability", defined as the persistence of an inter-firm relationship beyond the initial agreement, pursuing the same technological goals.

Keeping in mind both definitions of stability (short-term stability of a single alliance versus long-term stability of a global relationship), our main propositions about alliance dynamics may be summarized as follows. As emphasized in the literature, the rate of failure of collaborative agreements is generally high, because of high internal and external uncertainty and because of the dangers associated with a potentially opportunistic partner. In other words, technological collaborations are not characterized by short-term stability. But if short-term stability occurs, or more precisely, if and when valuable specific assets are created, a kind of virtuous circle of success and increasing commitment may arise. In this perspective, a research agreement is seen as a kind of real option that will be exercised, only in the case of success, by way of further specific commitments, for instance, a second research agreement or an investment in a more formal structure, or an equity agreement such as a joint venture. Several authors[16] mention this idea of an escalation of commitment and satisfaction. It is a good example of what is called "long-term stability" of an inter-firm relationship.

Section 10.2 was devoted to underlining relevant theoretical results from the rich and diverse literature about strategic inter-firm alliances. Most of the papers cited focus implicitly on spontaneous collaborations, i.e., collaborations that are not fostered by government policies. Section 10.3 challenges these theoretical propositions with empirical information concerning the particular case of EU sponsored R&D collaborations.

10.3 The Specificities of EU Sponsored Collaborative Projects

We use the conceptual framework elaborated in the previous section to locate and characterize a particular case of collaboration: R&D agreements sponsored by the European Framework Programmes (FWP). More precisely, this section tries to identify the specific incentive, learning, and coordination properties of Brite-Euram projects, and to confront them to the main analytical results in Section 10.2.

---

[16] Cf. Doz 1996; Doz et al. 2000; Bureth et al. 1997; Ring and Van de Ven 1994; Wolff 1992

Our empirical material is mainly qualitative information obtaine through official information channels and through numerous in-depth in terviews[17] covering a representative sample of 50 Brite-Euram projects (cf Bach et al. 1995). The interviews were conducted at the beginning of th 1990s. Although they were not conducted for the specific purpose of thi chapter, they yielded a lot of information that is relevant here, for instance about the stability of collaboration, the creation of new technological as sets, etc. We do not try to use our empirical information to give statistica support to a theoretical hypothesis. We simply use it to make inference and to elaborate several theoretical propositions, or stylized facts, accord ing to an inductive research process.

Our intention was to highlight certain outstanding features specific t these EU sponsored collaborative projects, at the individual level. We pro ceed by contrasting two collaborative patterns: that of a typical EU spon sored R&D collaboration versus that of an idealized, pure form of sponta neous collaboration. We develop the idea that these two collaborativ patterns exhibit sharp differences. These differences are mainly due to th presence of the rules imposed by and the characteristics of EU FWPs. W suggest that the main determinant of these differences is related to the stra tegic importance of the collaborative research from the point of view of th firm: the research undertaken in a spontaneous collaboration is presumabl closer to the firm's core competence than the research undertaken in a gov ernment sponsored R&D project.

10.3.1 Incentives to Form EU Sponsored R&D Collaborations

The motives or incentives for collaborating described in the literature (cf 10.2.1) probably also apply to EU sponsored and spontaneous collabora tions. Policy makers often emphasize cost sharing motives. However, ou observations reveal that complementary, dissimilar partnerships are much more frequent in the case of Brite-Euram projects, than collaboration be tween partner firms that are similar and whose sole objective is to achiev critical mass. So, the skill-sharing motive probably prevails in the Brite Euram case. Sakakibara (1997a) in his study of Japanese R&D consorti also stresses the importance of the "skill-sharing" as opposed to the "cost sharing" motive. Hence, we can formulate the following proposition:

- Proposition 1: In the case of EU sponsored collaborations, "skill sharing" motives and access to complementary knowledge probabl

[17] R&D managers in each partner firm were interviewed.

prevail in terms of incentives over "cost-sharing" motives and achieving a critical mass.

However, the necessity for firms to disclose information, and the existence of subsidies in EU sponsored collaborations, can induce specific incentives. The strategic importance of the project, that is to say its distance from the "core competences" of the firm, is a relevant development in this respect.

Information disclosure: combining complementary resources for peripheral activities and gaining reputation

Participation in a public programme means that certain information concerning the project becomes publicly available. In the case of Brite-Euram and of other European programmes, the EU Cordis database provides free access to a list of funded projects, including a summary of the research objectives and the different partners. From the strategic point of view of a specific company, several assumptions can be made about this information disclosure:

- the firm does not mind revealing this kind of information because the project is not critical for it, in the sense that it is one of its peripheral activities;
- the firm wants to reveal this information as a signalling strategy. For instance, it wants to signal to the external world a specific technical competence, its willingness to cooperate, or intention to enter a new research area, etc. This signalling strategy is *a priori* compatible with the hypothesis that the project is connected with peripheral activities, insofar as competitors, clients and suppliers would generally be familiar with the core activities of the firm[18]. Thus, signalling of a core activity seems to be less important for the firm than signalling other domains of interests that might attract new partners;
- the signalling strategy may also be linked to reputation effects. The diffusion of information – concerning reliability – to the microcosm of firms participating in a public programme will play an important role in creating and/or maintaining a good reputation. This good reputation can be considered as a strategic asset, attracting new "public" research contracts, for instance.

---

[18] Of course, we recognize that core activities may be blurred, complex and evolving. In this case, signaling peripheral competences may also be a means to provide information about potential, future core competences.

The specificity of EU sponsored collaborations at this stage can be summed up by proposition 2:

- Proposition 2: In the case of EU sponsored collaborations, the imposed disclosure of information also suggests that gaining reputation, networking, and/or signalling are among the basic motives for collaborating.

Spontaneous research collaborations represent an alternative means to access external complementary knowledge, without necessarily disclosing this strategy publicly. In other words, partners can choose to advertize or not the existence of their collaboration. If the R&D project is closely connected to the core activity of the firms, they may prefer to keep the cooperation secret[19]. Spontaneous agreements are thus more compatible with the preservation of secrecy, often required for highly strategic activities. This leads on to the next subsection, concerning the strategic importance of government sponsored compared to spontaneous collaborations from the point of view of the firm.

Subsidy: a means used to explore new technological options

Government-sponsored agreements benefit from public subsidies[20], which, from a social point of view, should not be seen as a substitute for private funds, but as being complementary [21]. Let us consider that firms, at least big ones, generally do not pursue only a single research project, but manage a portfolio of projects. From this perspective, the opportunity to benefit from external subsidies can be analyzed from the point of view of the company in different ways:

- the company "free rides" and uses public money to do what it would have done without it. This corresponds to a case of pure substitution where the policy is *a priori* useless. Nevertheless, the company can allocate the money saved to other research projects. In this sense, the subsidy allows the firm indirectly either to open up a new research project or to expand an existing one;
- the company takes advantage of the subsidy to carry out a project it would not have undertaken, or not to such an extent. The subsidy allows

[19] We do not mean that all spontaneous agreements imply secrecy. At a development stage, or subsequent to promising preliminary results, firms may have an interest in disclosing the cooperation for competitive reasons: to patent a technological innovation, to be the first to innovate, to advertize their comparative advantage, etc.

[20] In the 5th EU FWP, for example, the financial contribution of the Community represents 50% of the eligible project costs.

[21] Cf. David et al. 2000; see also Bozeman and Dietz 2001.

the company directly to open up a new research project or to carry it out in a more ambitious way (broader objectives, with more partners, time-saving, etc.).

The decision to finance research collaboration (using public or internal private funds) is connected with the way companies manage their knowledge. The more strategic (i.e. close to the core competences) the knowledge, the more the company will be prompted to invest on its own. Here, again, we find the idea that the spontaneous strategy is more likely to be related to the management of critical, "close to core" knowledge. More precisely, the subsidy may be considered as a direct or indirect means to open up new technological options, that is to say, to broaden the scope of exploration[22]. It is presumed that the new options are located at the periphery of the firms' activities. They might, of course, eventually become central. By contrast, private funding explores previously selected options, considered as having strategic priority and connected to the – future – core activities of the company. To some extent, government-sponsored collaboration tends to be more “exploratory” and private funded partnerships more "exploitation" oriented. The above discussion can be synthesized in the following proposition:

- Proposition 3: In the case of EU sponsored collaborations, the imposed disclosure of information and the presence of a subsidy mean that:
  (i) firms tend to collaborate mainly on projects connected to peripheral activities;
  (ii) collaborative projects can be considered as highly exploratory means to open up new technological options.
- Proposition 3bis: By contrast, spontaneous research agreements seem to focus on:
  (i) "close to the core" competences;
  (ii) collaborative projects that aim at exploring pre-selected technological options.

Supporting our view, Sakakibara shows for Japan that:
"support for R&D consortia by the Japanese government is modest and declining, and there is no clear link between the existence of R&D consortia and industry competitiveness. R&D consortia participants perceive sharing complementary knowledge to be the single most important objective of R&D consortia. […] R&D consortia work as a complement of private R&D. The overall subjective evaluation of the typical project's success is modest, and participants do not perceive R&D consortia to be

[22] Exploration in the sense of March (1991).

critical to the establishment of their competitive position" (Sakakibara 1997b, p. 449).

10.3.2 Learning in EU Sponsored Collaboration: the Predominance of Unilateral Learning

Although exploration is obviously an ingredient of both types of agreement, we want to underline some distinctive features of "public" partnerships. In the previous section, we suggested that the object of the publicly funded joint research is usually rather distant from the current core competences, so that learning is of a more exploratory nature than in the case of private collaboration in general. If we exclude a few cases of (very) small companies with restricted resources, the strategic intent behind Brite-Euram was generally not to create or to move quickly towards new core competences. From our investigation it seemed to us that companies cooperated mainly in order to explore possible options and to screen a given technological area.

Section 10.2.2 developed two notions of learning relevant to the case of technological collaboration: unilateral and interactive learning. Actually, we found that the Brite-Euram projects exhibited important effects in terms of unilateral, individual learning, but no strong effects in terms of interactive collective learning. One of the prevailing indirect effects identified in Bach et al. (1995) is "technological effects"[23]. By definition, indirect technological effects result from the partial redeployment by one party, of the technology acquired through the European project. Such effects are considered to be indirect in the sense that the new technological knowledge is applied (transferred) to an activity that was not involved in the joint research. Almost all indirect technological effects typically correspond to the creation of knowledge assets, which are individually redeployable by a particular partner[24]. Thus they can be considered to be a manifestation of positive unilateral learning. Participation in a Brite-Euram project also substantially improved the "learning to learn" abilities of the participants. Participants often recognized that they had gained considerable experience in the course of managing multipartite, EU sponsored R&D collaborations. This activity was perceived as facilitating subsequent contributions to such programmes, or to other types of joint research pro-

[23] In the frame of the particular evaluation methodology used by the authors, technological effects represent 50% of all indirect effects (the latter include technological, but also network, organizational, and critical mass effects).

[24] Indirect technological effects due to collective redeployment of knowledge are few.

jects. In terms of opportunistic learning, we found no obvious cases of learning races leading to effective hold ups in partners' competences. Nevertheless, free riding, in the sense of adopting "wait and see" positions and lack of true commitment, did occur.

We now turn to the role of interactive learning, i.e., the creation of collective competences and specific indivisible assets that cannot be individually appropriated and redeployed by one party. Whereas spontaneous R&D collaborations sometimes conclude with an explicit intention of creating such specific collective assets – from the perspective of a long-term relationship – this was never the case in Brite-Euram projects (or at least, in our representative sample of 50 projects). However, we might qualify this argument by recognizing that partners often have to learn together how to integrate the different contributions in the final product or process. Interactive learning may thus occur in this weaker form.

- Proposition 4: In the case of EU sponsored collaborations, learning is basically unilateral: most of the assets created through these projects are largely redeployable, separable and appropriable individually by each partner.
- Proposition 4bis: By contrast, spontaneous research agreements seem to favour interactive learning and the creation of non-redeployable assets.

So, the newly created assets are generally not of the collective, non-redeployable kind described above. As a consequence we may advance the notion that long-term stability, and particularly the rationale of increasing commitment emphasized by several authors (cf. 10.2.4), is not a typical characteristic of publicly sponsored agreements. This idea is partly supported by our empirical material: none of our sample of Brite-Euram projects gave rise to subsequent collaboration with increasing commitment in the same field[25], such as a manufacturing equity joint venture, for instance. In other words, we found no direct effects (i.e., effects in the same technological field) linked to the creation of a new agreement. This does not mean that indirect networking effects did not exist; on the contrary, firms established new business links (commercial activity, research contract sponsored, or otherwise) thanks to the sponsored collaboration. But these networking effects, more often than not, consisted of increased numbers of contacts or exploring a new technology with a previous partner, rather than any deepening of the research area based on a rationale of increasing commitment. By spontaneous inter-firm collaborations, we do not mean that

[25] A complementary explanation resides in the fact that antitrust pressures and internal regulation of R&D programmes usually discourage repetitive contracting in the same field and with the same partners.

that firms always exhibit long-term stability: effective genesis of specifi assets requires short-term stability and is typically difficult to obtain Rather, we simply assert that, if and when specific assets have been cre ated, private partnerships tend to promote longer term relationship (through a sequence of increasing commitment with the same partners an in the same field) than government-sponsored agreements.

It is worth mentioning that the specificities of learning through sponta neous versus EU sponsored collaborations are consistent with our proposi tions concerning the incentive dimension in Section 10.3.1. In order to cre ate valuable, collective, specific assets, the partners to a spontaneou collaboration must have clear prospects of developing pieces of knowledg that are relatively close to their core competences and long-term horizons Obtaining peripheral knowledge is consistent with the fact that govern ment-sponsored collaborations are triggered by an external organization (government agency), and not due to an "internal" awareness of environ mental pressures or opportunities[26].

10.3.3 Coordination of Activities in EU Sponsored Collaboration: pre-Defined Rules and Arbitration as a Short-Term Stabilizing Factor

In this section we will assume that inter-firm technological collaboration are governance modes that offer decisive advantages, in terms of flexibil ity, compared to hierarchical governance structures. We will focus on th rules of coordination operating inside a given collaboration. The aim o coordination is to provide compatibility and coherence to individual ac tions, as well as to decentralized learning processes, in order to reach global objective. Moreover, any collaborative agreement has to resolve th problem of the distribution of the roles and tasks of the partners, as well a their articulation. Here, again, our Brite-Euram collaborative projects show strong particularities – some imposed by the design of the European pro gramme itself.

Pre-defined rules: stability vs. potential learning rigidities

Very often in the framework of a technological policy, the public organiza tion in charge of the management and the control of the programme re quires certain information and imposes certain (minimal) rules that mus be respected by all partners. For instance, the research contracts signed by partners to the Brite-Euram Programme contain provisions about the allo

---

[26] For an interesting development concerning "triggering entities" in connection with "engineered networks" of collaborations, see Doz et al. (2000).

cation of funds and budget between the partners, the duration and milestones of the agreement, the contribution to be made by each partner, and the objectives of the project. European programmes also impose certain minimal inter-partner coordination rules in terms of allocation of work: *ex ante* definition of work packages, organization of a minimum number of meetings, etc. Moreover, this kind of programme usually requires that the partners agree on the results and/or property rights. This pre-defined framework eases the coordination of the partners in terms of allocation of resources (money, competences, tasks, property rights), and contributes to building the channels used to communicate and exchange research results. In other words, it helps to set the initial conditions of the collaboration and constitutes an important stabilizing element.

Nevertheless such *ex ante* specification of rules also introduces potential rigidities. More importantly, it may confine learning to specific zones and types. We propose that it primarily favours unilateral learning to the detriment of true interactive learning. To illustrate this point, we observed that in most cases of Brite-Euram partnerships, the organization of work between the partners often consisted of a clear *ex ante* separation of the tasks between the parties. Actual organization of the work seemed to correspond more or less to the minimal EU requirements and/or guidelines (in terms of defining work packages and number of meetings). These rules, when strictly applied, constitute a very poor framework for stimulating interactive learning. More intensive exchanges, or even the creation of a common research facility, may be necessary to stimulate effective processes of collective knowledge creation.

- Proposition 5: In the case of E.U. sponsored collaborations, the existence of pre-defined rules favours short-term stability, but also brings potential rigidities and impediments to interactive learning.

In a spontaneous collaboration strategy, the partners have first to elaborate and agree upon the kinds of rules to be applied before specifying their content. Spontaneous agreements, therefore, have to overcome an additional problem, which is to define the "borders" of the contract. In the negotiation phase, partners must learn which rules should be created, and also how to implement and manage them. This first step confers on the partners a higher degree of freedom and flexibility compared to a European sponsored partnership. To maintain or even extend this flexibility the partners may well be induced to formalize and codify fewer rules and contractual safeguards than in a government-sponsored agreement. But flexibility and informal coordination have a counterpart. This greater freedom may entail more hazards and misunderstandings, which result in the well documented, high failure rates, i.e., to the premature end of cooperation.

Both spontaneous and sponsored collaborations do, in most instances, specify the duration of the partnership. Respecting the planned time duration may be an indication of the stability of the relationship. We saw in section 10.2.4 that a high rate of failure characterizes most spontaneous technological collaborations. Interestingly though, from our evaluation of Brite-Euram agreements[27] this was not the case. Our evidence suggests that the rate of break-up is very small: among 50 statistically representative selected agreements only one failed at the beginning. It might be concluded that short-term stability seems to be an important characteristic of EU sponsored collaborations. This observation is consistent with the assumed stabilizing properties of the pre-defined coordination rules imposed by European R&D FWPs. The "stability" argument is further reinforced by the discussion in the next subsection.

The existence of an arbitrator and reputation effects: reinforced stability

This aspect is linked to the previous one. In a public policy, the agent in charge of the management and control of the programme can play the role of arbitrator, in the sense of Williamson (1979), should a conflict arise. For instance, in the case of non-enforcement of the agreed rules by one of the partners, the public supervisor imposes some credible threat (no more subsidy payments, no reimbursement). The supervisor can exclude one of the partners and help the remaining group to stabilize[28]. Another factor, that has already been mentioned, might explain the relatively higher stability of EU sponsored collaborations: the strategic importance of networking and reputation effects in the microcosm of firms participating in a European research programme. These two elements lead us to the following proposition:

- Proposition 6: In the case of EU sponsored collaborations, the conjunction of a public supervisor playing the role of an arbitrator and strong reputation effects favours short-term stability.

In spontaneous agreements, there is no official arbitrator and the partners have to define their own solutions to conflicts. Asking a third party (very often a lawyer) to intervene is usually an expensive alternative that does not preserve the continuity of the relationship and is generally used only if the damages are significant. The absence of an arbitrator may be

[27] Cf. Bach et al. (1995).

[28] For a more general discussion of the role of government agency in discouraging opportunistic behaviour in collaborative R&D, see Tripsas et al. (1995). A complementary view recommends the presence of a “principal” as a good way to coordinate a group of agents (cf. Picard and Rey, 1988).

another reason for the relatively higher percentage of failure in spontaneous agreements compared to government-sponsored ones.

10.3.4 Two Contrasted Scenarios of Evolution

The bi-polar characterization of inter-firm research agreements described above leads us to assemble incentive, coordination and cognition features in such a way as to exhibit a strong internal coherence between the characteristics of one given configuration. For instance, the secrecy aspect put forward as a characteristic of private R&D cooperation is consistent with the idea of close connection to the core activity of the partners, which, in turn, seems to be highly compatible with greater commitment and more "exploitation oriented" learning. Also, it is consistent with the fact that the partners prefer to preserve their strategic scope of action, which includes, among other things, the free specification of cooperative rules. In our view, internal organizational coherence is also critical to understanding agreement dynamics – in particular, the stability, success, and overall logic of a given relationship. Keeping in mind our two definitions of stability, we can now combine the incentive, learning, and coordination features of agreements identified in the previous section in order to build two contrasting scenarios of evolution.

Our main propositions at this stage can be formulated as follows.

- Proposition 7: Government-sponsored R&D agreements are most often associated with strong "short-term" stability, but they do not necessarily favour increasing commitment and longer-term relationships.
- Proposition 7bis: By contrast, spontaneous research agreements seem to be characterized by a much higher degree of instability and failure in the short term, whereas they promote long-term relationships in case of a first success.

Let us first consider the case of publicly funded agreements. As mentioned previously, the mere existence of pre-defined coordination rules, combined with the presence of a third party able to arbitrate in case of conflicts, should promote stability of the research project. This "short-term stability" argument is further reinforced by the reduced strategic weight linked to a peripheral activity as compared to a core business. Most importantly, the signalling strategy associated with potentially strong reputation effects inside the whole network of firms participating in the government programme may exert very strong, although indirect, incentive pressures. Willingness to create or maintain a good reputation should impede or at least discourage opportunistic behaviour; it adds to the incentive to get things done, that is to say, to complete previously announced R&D projects. Unilateral learning can be considered as a normal by-product. But the newly created assets are not of the collective, non-redeployable kind described earlier. Hence, long-term stability, and particularly the rationale of increasing commitment, are not likely to occur very often, despite the higher probability of short-term stability and "success". However, it may be that experience gained in the management of government-sponsored projects leads to subsequent contracting in another field and/or with other companies (networking dynamics).

Let us turn now to our pattern of idealized spontaneous collaborations. Such collaborations, because of their proximity to the partners' core business (perceived as critical for companies' survival) or, more generally, because of the strong motivations and expectations that triggered them, run higher risks of failure in the short run. However, if specific, non-appropriable assets are generated as the result of an effective interactive learning process, then a virtuous escalation of commitments and successes may occur. These are the reasons why we – somewhat paradoxically – stress both the short-term instability of spontaneous collaborations and their "long-term stability". Needless to say, we recognize that short-term stability is an obvious prerequisite for long-term stability, but not a sufficient condition.

Our bipolar characterization and our evolution scenarios are summarised in Table 10.1.

Of course, Table 10.1 relates to two idealized cases, simplified for analytical purposes. The real world is more complex and may display a set of "intermediate cases", which combine the characteristics of EU-sponsored and spontaneous collaborations. For instance, social norms and predefined rules may also emerge in spontaneous cooperation networks open to multiple membership, such as R&D consortia. Also, as mentioned in section 10.3.2, a weak form of a specific asset can sometimes emerge during an EU-sponsored collaborative process (for example, the phase of integration of the various contributions). So, the idea that EU-sponsored collaborations do not produce "increasing commitment" needs to be qualified. Although in the 50 Brite-Euram projects we surveyed this evolution was not observed, we cannot exclude the possibility that promising results (high potential, future benefits), and good inter-firm relationships might induce the partners to continue the collaboration, but not as part of a government sponsored programme and without any public subsidy. This behaviour somehow assumes that the continuing collaboration involves a "close to the core" rather than a "close to the periphery" to activity. At this point, the firm is embarking on the "spontaneous collaboration" pathway described above, with the advantage of a first common stable experience.

Table 10.1. Government sponsored versus spontaneous agreements: two idealized collaborative patterns

	Government sponsored agreement	Spontaneous agreement
Incentive	Subsidies	Private funding
	Public info/signalling strategy	Secrecy
	Reputation effects	Trust
	Develop peripheral competences	Develop future core competences
Coordination	Pre-defined rules	Rules to be created/flexibility
	Arbitration by a third party	Self resolution of conflicts
Learning	Individual learning	Towards interactive learning
	creation of redeployable assets	creation of endogenous specific asset
Evolution	Short-term stability	Short-term instability
	No long-term duration of the global relationship between the partners	Long-term duration of the relationship in case of success i.e. increasing commitment
	"Networking" rationale	"Increasing specificity" rationale

10.4 Policy Implications

If the government-sponsored collaborations (like spontaneous collaborations) obey specific motivation, learning, coordination, and evolution features, then it becomes necessary to revisit the rationales behind certain technology policy instruments, such as the European research FWPs. This is the subject of this final section.

10.4.1 Revisiting the Rationale of EU Research Programmes in the Light of Firms' Incentives to Collaborate

As mentioned in section 10.1.1, cost-sharing is one of the motives emphasized by policy makers and assumed in a large part of the neoclassical literature (Katz 1986; D'Aspremont and Jacquemin 1988; Suzumura 1992, Salant and Shaffer 1998, etc.). This literature considers that R&D cooperative agreements are a means of reducing market failures induced by the characteristics of the technology. It typically assumes that firms are symmetrical in terms of skill and knowledge and analyzes the conditions under which R&D cooperation brings social benefits. State intervention in areas where firms would not cooperate spontaneously, but where R&D collaboration would benefit the society, is justified. It should be noted that this assumes that: 1) government-sponsored agreements are equivalent to spontaneous ones; 2) the sponsored firms have similar resources and cost-sharing is the main motive for collaboration; 3) the partners produce only one good and are active in the same market. The validity of these assumptions is questionable on at least three counts: firms need to access complementary knowledge; they collaborate to improve their reputation; they generally manage a portfolio of R&D projects, among which spontaneous agreements and sponsored agreements should be considered as quite separate in strategic terms.

From "cost sharing" to complementarity

Sakakibara (2003) shows in a three-stage model that R&D consortia that include firms with complementary knowledge are welfare enhancing. He underlines that the consortium Sematech, comprising a narrow set of industries with, therefore, little complementarity between participants, were able to reduce the cost of R&D in the semiconductor industry, because of the dominance of cost-sharing effects among partners. In contrast, Japanese R&D sponsored consortia generally involve firms from much more diverse industries. Thus, there is a higher probability of exchange of complementary knowledge and positive learning effects. Sakakibara concludes that "governments should take the organization of R&D consortia into account when deciding on the types of consortia they want to promote" (Sakakibara 2003, p. 126).

Similarly, Bach et al. (1995) found that firms associated with universities – i.e., with clearly different and complementary organizations – generated more indirect effects and were more effective at achieving the objectives of the research project. The results were the same for consortia where complementary partners, such as users and producers, were associated, compared with consortia involving users or producers only. Moreover, Sakakibara and Brandstetter (2003) confirm empirically the theoretical prediction of Katz (1986), according to which product-market proximity has a negative impact on the outcomes of public consortia. Policy makers should consider knowledge complementarity as being critical, with cost-sharing a secondary issue. The way that consortia are structured (firms from diverse industries with technological proximities, associating users and producers, and showing high potentialities for exchanging complementary knowledge) will influence outcomes and thus the impact of the public policy.

Taking networking into consideration

Bach et al. (1995) show that firms participating in Brite-Euram generated networking effects. Firms established new business links with some of their partners, or with the partners of their partners. Thus, networking, even if it does not take the form of a reinforced commitment, is an important outcome of public policy. Link et al. (2002) find that the US ATP (Advanced Technology Program) encouraged participating firms to join other research consortia. Broadening networks is certainly an important policy means to disseminate knowledge in the economy, to increase the coordination of complementary competences, and, thus, to promote technological innovation. This learning argument, where networks are considered as privileged *loci* for innovation, is especially important in the evolutionary economics rationale[29] for publicly subsidized research programmes. However, what is the economic justification when networking becomes the primary objective of the firm rather than a means of exchanging and creating new valuable knowledge? Proposition 2 states that the acquisition of network assets represents – at least – an important motive for collaborating. But, what happens if it is the only motive? In cases where networking, signalling, or adding to reputation are a company's only motives for collaborating and no real technological innovation is intended, or can occur, then the subsidy would not be justified. At a more general level, the need for and effectiveness of supporting a network over a prolonged period might also be questioned.

Consideration of the firm's portfolio of projects

The portfolio of R&D activities of a company, indicating how they manage their sets of competences, might influence a collaboration. The decision to enter into a certain type of collaboration for a given activity often depends on the distance of this activity from a firm's core competences. We would maintain that an activity closer to the core has a higher probability of being funded in house, whereas activities on the periphery will usually be managed by applying to government-sponsored programmes. In our analysis, where we want to stress the necessity to distinguish between sponsored and spontaneous collaborations, the existence of a portfolio specifically indicates the variety of projects carried out by a given firm. First, let us consider that firms are able to rank their projects in terms of strategic priority[30]. Subject to the money being available, firms know which projects to fund internally (i.e., those close to the core), and which will need public subsidy (i.e., those close to the periphery) because they are currently of less importance strategically, therefore more risky, although they are of importance for the future (i.e. new valuable options).

[29] See Metcalfe (1995) for an overview of technology policy rationales from an evolutionary economics perspective.

[30] Of course the projects are not independent and there are some overlaps in terms of the required knowledge base.

This has an important implication for the debate in terms of input (and output[31]) additionality vs. substitution. Input additionality exists when, provided the actions are welfare improving, the State sponsors actions which agents would not have resourced themselves. Substitution is a situation of "waste" where public funds are used to support actions that would have been carried out by the firm in any case. If sponsored and spontaneous collaborations differ in strategic terms (as mentioned above), then funding through EU programmes should be devoted explicitly to supporting valuable peripheral activities rather than highly strategic projects. By definition, firms with a portfolio of projects will not ask the EU to fund their "close to the core" strategic projects and the State should not offer to fund them. In other words, substitution (between spontaneous and publicly funded projects) should not be a major issue, since spontaneous projects tend to be privately funded – at least by large companies. Thus, if a substitution issue exists, it is in relation to peripheral activities. Waste and free-riding occur when firms deliberately apply to public programmes for funding to carry out unimportant, minor projects under a "wait and see", or pure networking strategy. Such waste also occurs when firms receive funds for peripheral activities that they would have funded themselves (in wealthy periods). In other words, the policy maker's problem of substitution vs. additionality should be turned into "opening new valuable options" vs. "sustaining minor activities". Of course, this supposes that firms have a clear vision about the "good" options to open and are able to differentiate between peripheral activities that might be considered as important for the future, and minor activities that will never be of any importance. It also supposes that the State is able to select the technological options with a high "public" value (in the social interest).

In other words, we would suggest that cooperative policies should take care of the strategic positioning of the project inside the company. We contend that analysis of the strategic weight of the collaborative project for each partner could usefully complement the current pre-competitiveness criterion of selection. If the stage at which cooperation takes place (i.e. pre-competitive R&D) is the unique criterion, spontaneous and subsidized collaborative R&D seem to be equivalent and could be considered to be substitutable. However, when strategic considerations at the firm level are taken into account the picture changes radically.

[31] Output additionality answers the question: would we have obtained the same outputs without policy intervention?

10.4.2 Toward Revisiting the Rationale of EU Research Programmes in the Light of Specific Inter-Firm Learning and Coordination Mechanisms

Propositions 3, 5 and 6 express the following ideas: learning in sponsored agreements is exploratory due to the distance of the project from current core competences; the existence of pre-defined rules induces short-term stability even if it entails potential rigidities in terms of interactive learning; short-term stability is reinforced by the presence of an arbitrator (the public agency) and reputation effects.

Revisiting the rules concerning the division of labour between the partners

The existence of pre-defined rules, especially in relation to the project's research agenda (work packages, milestones, partner contributions, theoretical background, technological options, etc.), might strongly influence the direction of learning and the actual learning mechanisms. For instance, the application forms themselves impose a precise division of labour and a minimum number of meetings. The project design thus implicitly assumes that each participant will work on its contribution in-house and will learn about the results of other partners during meetings, or in a subsequent integration phase. In our view this resembles a kind of modular organization, which presupposes an efficient pre-defined architecture that makes it possible to build each module (produce each work package) independently of the others. The relevance of an *a priori* rigid architecture in the case of highly exploratory projects is questionable. In such an uncertain context, much more integrative work, and especially *ex ante* integrative work, is likely to be necessary if any technological value and success is to be achieved. In other words, interactive learning is likely to be involved. However, the imposed design and coordination of projects enforces a high level of unilateral learning and enables little learning by interacting.

The project design should foster interactive learning and, thus, the possibility to create new knowledge. Formal rules and procedures rarely trigger learning. Extrinsic incentives (here, very restrictive pre-defined rules) cannot counter weak intrinsic learning motivation (cf. Kreps 1997). Rules that are too rigid can be counter-productive and even impede learning and/or innovation. Thus, rules should be seen as means of stabilizing the initial condition and not as means of monitoring or inducing learning; flexibility in the design and the monitoring of projects, therefore, is important. In addition, the administrative burden for firms should not be too heavy, such that it hinders learning. Finally, "wait and see" behaviour, amounting to lack of involvement and commitment from any of the partners (i.e., free-riding) should be eliminated by the supervisor of the public programme dealing out appropriate punitive actions.

Networking and the danger of lock-in effects

Networking, and more precisely increased numbers of business contacts seems to be an important motive for both firms and policy makers, whose intention is to stimulate creative interactions and technological innovations. But policy makers must be aware of the danger of lock-in effects inside the created networks. In the longer term, once the networks have been set up, negative path dependencies, in the sense that participants become too close to one another to be able to create new valuable knowledge, must be avoided. If it is evident that the group of companies benefiting from subsidies is the same for two or more projects, that is to say, when the network of interacting firms has stopped expanding, then the positive networking effects have reached their limit. When this occurs, continuing to award funding runs the risk of reinforcing the position of the initial contributors and blocking entry of possible new partners.

10.5 Conclusion

In this chapter we have tried to characterize the organizational properties and the micro-mechanisms of a particular case of inter-firm technological collaboration: the EU-sponsored collaborative projects. In order to achieve this goal, we used a three dimensional analytical grid in terms of incentive to cooperate, and learning via collaboration and coordination devices. A first application of this grid led us to the literature relevant to strategic alliances in general. We then complemented this with empirical material drawn from numerous Brite-Euram projects.

Our main results are of a theoretical nature. They consist of the elaboration of stylized facts, i.e., two contrasted, idealized collaborative patterns EU-sponsored research collaborations and spontaneous research collaborations. Our findings can be summarized as follows:

- government-sponsored collaborations generally focus on peripheral competences, submit to pre-defined rules, and favour exploratory, unilateral learning;
- spontaneous collaborations can also concern the creation of more critical knowledge (i.e., closer to core competences); they need to define their own operating rules, and they sometimes trigger an interactive learning process that generates valuable collective specific assets;

– evolution paths differ: government-sponsored collaborations seem to be more stable in the short run (no premature end), but less persistent in the long run in the case of success.

From the identification of the specificities of EU-sponsored technological collaborations, we were able to derive some important policy considerations concerning the potential rigidities of European programmes (lack of flexibility in the organizational design of the projects, learning impediment, lock-in effects in the long run). We also tried to revisit the debate concerning additionality versus substitution in sponsored R&D projects, by examining in depth the specific motivations behind such partnerships from the micro-strategic point of view of the firm. Clearly, the evidence presented in this chapter has some limitations, one of the most important being the lack of a wide-scale, comparative empirical analysis. The next step in this exploratory research would be to carry out a survey in order to provide some empirical support for our theoretical hypothesis.

10.6 References

Amesse F, Cohendet P (2001) Technology transfer revisited from the perspective of the knowledge-based economy. Research Policy 30: 1459-1478.

Ancori B, Bureth A, Cohendet P (2000) The economics of knowledge: the debate about codification and tacit knowledge. Industrial and Corporate Change 9(2): 255-288.

Aoki M (1988) Information, incentives and bargaining in the Japanese economy. Cambridge University Press, Cambridge.

Argyris C, Schön DA (1978) Organizational learning: a theory of action perspectives. Addison Wesley, Reading.

Arrow KJ (1962) The economic implication of learning by doing. Review of Economic Studies 26: 155-173.

D'Aspremont C, Jacquemin A (1988) Cooperative and non-cooperative R&D in duopoly with spillovers. American Economic Review 78: 1133-1137.

Avadikyan A, Llerena P, Matt M, Rozan A, Wolff S (2001) Organizational rules, codification and knowledge creation in inter-organization cooperative agreements. Research Policy 30: 1443-1458.

Bach L, Ledoux MJ, Matt M, Schaeffer V (1995) Evaluation of the economic effects of Brite-Euram programmes on the European Industry. Scientometrics 34(3): 325-349.

Bozeman B, Dietz JS (2001) Strategic research partnerships: constructing policy-relevant indicators. Journal of Technology Transfer 26: 385-393.

Bureth A, Wolff S, Zanfei A (1997) The two faces of learning by cooperating: the evolution and stability of inter-firm agreements in the European electronics industry. Journal of Economic Behaviour and Organization 32(4): 519-537.

Butler R, Carney MG (1983) Managing markets: implications for the make-buy decision. Journal of Management Studies 20(2): 213-231.
Child J (1997) Learning trough inter-organizational cooperation. In: Proceedings of EMOT program Conference, Stresa, Italy, 11-14 September.
Ciborra CU (1991) Alliances as learning experiences: cooperation, competition and change in the high-tech industries. In: Mytelka L (ed.) Strategic partnerships and the world economy, Pinter, London, pp. 51-77.
Cohen WM, Levinthal DA (1989) Innovation and learning: the two faces of R&D. Economic Journal 99: 569-610.
David PA, Hall BH, Toole AA (2000) Is public R&D a complement or a substitute for private R&D? A review of the econometric evidence. Research Policy 29: 497-530.
De Bondt R (1997) Spillovers and innovative activities. International Journal of Industrial Organization 15(1): 1-28.
Dosi G (1988) Sources, procedures and microeconomic effects of innovation. Journal of Economic Literature 26: 1120-1171.
Doz YL (1996) The evolution of cooperation in strategic alliances: initial conditions or learning processes? Strategic Management Journal 17: 55-83.
Doz YL, Olk PM, Ring PS (2000) Formation processes of R&D consortia: which path to take? Where does it lead? Strategic Management Journal 21: 239-266.
Dyer JH, Singh H (1998) The relational view: cooperative strategy and sources of interorganizational competitive advantage. Academy of Management Review 23(4): 660-679.
Hagedoorn J (2002) Inter-firm R&D partnerships: an overview of major trends and patterns since 1960. Research Policy 31: 77-492.
Hagedoorn J, Schakenraad J (1993) A comparison of private and subsidized R&D partnerships in the European information technology industry. Journal of Common Market Studies 31(3): 373-390.
Hagedoorn J, Schakenraad J (1992) Leading companies and networks of strategic alliances in information technologies. Research Policy 21: 163-190.
Hamel G (1991) Competition for competences and inter-partner learning within international strategic alliances. Strategic Management Journal 12 (Special Summer Issue): 83-103.
Katz MJ (1986) An analysis of cooperative research and development. RAND Journal of Economics 17: 527-543.
Katz MJ, Ordover JA (1990) R&D cooperation and competition. Brookings Papers on Economic Activity: Microeconomics, pp. 137-203.
Kogut B (1991) Joint ventures and the option to expand and acquire. Management Science 37: 19-33.
Koza MP, Lewin AY (1998) The co-evolution of strategic alliances. Organization Sciences 9(3): 255-264.
Kreps DM (1997) The interaction between norms and economic incentives. Intrinsic motivation and extrinsic incentives. American Economic Review 87(2): 359-364.

Larsson RL, Bengtsson L, Henriksson K, Sparks J (1998) The interorganizational learning dilemma: collective knowledge development in strategic alliances. Organization Sciences 9(3): 285-305.

Llerena P, Matt M (1999) Inter-organizational collaboration: the theories and their policy implications. In: Gambardella A, Malerba F (eds.) The organization of economic innovation in Europe. Cambridge University Press, Cambridge, pp. 179-201.

Link AN, Paton D, Siegel DS (2002) An analysis of policy initiatives to promote strategic research partnerships. Research Policy 3: 1459-1466.

Lundvall BA (1988) Innovation as an interactive process: from user-producer interaction in the national systems of innovation. In: Dosi G, Freeman C, Nelson R, Silverberg G, Soete L (eds.), Technical change and economic theory, Pinter Publishers, London, pp. 349-369.

Macneil IR (1974) The many futures of contract. Southern California Law Review 47: 691-816.

Malerba F (1992) Learning by firms and incremental technical change. The Economic Journal 102: 845-859.

March JG (1991) Exploration vs. exploitation in organizational learning. Organization Science 2(1): 71-87.

March JG, Simon H (1993) Organizations revisited. Industrial and Corporate Change 2(3): 299-316.

Mariti P, Smiley R (1983) Cooperative agreements and the organization of industry, Journal of Industrial Economics 4: 437-451.

Matt M, Wolff S (forthcoming 2004) Incentives, coordination and learning in government-sponsored vs. spontaneous inter-firm research cooperation, International Journal of Technology Management.

Metcalfe S (1995) The economics foundations of economic policy: equilibrium and evolutionary perspectives. In: Stoneman P (ed.) Handbook of the economics of innovation and technological change. Blackwell Handbooks in Economics, Oxford UK and Cambridge USA, pp. 409-512.

Ouchi WG (1980) Markets, bureaucracies and clans. Administrative Science Quarterly 20: 129-141.

Palay TM (1984) Comparative institutional economics: the governance of rail freight contracting. Journal of Legal Studies 13(June): 265-288.

Parkhe A (1991) Interfirm diversity, organizational learning, and longevity in global strategic alliances. Journal of International Business Studies 22: 579-601.

Picard P, Rey P (1988) Recherche-Développement, Incitation et Coopération. In : Mélanges économiques, Essai en l'honneur de Edmond Malinvaud, EHESS, Economica, Paris, pp. 315-342.

Prahalad CK, Hamel G (1990) The core competence of the corporation. Harvard Business Review 68(May/June): 79-91.

Powell WW, Koput KW, Smith-Doerr L (1996) Interorganizational collaboration and the locus of innovation: Networks of learning in biotechnology. Administrative Science Quarterly 41: 116-145.

Richardson GB (1972) The organization of industry. Economic Journal 82: 88.
896.

Ring PS, Van de Ven AH (1994) Developmental processes in cooperative intero
ganizational relationships. Academy of Management Review 19: 90-118.

Rosenberg N (1982) Inside the black box. Cambridge University Press, Car
bridge.

Sakakibara M (1997a) Heterogeneity of firm capabilities and cooperative researc
and development: an empirical examination of motives. Strategic Manag
ment Journal 18: 143-164

Sakakibara M (1997b) Evaluating government-sponsored R&D consortia in Japa
who benefits and how? Research Policy 26: 447-473.

Sakakibara M (2003) Knowledge sharing in cooperative research and develo
ment. Managerial and Decision Economics 24: 117-132.

Sakakibara M, Brandstetter L (2003) Measuring the impact of US research conso
tia. Managerial and Decision Economics 24: 51-69.

Salant SW, Shaffer G (1998) Optimal asymmetric strategies in research joint ve
ture. International Journal of Industrial Organization 16: 195-208.

Suzumura K (1992) Cooperative and non-cooperative R&D in an oligopoly wi
spillovers. American Economic Review 82: 1307-1320.

Teece DJ (1992) Competition, cooperation and innovation: organizational a
rangements for regimes of rapid technological progress. Journal of Econom
Behaviour and Organization 18: 1-25.

Teece DJ (1986) Profiting from technological innovation: implications for integr
tion, collaboration, licensing and public policy. Research Policy 15: 285-305

Tripsas M, Schrader S, Sobrero M (1995) Discouraging opportunistic behavior
collaborative R&D: a new role for government. Research Policy 24: 367-389

Vonortas NS (2000) Multimarket contact and inter-firm cooperation in R&I
Journal of Evolutionary Economics 10: 243-271.

Williamson OE (1989) Transaction cost economics. In: Schmalensee R, Will
RD (eds.) Handbook of industrial organization Vol. 1. New York, North Ho
land, pp. 136-182.

Williamson O E (1985) The economic institutions of capitalism. Free Press, Ne
York.

Williamson OE (1979) Transaction costs economics: the governance of contra
tual relations. Journal of Law and Economics 22(2): 233-261.

Wolff S (1992) Accords inter-entreprises et flexibilité: elements théoriques et a
plication au secteur des telecommunications, PhD Thesis, Université Lou
Pasteur, Strasbourg.

11 How International are National (and European) Science and Technology Policies?

Jakob Edler and Frieder Meyer-Krahmer

FhG-ISI, Karlsruhe, E-mail: j.edler@isi.fraunhofer.de
BETA, Strasbourg and FhG-ISI, Karlsruhe, E-mail : Frieder.Meyer-Krahmer@isi.fhg.de

11.1 Introduction

The internationalization of research and development is a fact. Numerous indicators point to a strong tendency towards intensified international activities as regards the generation and exploitation of new scientific knowledge and technologies. At the same time, the corporate strategies designed to internationalize R&D activities follow ever more differentiated rationales: there is apparently no one single reason, but a whole range of different motivations for international collaborations.

The consequences for science and technology (S&T) policy should be obvious: if more and more of the activities within a given country or region are conducted by foreign-based actors, and if the home-based actors reach out further and further beyond the borders of their home country (or region), it should be self-evident that national or regional policies have reacted – one way or the other.

In this chapter we concentrate on the international activities of multinational companies (MNC) and provide some evidence for the empirical trends both as regards the technological activities of MNCs and national policies to exploit these tendencies. It will be shown that – despite the obvious trends – national policy-makers have not fully understood the necessities to tailor appropriate, comprehensive approaches and that the contexts and existing activities in different countries vary considerably. Finally, we derive a couple of principal lessons to be considered by national and European policy-makers that seek to foster and take advantage of the global optimization of R&D portfolios.

11.2 Patterns of International R & D

11.2.1 Scale and Scope of International R&D

Recent trends. There are three different dimensions to the internationalization of science and research: international exploitation of nationally generated knowledge and technology; international technological and scientific cooperation and exchange; and the international generation of knowledge and innovation (Archibugi and Michie 1995; Archibugi and Iammarino 1999; Meyer-Krahmer et al. 1998). Current data indicate a robust tendency towards growing internationalization of science and research in all three dimensions (OECD 1997a, 1998a,b,c, 1999; UNCTAD 1996; UNESCO 1998; Narula 1999). Although certain aspects of this internationalization trend are well documented, and some effects can be quantified, the overall processes are extremely complex and the outcomes/impacts are highly uncertain. The existence of the phenomenon is generally accepted, but its importance and the trends are currently the topic of a lively debate (see Kuhlmann and Meyer-Krahmer 2001). Since the early 1980s, the international generation of innovation has increased, and affected the internationalization of research and development (R&D). During earlier periods of international expansion (the 1960s and 1970s), MNCs first built up their sales, distribution and assembly operations in foreign countries. In later phases (late 1970s/early 1980s), efforts were directed towards supporting foreign subsidiaries with corresponding capacities in application engineering and applied R&D. Although initially the tasks of development departments abroad were limited to adapting product and process technologies from the home country to local production and market requirements, there was a clearly recognizable trend, from the late 1980s, towards strengthening R&D in foreign countries and extending the global competence portfolio. Increasingly, research became established at a high level in foreign locations.

If the situation up to the end of the 1970s was largely characterized by the dominance of a world centre for research and innovation (the United States in many important fields of technology, and western Europe in individual fields, such as chemistry), it is now true to say that, for the most important fields, several centres are crystallizing out within the Triad countries – and in some instances even beyond. These are in fierce competition with one another, and, from time to time, very rapid changes in ranking take place. Because of this development, enterprises that are leading performers of R&D have to demonstrate a presence in several locations at the same time, establish sufficiently competent and extensive structures there, and react as quickly as possible to dynamic changes in relative location advantages.

For this reason, R&D centres and product development capacities were established within the same corporation at several different Triad locations as a part of entrepreneurial integration strategies. At the same time, attempts are being made, through R&D cooperations and strategic technology alliances, to form networks as quickly and as flexibly as possible between institutionally and regionally scattered centres of competence. Empirical evidence has shown that since the 1980s the number of newly established strategic technology alliances has increased considerably (see Hagedoorn and Schakenraad 1990, 1993; Narula 1999), especially in the most dynamic technology fields, such as biotechnology, new materials and, above all, information technologies.

Complex mix of motives and the role of lead markets

Obviously, in a somewhat simplified typology globalization follows *different paradigms in different entrepreneurial functions* (Gordon 1994): (1) the internationalization of markets is determined by the search for markets with high income and low price elasticities of demand in conditions of free world trade, (2) the transnationalization of production locations is driven by the regime of production possibilities (qualified workforce, supplier-producer networks, costs, other comparative advantages, closeness to market), and lastly, (3) globalization is characterized by the pursuit of system competence through global "R&D sourcing" and the orientation towards the excellence of (national) innovation systems and related institutions[1].

However, there is growing evidence that the "three worlds" postulated in the above "three-different-paradigms" approach repeatedly impinge on one another, so that the various paradigms merge again to some extent – markets and the excellence of innovation systems are taken into consideration *together*: recent studies on determinants of location factors of the internationalization of research and development (Reger et al. 1999; Jungmittag et al. 1999) show that in different key technologies the three paradigms play varying roles. Differences between sectors regarding the degree of liberalization of international trade, the regulation of streams of direct investments, specific features of regional demand, economies of scale in production and the internationalization of technological knowledge, result in different levels of internationalization. Surveys in three selected technology fields indicated that the internationalization of R&D is mainly influenced by three factors, namely:

---

[1] In management theories these terms „transnationalization“ and „globalization“ are used the other way round (see, e.g., Bartlett and Ghoshal 1989).

- early linkage of R&D activity to leading, innovative clients ("lead users") or to the "lead market";
- early coordination of the enterprises' own R&D with scientific excellence and the research system;
- close links between production and R&D.

Different patterns within Europe
Within Europe, the degree of internationalization varies considerably. The roles of a home for international expansion and a host for foreign R&D laboratories are mostly concentrated in a few countries – mainly Germany France and the United Kingdom. Within the European Union, three main clusters of countries – which are related to the companies' strategies – can be identified:

- *small, highly developed European countries*, such as Belgium, Sweden or the Netherlands (also Switzerland), where global players perform up to more than half of their R&D activities outside their home country These countries have a relatively small pool of domestic R&D resources; firms therefore invest heavily in the international generation of innovations;
- *large European countries* with large technology bases and markets, such as Italy, Germany, and France, and where "their" MNCs perform between one-fifth to one-third of their R&D activities abroad. Nevertheless, many large enterprises in these large European countries, particularly in the machinery, transportation and electrical engineering sector tend to concentrate a significant part of their research in the country of origin;
- *"intermediate countries"*, such as Spain, Portugal and Ireland, participate somehow differently in the new international division of labour These countries lack well-equipped technological infrastructures and resources and are characterized by high foreign inward R&D investment and very low outward R&D investment. On the one side, MNCs contribute, quantitatively and qualitatively, to a great extent to the technological efforts of these countries. On the other side, there are, firstly, a considerable number of innovative domestic companies which do not internationalize (neither via exports nor foreign direct investment (FDI)). Secondly, most domestic firms operating in international contexts use exports as the basic and almost unique way of internationalization.

The (decreasing) weight of Europe as host to international R&D
The significance of the big European countries playing host to international R&D is qualified if one compares their weight as host to that of other regions in the world, mainly the US. In the course of the 1990s the attractiveness of Europe as a location for foreign companies declined. Of all R&D expenditure *under foreign control* in the manufacturing industry *within the OECD countries*, the share that is spent in the US has grown from 45.3% to 55.5%, the share of Japan has grown from 2.8% to 4.1%, while the share of Germany, France, the UK and the rest of the OECD countries has declined (OECD 2001; p. 26, fig. 10).

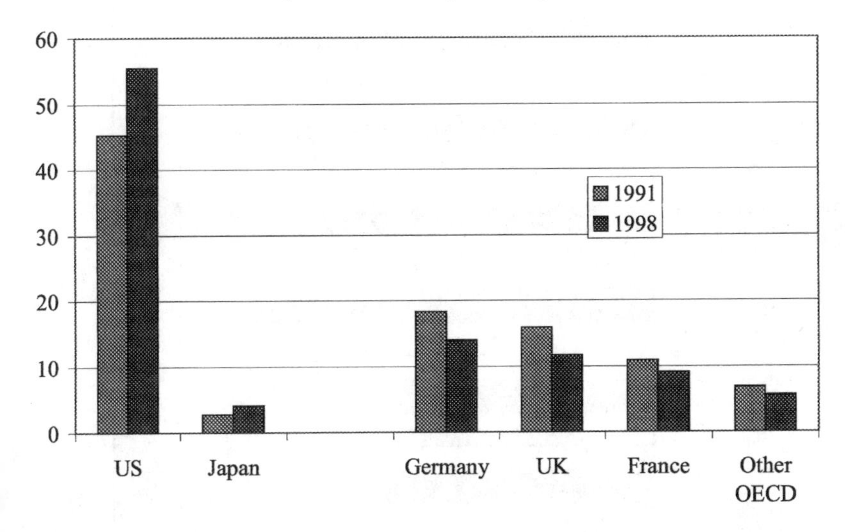

Source: OECD (2001).

Fig. 11.1. Share of R&D expenditure under foreign control in the manufacturing industry in selected OECD countries

11.2.2 Lessons on Location Factors of MNCs

The multitude of studies that have been produced in the last five to ten years has shown that R&D is motivated by a very broad variety of factors[2]. For example, Edler et al. (2002) have identified a whole range of reasons that drive internationalization.

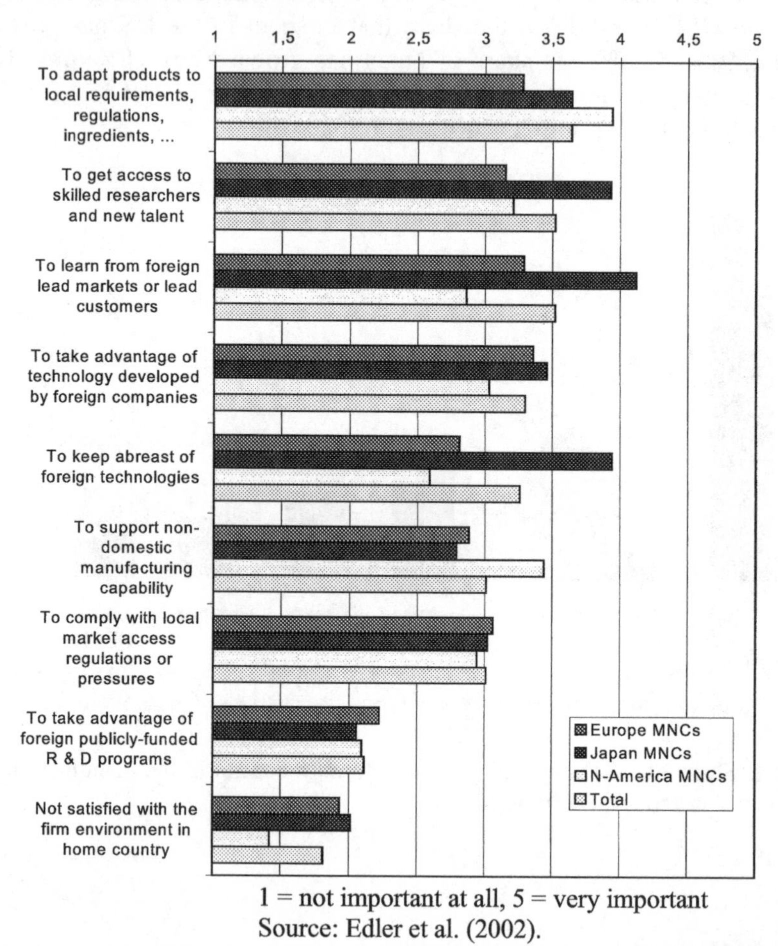

1 = not important at all, 5 = very important
Source: Edler et al. (2002).

Fig. 11.2. Motivations for MNCs from the Triad regions to invest in R&D abroad

[2] See Criscuolo et al. (2001) for an overview.

Figure 11.2 shows, that despite this variety, the most important reasons can be grouped into two basic motivations: *knowledge exploiting* vs. *knowledge augmenting*[3].

Knowledge exploitation (R&D abroad to meet the peculiarities of foreign markets)

- *Mode*: Knowledge exploitation encompasses all motives that are related to R&D work done in order to adjust the existing technologies, products, and processes to meet the needs of local demand, supply, regulation (standards, etc.). In this mode, the major knowledge is generated in the home country, and, in a second step, exploited abroad by fine-tuning technological developments towards different needs and to support foreign production and marketing.
- *Empirical evidence*: For most companies, the bulk of activity is still the support of local production and marketing abroad (Cantwell 1995; Cantwell and Kosmopoulou 2001; Patel and Vega 1999; Patel and Pavitt 2000; Serapio and Dalton 1999). For example, for German MNCs various studies (Legler et al. 2000; Beise and Belitz 1998; Belitz 2002; Edler 2003) have shown that the technological areas developed abroad are very similar to those at home, indicating that companies mainly adapt what they have developed at home. This also means that for the bulk of R&D investment in a given host country the characteristics of its internal market (size, advanced users, advanced suppliers, product or process regulations) are more important than the quality of the science and research system.

Knowledge augmenting

- *Mode:* Knowledge augmenting, on the other hand, means that the international arena is used to generate new knowledge. Innovation is more and more knowledge- and speed-driven, MNCs are forced to be quick and excellent at the same time. Therefore, MNCs have perfectly genuine motives to tap into existing, forefront knowledge centres of excellence abroad and to take advantage of them. This might be done through simple monitoring activities, through integrating into existing scientific networks, or through employing scientific talent. In this paradigm of in-

[3] This is confirmed by a very broad range of recent studies, see, e.g., Dunning andNarula (1995), Meyer-Krahmer et al. (1998); Edler et al. (2001); Meyer-Krahmer (1997); Niosi (1999); OECD (1999); Pearce (1999); Pearce/Singh (1992); Criscuolo et al. (2001); Boutellier et al. (1999), Kuemmerle (1999) and Kumar (2001).

ternational R&D, the search for excellent research centres around the globe and the build up and re-transfer of knowledge are major tasks for "globally learning companies" (Meyer-Krahmer et al. 1998).

- *Empirical evidence:* a growing number of studies finds the knowledge augmenting mode becoming more and more important (Florida 1997; Koopmann and Münnich 1999; Boutellier et al. 1999; Cantwell 1995; Edler et al. 2001; Dunning and Wymbs 1999; Granstrand 1999; Pearce and Singh 1992; Pearce 1999; Criscuolo et al. 2001; Narula 1999). It has been shown that knowledge seeking and generating abroad correlates with
  - the knowledge intensity of the technological area[4],
  - the intensity and scope of the corporate R&D,
  - the perception of researchers and managers that the knowledge base of the host country is more advanced.

Secondary location factors

Beyond the two major strands of motivations a set of other location factors can be defined, the meaning of which, as identified in surveys, is in most cases lower than the meaning of the two sets of reasons just discussed.

- Vertical *cooperation* with local partners (suppliers, (lead) customers). The importance of vertical integration has grown, and so has the inclination of MNCs to locate parts of their R&D close to their most important suppliers and/or customers (Just 1997).
- *Efficiency (research costs)*: in some areas of research, the actual costs of performing it play a major role. It has been shown that if the level of expertise needed is available for less cost, and if the infrastructure limits transaction costs, research also follows efficiency (Gerybadze et al. 1997).
- *Follow the competitor*: in some cases, mainly in oligopolistic markets, a "follow the leader" effect has been observed even for R&D activities (Kumar 2001; Pearce and Singh 1992).
- *General political and financial framework conditions*: in none of the recent studies does the general political framework or a different financial

[4] A recent study on German MNCs abroad and foreign MNCs in Germany has shown a clear correlation between knowledge intensity of a technological field (average number of cited publications in a patent) and degree of international activity (Edler 2003).

framework (venture capital etc.) play a significant role as a driving force for R&D investment[5].

- *Public RTD policy*: all studies reviewed (including especially Kumar 2001; Edler et al. 2002 and Edler et al. 2003) find that public policy (R&D programmes, patent regime, some kind of indirect supporting schemes, etc.) are *not* a major determining factor, they are far less important than market size and requirements, or the various forms of knowledge supply.
- *"Side effects"*: finally, in interpreting data on R&D internationalization it is important to note that the build up of foreign research capacity is very often the by-product of merger and acquisition activities that are not driven by R&D considerations (Boutellier et al. 1999). As Archibugi (2000) has stressed, and as has recently been shown for German MNCs abroad (Edler et. al. 2003), the R&D intensity of foreign subsidiaries tends to be smaller than that of domestic companies. Therefore, it is true that external growth in R&D capacity by foreign firms through "side effects" increases the share of international R&D activities, however, in the medium and long run, such "side-effect" R&D capacities are often reduced if not closed down. Moreover, there are instances where existing linkage to the local knowledge base dissolves after post acquisition re-organizations[6].

11.3 Existing Policy Activities for Internationalization

Governments act on the increasing internationalization of science and technology, albeit slowly. Edler and Behold (2001) studied national public policies to exploit international science and industrial research. They analyzed strategies and initiatives set in motion by key actors in science and technology policy in eight countries (USA, Japan, France, United Kingdom, Netherlands, Switzerland, Malysia, South Korea)[7]. These activities could be directed at attracting or absorbing foreign knowledge or carriers

[5] However, in most cases where big MNCs are analyzed, different cultures of venture capital provision may play a very different role for small and young companies.

[6] An important issue for future research on foreign R&D investment will be what kind of *negative* effects might occur for local or regional innovation systems in the long run.

[7] Unfortunately, Germany was excluded from the empirical analysis that was conducted on behalf of the German Ministry for Education and Research.

of knowledge. The study found that, although there is awareness of a general leverage for policy, few countries develop an integrated policy strategy to address these issues. Nevertheless, the study identified a number of interesting initiatives that can serve as "benchmarks" for policy-makers elsewhere.

Following Meyer-Krahmer and Reger (1999), Edler and Behold see the rhetoric on internationalization as aiming either at enhancing the inward activities of foreign actors (attractiveness) or the outward activities of national actors to exploit knowledge generated abroad or to contribute to this generation (absorption). The following problems are typically identified:

- lack of attractiveness or insufficient national supply of scientists and industrial R&D in certain fields;
- brain drain; and/or
- isolation from, or lack of integration into, processes of international generation of knowledge.

From the strategic documents[8], five principal strategies were identified, which, without being all-embracing, cover the major strategic efforts of national administrations:

1. attraction of foreign scientists;
2. attraction and integration of foreign industrial R&D;
3. improvement of access to foreign knowledge and to technological lead markets;
4. targeted learning from administrative and management practices abroad; and
5. support for the international networking efforts of firms and scientists. On the basis of these three problem dimensions and five strategic trajectories, Table 11.1 shows the characteristics of the countries.

[8] These are principal policy papers coming from research or economy ministries or state agencies. For an illustration, see, in the case of Japan e.g. Science and Technology Agency (Hg.) (1995) or Ministry of International Trade and Industry (MITI) (annually), or in the case of the USA e.g. Department of Commerce (1998, 1999).

Table 11.1. Strategic country characteristics

	USA	UK	CH	F	J	NL	SK	MY
Problem areas								
Lack of attractiveness for international scientists and industrial R&D capital	+	–	–	++	++	++	++	++
Brain drain	–	(+)	(+)	+	–	+	++	+
Insufficient integration into and exploitation of the global generation of knowledge	+	–	++	++	+	+	+	–
Strategic trajectories								
Attraction of foreign scientists	+	–	+	+	+	+	(+)	+
Attraction and integration of foreign industrial R&D capacities	+*	++*	–	–	–	+	(+)	++
Better access to foreign knowledge and technological lead market	+	++	++	–	+	+	++	–
Targeted learning from foreign administrative and management practices	+	+	–	–	+	++	–	–
Support of individual firms and scientists in international networking efforts	–	–	++	++	++	++	+	+

– no problem awareness/no strategic measures, (+) some problem awareness/no explicit measures yet, + some problem awareness/some measures taken, ++ high level of problem awareness/targeted strategic measures, * in the USA and in the UK attraction policies are mainly undertaken by the states and regions
Source: Edler and Boekholt 2000

In principle, three types of countries with similar strategic positioning can be defined, with the Netherlands being a special case. First, Switzerland, the USA, and the UK can be considered as a group of countries where policy-makers and administrators alike are convinced that their home country is attractive for foreign knowledge seekers. The overarching strategic orientation in these countries is towards the established strengths of the national innovation system, where world-class excellence will generate sufficient attractiveness. These countries concentrate on absorption. Switzerland, albeit developing a foreign science policy, has no explicit public strategy to attract foreign researchers, especially for industrial R&D. Internationalization of academia is largely based on decentralized initiatives.

Second, despite their obvious differences, Japan and France have shown a lack of attractiveness and in addition even face a brain drain. In both countries, administrations have recognized that foreign scientists face a range of disincentives to integrate in their national innovation systems. Incentives for immigration are seen thus as strategically important.

Third, Malaysia and South Korea face the key challenge of catching up with the OECD economic and innovation systems. However, their strategies to exploit international science and research (S&R) are different. Apart from following a strategy of imitating foreign-developed technologies and products, South Korea has put little emphasis on international S&R, but has rather supported students and senior scientists in going abroad. Malaysia has chosen to concentrate on attracting foreign industry.

The Netherlands are a special case. The national economy and the innovation system have traditionally been very open and internationally integrated. Also, research policy and industrial policy have traditionally been linked. Consequently, the internationally oriented activities of the country mainly aim at pushing for more international integration of the national industry, both within the country (attraction schemes) and abroad, and the country has indicated its willingness to import best organizational practice to that end.

11.4 Consequences and Issues for Technology Policy in Europe

What does all this mean for a more appropriate technology policy in Europe? As a general consequence of this situation, the premise of national science and technology policy encountered in many countries that the main benefit from the public allocation of resources in this policy area flows into the national economy, is progressively dissolving. Not only the know-how produced in the national innovation system, but also other public investments, for instance in training and education, are increasingly being swept into the stream of the international exchange of knowledge. This development enlarges the focus of policy: it is not simply the appropriation of nationally generated knowledge that is involved, but the strengthening of a generally beneficial, interactive transactional exchange of knowledge. It is possibly as important to absorb knowledge that has been generated worldwide, as it is to support the production of knowledge in one's own country. This statement is very important for technology policy on the national as well as the European level.

As a consequence of this analysis of the changes in the innovation strategies of large MNCs there are at least four dimensions that should be kept in mind by national and European policy-makers (Meyer-Krahmer and Reger 1999):

- strengthening European absorptive capacities and cooperating with non-European countries;
- attracting innovative companies from non-European countries;
- lead markets and learning for the mastery of complex innovations;
- integrating different policies towards an innovation policy in Europe.

To meet the challenges that arise within these four dimensions, policy-makers need to draw the appropriate conclusions from analysis of the location factors identified above. After all, policy needs to appeal to the decision-makers within these corporations. For heuristic reasons, the distinction between "market adaptation" and "knowledge creation" is initially maintained, while it is clear that a combination of measures to foster both modes at the same time are the most promising.

11.4.1 Policy Challenges Stemming from the Market Adaptation Mode

- Companies can be driven into investing in R&D if they sense a market to be a lead market requiring R&D presence alongside production or sales. This can be caused simply by different local demand (taste, tradition, etc.), by technologically advanced public or private demand, by advanced regulation or future oriented standards. If – in addition – a market is of a certain critical size, the adaptation to those local conditions triggers R&D investment. Therefore, European policy should identify and foster possible lead market areas, i.e., areas in which the end-user market is regarded as a trendsetter internationally. Especially in these markets, standards-setting regulations will drive European and non-European companies into research and development activities (Meyer-Krahmer and Reger 1999). Candidates for such lead markets in Europe could be in the field of pharmaceuticals, mobility (especially cars), (mobile) communication, and "sustainable" products and energy (especially fuel cell) (see Meyer-Krahmer (2004)).

- In this market adaptation – and similarly in the lead market oriented mode – direct policy measures seem to be less effective. Nevertheless, policy must ensure that those foreign companies that are willing to exploit a lead market and learn within lead markets – thus creating value within a host country – have access to cooperation partners, especially lead users. Here, public support schemes – including public procurement to trigger innovation – should not be discriminatory.

11.4.2 Policy Challenges in the Knowledge Creation Mode

- The greatest challenge, obviously, is to make a country or region scientifically or technologically attractive. Attractive locations for MNCs investing in the generation of new forefront knowledge are characterized by an excellence science system (excellent human capital, especially talent) that is accessible to foreign companies. European policy must foster the existence of, and accessibility to, scientific excellence and scientific-technological networks, including eagerness of universities and institutes to cooperate with (foreign) MNCs, including long-distance cooperation.
- As the necessity to integrate knowledge from very diverse technological areas into the industrial R&D process increases, and the absorption of knowledge from neighbouring fields becomes more important, locations that can offer accessibility to a wide scope of scientific and technological activities will become more attractive in the future. To ease the access to this wide scope within Europe would be an important element of any scheme to attract foreign MNCs to Europe.
- The enabling infrastructure, most importantly ICT networks, must be excellent, as coordination with the headquarter for reasons of integrating knowledge globally is crucial in this mode.

For the sake of analysis, the market adaptation and knowledge creation modes are mostly dealt with separately. The most important policy advice to be given, however, is that in order to attract foreign industrial R&D it is increasingly necessary to develop policy schemes that integrate the requirements of both modes, as the combination of advanced conditions to generate knowledge that feeds into innovation with a market that is able to absorb these innovations (lead markets) obviously has the greatest – and most likely sustainable – attractiveness for multi-national enterprises (Manes). At the same time the host country benefits, not only from advanced R&D activities and related networking, but also from value adding production activities.

Furthermore, some additional policy principles can be derived from the secondary location factors. As companies tend to follow the technological leader, the attraction of the prime players in the market should be a major goal, as this might more easily lead to agglomeration effects within a given region. In addition, as vertical cooperation is a major reason for many companies, it is important to ensure that access to local industrial clusters is not hindered for foreign MNCs. Finally, as public monetary incentives are of relatively minor importance for foreign MNCs, the issue is not whether the MNCs are eligible to receive additional money, but whether existing support schemes – and their regulations (exploitation, appropriation) – diminish the possibility of foreign MNCs to enter into technological or scientific cooperation schemes. However, from the importance that foreign MNCs attribute to integration into the local innovation systems, public policy can have a detrimental effect if it leads to the exclusion of foreign MNCs from certain cooperation schemes. This is perhaps the greatest problem with RTD support schemes, which are not open to companies from other countries, public policy should re-consider their openness. Therefore, it would be of major importance to have a new look at the openness issue, as the last comprehensive overview was OECD (1997b).

11.4.3 Limitations and Counterproductive Tendencies

The conclusions derived in this chapter are based on the empirical evidence of international R&D and the premise that policy needs to be tailored to meet the needs of internationally active MNCs. However, in order to avoid a somewhat naïve approach to appropriate policy-making, we conclude by pointing towards potential dangers.

Firstly, MNCs do not seem to rate direct policy measures geared towards R&D investment in their potential host countries as being a mature location factor. By far more important are the market and/or knowledge generation conditions. This means that expensive schemes seeking to attract companies simply by providing some kind of monetary or fiscal incentives might be a waste of money.

Second, there is tendency for winners to win and losers to lose and for attraction schemes in any form to create even more concentration and agglomeration within Europe, conflicting with cohesion policies. Therefore, it seems sensible that centres of excellence in Europe should rather be virtual, and characterized by new schemes of long-distance cooperation. The logic of the European Research Area (ERA), to combine complementary excellence throughout Europe, could result in a new mode of attractive ness. The idea of building up a critical mass of excellence in any give technological area through having competence centres cooperate and ex change, as is the major goal of the ERA, might be a major step in that di rection. The question, however, would be how accessible these (virtual centres would be foreign MNCs.

Third, making European companies re-locate existing R&D capacit back to Europe through some kind of incentive schemes could backfir Either to accompany local production or tap into knowledge structure abroad, the reasons for locating R&D capacity in foreign markets are rea sonable and, after all, serve the needs of the parent company in Europ Rather than thinking of attraction schemes, Europe has to make sure tha especially for knowledge centres and lead markets, it builds up more rea sons for MNCs to come back simply for their own good, not to assist pub lic policy.

For the time being, one might even help companies, even smaller one to integrate into international knowledge creation structures, since the ger eration of forefront knowledge fosters competitiveness at home as we (absorption). For example, policy could assist companies to install know edge management practices that ensure re-integration of knowledge.

11.5 References

Archibugi D, Michie J (1995) The globalization of technology: a new taxonom Cambridge Journal of Economics 19: 121-140.

Archibugi D, Iammarino S (1999) The policy implications of the globalisation o innovation. In: Archibugi D, Howells J, Michie J (eds.) Innovation policy in global economy. Cambridge University Press, Cambridge, pp. 242-271.

Archibugi D (2000) The globalisation of technology and the European innovatio system. Paper prepared as part of the EU Project: Innovation Policy in Knowledge Based Economy.

Bartlett CH, Ghoshal S (1989) Managing across borders: the multinational solu tion. Bartlett CA; Doz Y.; Hedlund G (eds.) Managing the Global Firm Routledge, London.

Beise M, Belitz H (1998) Trends in the internationalisation of R&D – the Germa perspective. Vierteljahreshefte zur Wirtschaftsforschung des DIW 67(2): 67 85.

Belitz H (2002) Deutschland als Forschungsstandort multinationaler Unterneh men, DIW-Wochenbericht 16/02.

Boutellier R. Gassmann,O. von Zedtwitz,M.(1999) Managing global innovatio Uncovering the secrets of future competitiveness. Springer, Heidelberg.

Cantwell JA (1995) The globalisation of technology: what remains of the product cycle model? Cambridge Journal of Economics 19: 155-174.

Cantwell J, Kosmopoulou E (2001) What determines the internationalisation of corporate technology. In: Forsgren M, Havila V, Håkansson, H (eds.) Critical Perspectives on Internationalisation. Pergamon, Amsterdam, pp 305-334.

Criscuolo P, Narula R, Verspagen B (2001) Measuring knowledge flows among European and American multinationals: a patent citation analysis. Paper presented at the The Future of Innovation Studies conference, Eindhoven.

Dunning JH, Wymbs C (1999) The geographical sourcing of technology-based assets by multinational enterprises. In: Archibugi D, Howells J, Michie J (eds.) Innovation policy in a global economy. Cambridge University Press, Cambridge, pp. 185–224.

Dunning J, Narula R (1995) The R&D activities of foreign firms in the United States. International Studies of Management & Organization 25(1-2): 39-73.

Edler J, Boekholt P (2001) Internationalisierungsstrategien in der Wissenschafts- und Forschungspolitik, Study for the German BMBF, Bonn.

Edler J, Meyer-Krahmer F, Reger G (2001) Managing technology in the top R&D spending companies worldwide - results of a global survey. Special Issue of the Engineering Management Journal, Managing High Technology Research Organizations 13(1): 5-11.

Edler J, Meyer-Krahmer F, Reger G (2002) Changes in the strategic management of technology - results of a global benchmarking study. R&D Management Special issue 32(2): 149-164.

Edler J, Döhrn R, Rothgang M (2003) Internationalisierng industrieller Forschung und grenzüberschreitendes Wissensmanagement. Eine emirische Analyse aus der Perspektive des Standortes Deutschland, Karlsruhe.

Edler J (2003) International research strategies of multinational corporations: A German perspective. Technological Forecasting and Social Change 71: 599-621.

Florida R (1997) The globalization of R&D: results of a survey of foreign-affiliated R&D laboratories in the USA. Research Policy 26: 85-103.

Gerybadze A, Meyer-Krahmer F, Reger G (1997) Globales Management von Forschung und Innovation, Schaeffer-Poeschel, Stuttgart.

Gordon R (1994) Mastering globalisation. Seminaron The Future of Industry in Europe, Brussels.

Granstrand O (1999) Internationalization of corporate R&D: a study of Japanese and Swedish corporations. Research Policy 28: 275-302.

Hagedoorn J, Schakenraad J (1990) Inter-firm partnerships and co-operative strategies in core technologies. In: Freeman C, Soete L (eds.) New explorations in the economics of technical change. Pinter Publishers, London, pp. 3-37.

Hagedoorn J, Schakenraad J (1993) Strategic technology partnering and international corporate strategies. In: Hughes K (ed.) European competitiveness. Cambridge University Press, Cambridge, pp. 60-86.

Jungmittag A, Meyer-Krahmer F, Reger G (1999) Globalisation of R&D and technology markets - trends, motives, consequences. In: Meyer-Krahmer F

(ed.) Globalisation of R&D and technology markets: consequences for national innovation policies. Physica-Verlag, Berlin, pp. 37-78.
Just RW (1997) Die Internationalisierung der Unternehmensbereiche Forschung und Entwicklung: Theoretische Erklärungsansätze, Determinanten und Indikatoren unter Berücksichtigung des Wirtschaftsstandortes Deutschland. Lang, Frankfurt.
Koopmann G, Münnich F (1999) National and international developments in technology. Intereconomics 34(6): 267-278.
Kuemmerle W (1999) Foreign direct investment in industrial research in the pharmaceutical and electronics industries – results from a survey of multinational firms. Research Policy 28: 179-193.
Kumar N (2001) Determinants of location of overseas R&D activity of multinational enterprises: the case of US and Japanese corporations. Research Policy 30: 159-174.
Kuhlmann S, Meyer-Krahmer F (2001) Internationalisation of innovation, interdependence and innovation policy for sustainable development. In: Sweeney G. (ed.) Innovation, economic progress and the quality of life. Edward Elgar Publishing. Ltd., Brookfield, pp. 86-110.
Legler H, Beise M, Gehrke B, Schmoch U, Schumacher D (2000) Innovationsstandort Deutschland. Chancen und Herausforderungen im internationalen Wettbewerb. Moderne Industrie, Landsberg/Lech.
Meyer-Krahmer F (1997) Konsequenzen veränderter industrieller FUE-Strategien für die nationale Forschungs- und Technologiepolitik. In: Gerybadze A, Meyer-Krahmer F, Reger G (eds.) Globales Management von Forschung und Innovation. Schäffer-Poeschel, Stuttgart, pp. 196-215.
Meyer-Krahmer F, Reger G (1999) New perspectives on the innovation strategies of multinational companies: lessons for technology policy in Europe. Research Policy 28: 751-776.
Meyer-Krahmer F (2004) Vorreiter-Märkte und Innovation. Ein neuer Ansatz der Technologie- und Innovationspolitik. In: Steinmeier F-W (ed.) Made in Germany '21. Hoffmann & Campe Verlag, Hamburg, pp. 95-110.
Meyer-Krahmer F, Archibugi D, Durand T, Molero J, Pavitt K, Soete L, Sölvell Ö, Reger G (1998) Internationalisation of research and technology: trends, issues, and implications for science and technology policies in Europe. ETAN Working Paper.
Ministry of International Trade and Industry (MITI) (published annually) Trends and future tasks in industrial technology (White Paper).
Narula R (1999) Explaining the growth of strategic R&D alliances by European firms. Journal of Common Market Studies 37(4): 711-723.
Niosi J (1999) The internationalisation of industrial R&D: from technology transfer to the learning organisation. Research Policy 28: 107-117.
OECD (1997a) National Innovation Systems. OECD, Paris.
OECD (1997b) Access of foreign firms to publicly-funded European RTD programmes (OECD/GD(97)). OECD, Paris.
OECD (1998a) Facilitating technology co-operation: proceedings of the Seoul conference. OECD, Paris.

OECD (1998b) Internationalisation of industrial R&D. Patterns and trends. OECD Paris.
OECD (1998c) Main science and technology indicators. OECD, Paris.
OECD (1999) Globalisation of industrial R&D - policy issues. OECD, Paris.
OECD (2001) Measuring globalisation. The role of multinational in the OECD economies. OECD, Paris.
Patel P, Pavitt K (2000) National systems of innovation under strain: the internationalisation of corporate R&D. In: Barrell R, Mason G, O'Mahony M (eds.) Productivity, innovation and economic performance. Cambridge University Press, Cambridge, pp. 217-235.
Patel P, Vega M (1999) Patterns of internationalisation of corporate technology: location vs. home country advantages. Research Policy 28: 145-155.
Pearce RD, Singh S (1992) The globalization of research and development, London.
Pearce RD (1999) Decentralised R&D and strategic competitiveness: globalised approaches to the generation and use of technology in multinational firms. Research Policy 28: 157-178.
Reger G, Beise M, Belitz H (1999) Internationalisierung technologischer Kompetenzen. Trends und Wirkungen in den Schlüsseltechnologien Pharmazeutik, Halbleiter und Telekommunikation. Physica, Heidelberg.
Science and Technology Agency (Hg.) (1995) Kagaku Gijutsu Kihonhô ni tsuite: The Science and Technology Basic Law, Law No. 130 of 1995, Tokyo.
Serapio M, Dalton D (1999) Globalization of industrial R&D: an examination of foreign direct investments in R&D in the United States. Research Policy 28(2-3): 303-316.
US Department of Commerce (Technology Administration) (1998) International plans: the global context for U.S. technology policy. Department of Commerce, Washington, DC.
US Department of Commerce (Technology Administration) (1999) Globalizing industrial research and development. Department of Commerce, Washington, DC.
UNESCO (1998) United Nations Educational, Scientific and Cultural Organisation, World science report 1998. United Nations, New York.
UNCTAD (1996) United Nations Conference on Trade and Development, World investment report 1996. United Nations, New York/Geneva.

12 Universities Specificities and the Emergence of a Global Model of University: how to Manage These Contradictory Realities

Chantale Mailhot and Véronique Schaeffer

HEC, Montréal, E-mail: chantal.mailhot@hec.ca
BETA, Strasbourg, E-mail: schaeffer@cournot.u-strasbg.fr

12.1 Introduction

Debate has been growing among policy-makers, scientists, and industrialists about the role of the university in modern society (Etzkowitz and Leydesdorff 2000, p. 109). How the missions of universities have evolved has depended on the link established between the economic dynamic and the production of knowledge at different periods of time and how science has been oriented by the policies in place in those periods.

Science and technology policies impact significantly on the activities involved in the creation and diffusion of knowledge in shaping the activities of scientific institutions and their relationships with other economic actors. The objectives of these policies have evolved in line with socio-economic change and different conceptions of the roles of scientific and economic institutions. Today, the governments in most industrialized countries tend to adopt convergent scientific policies, imposing constraints on universities in terms of the broadening of their mission in conformity with current economic constraints. This tends to increase the pressure on research activities

The transformation of universities is thus often seen from a deterministic point of view in which these institutions need only to adapt to the new mission being imposed by government. The requirement for continual performance improvement, in a context in which science is considered as a strategic economic resource and where higher education becomes more and more competitive, has produced an evolution in university management. In this chapter we examine the new necessity for management and consider how the role of the university has evolved.

Based on a strategic analysis model, we will demonstrate that universities cannot all respond to external pressures as firms would, and also that all universities cannot do so in the same way. Strategic analysis shows that the performance of a particular university does not depend on the adoption of a global model. We maintain that the diversity amongst universities is not taken into account at different political levels. The trajectories chosen by universities in the context of the evolution of their missions are specific to them and not compatible with the idea of a single university model based on the entrepreneurial university. Our aim in this chapter is to address some management and policy recommendations. Identification of the factors that characterize the ability of a university to move towards a proposed model of university is important in this context where the missions of universities are broadening because of their growing implications for economic development.

In the first part of the chapter we briefly describe the evolution of science and technology policy in order to underline the convergence of political orientation in the OECD countries and to characterize the global model of the university that accompanies it. The second part of the chapter will be devoted to the limitations of this global model. We will show that the existence of a global model does not correspond to reality in a world characterized by significant differences between universities, between countries and also between different universities within the same country. Finally, we will draw out some management and policy implications, especially in relation to the responsibilities of university management and the coordination with policy makers that will become necessary.

12.2 Science Policies and the Emergence of a New Global Model of University

Science, its activities, and its products are clearly linked to the social and economic domain. The activities of creation, diffusion and use of knowledge in the economy have gained in importance and the scientific systems and techniques involved have become more collective: "knowledge becomes an activity that is whole and openly multidimensional that must contribute simultaneously to the creation of certified knowledge, collective goods, competitive advantages, professional competencies but also to a culture of decisions shared by the majority" (Callon *et al.* 1995, p. 12). All the institutions involved in knowledge production activities, whether they are firms or universities, are now expected to produce all of these kinds of knowledge. This has not always been the case. Universities and firms once had very distinct roles and recognition of the link between the economy and the production of knowledge is relatively recent.

12.2.1 Science Policies and the Evolution of the Missions of the Universities

The explicit integration of the scientific, economic, and social domains dates back only to World War II, with the institutionalization of a policy *for* and *by* science. Since then, this policy has taken different directions that have been marked by the passage from measures based upon public financial support for universities with no conditions attached, to measures characterized by funding for specific projects (Vavakova 1998, p. 219).

Autonomy of universities

Universities for a long time were seen as being autonomous and as the places responsible for the creation and diffusion of knowledge. They were not ruled by any specific policies. The roles of universities and of businesses were quite distinct and can be summarized as follows. Universities were devoted to the production of knowledge, and firms and businesses to the creation and improvement of economic goods. Universities diffused knowledge across the socio-economic sphere through publications and education. The creation of knowledge was their starting point and determined industrial development and the creation of markets. The role of the government was to regulate exchanges, and to unconditionally finance universities in order that they could pursue their social objectives of knowledge creation and teaching.

The concept that fundamental knowledge delivered by universities is the origin of technical development in the industry for a long time justified the policies that were put in place. The "environment" of the university was characterized by very little regulation of its activities and a complete absence of competition in the transmission or creation of knowledge. Private funding for universities was only acceptable in the form of unconditional subsidies or donations.

Fundamental research was organized around universities and public laboratories that were structured within disciplines and were precisely specialized. In terms of their philosophy, or principles and rules of action, universities did not, for example, have to direct their research towards business objectives, which would have run contrary to the concept of public service. The fundamental driver of research was curiosity, and problems were defined within disciplinary matrices (OECD 1999, p. 46). This "contract" between science and society can be summarized as: "Government promises to fund the basic science that peer reviewers find most worthy of support, and scientists promise that the research will be performed well and honestly and will provide a steady stream of discoveries that can be translated into new products, medicines, or weapons" (Vavakova 1998, p. 210).

Accrued importance of socio-economic variables in research

The emergence of a second type of scientific policy, focused more on socio-economic demands than on producers of science, such as universities, can be linked to a new awareness from governments of their limits in relation to research and development (R&D) spending, to a decrease in research productivity, and to a questioning of its value (Vavakova 1998, p. 219).

The debate around science policies in the 1970s directly touches universities and the question of the relationship between universities and industry. The social contract between scientific institutions and governments was revised: "The changed world of modern science and modern government means that it is imperative to search for and begin to define a new contract, or series of contracts, between the institutions of democracy and the institutions of science. The scientific community needs to reach out and justify its claim on public resources by demonstrating where and how it is relevant in solving public problems. Science needs to earn the confidence of the public and the government and to enhance its contribution to the general welfare" (Guston and Keniston 1994, p. 32, cited in Vavakova 1998, p. 210).

For some, it is also the huge increase in scientific activity and the diffusion of results linked to massive investment that has been made in research and education, which has provoked significant transformations in the ways that knowledge is created and diffused[1]. New rationales of funding are being initiated in most OECD countries. Knowledge creating organizations are multiplying; scientific expertise is being redefined in the context of the links being established between the producers of knowledge, and scientific disciplines; the boundaries between them are becoming less well defined; professional bodies are being established; and scientific and technological systems are becoming global (Hamdouch and Depret 2001, p. 44).

Once integrated within a political and economic environment composed of competing organizations, universities are being forced to adopt more strategic styles of management. In this new setting, they have to reconcile environmental, economic, political, social and technological demands.

[1] In the United States, for example, new disciplines emerge from collaborations between universities and industrials in the context of large research programmes, like the Manhattan Project. Applied disciplines are institutionalized.

Global converging changes

The changes in R&D systems that have occurred since the 1990s have assumed such importance that many authors believe they are radical and global (Rip and Ziman 1990; Gibbons *et al.* 1994; Etzkowitz *et al.* 1998). Contemporary analysis of how economies have been transformed demonstrates that, from a global perspective, scientific and technical systems are being characterized by the transition towards a knowledge economy, with closer ties between science and technology, a tightening of relationships between companies and public laboratories, the development of global competition, the centralization of techno-science in international markets, and global intellectual property rights (OECD 1999, p. 47). These changes are taking place in a context of increasing research costs, and budgetary constraints at government level, both of which have had an impact on the funding of university research.

Currently, technology policies are largely aimed at promoting cooperation between the public and private sectors. Collaboration has become a management tool in government innovation programmes. This has resulted in new rationales in relation to funding, the promotion and organization of research, a reduction in public participation in the R&D effort, and the growing involvement of the private sphere within the public sector. In the early 2000s, in Japan and the United States, as in Europe, industry finances over 60 per cent of the R&D that is conducted – twice as much as at the beginning of the 1980s (Hamdouch and Depret 2001, p. 44).

As a result of this type of innovation policy, academic research has become strategic research[2], which is largely determined by economic and social needs and is evaluated by public authorities in terms of its contribution to national objectives. In the past, research was internally generated and assessed by peers. Innovation policy is closely linked, in the OECD countries, to a major reconfiguration of the role of universities in the creation, diffusion, and use of knowledge globally (Milot 2003, p. 68).

Universities are losing their monopoly over knowledge production to the benefit of numerous other institutions (research centres, industrial laboratories, consultancy firms, etc.). Science is no longer organized within disciplines and assessed by a limited community of peers. Solutions emerge from the context of application. The creation of knowledge has become a process that must be justified and legitimated against social values, the media, and political objectives (Gibbons *et al.* 1994). This also

---

[2] Rip (2002) quotes Irvine and Martin (1984) to define the concept of strategic research: "research carried out with the expectation that it will produce a broad base of know-edge likely to form the background to the solution of recognised current or future practical problems" (Rip 2002, p. 125).

means that universities are in competition with firms in the knowledge business in various aspects of its creation, diffusion, and promotion.

As a result, Rip describes a new regime of strategic science. This science combines preoccupations of relevance and excellence (Rip 2002: 123). The word "regime" is used to qualify the fact that these preoccupation today are legitimate. New rules of operation, organization, and legitimation of science are being established and are not limited only to the academic context (Rip 2002, p. 126).

12.2.2 The Emergence of an Entrepreneurial Model of University

Public policies concerning research aim at formalizing and orienting these new links between science and the economy. At the international level, there seems to be a convergence among the various measures adopted by many governments to direct the creation, diffusion, and promotion of science, towards the achievement of major national objectives.

Governmental policies in most of the OECD countries are based on collective innovation models (Wouters *et al.* 2002, p. 4) within which neither the academic institution nor the innovating firm is dominant (Gulbrandsen and Etzkowitz 1999; Gibbons *et al.* 1994; Callon *et al.* 1994). These models account for, reinforce, and legitimize the emergence of hybrid organizations, *multi or transdisciplinary,* which reside within the network of heterogeneous organizations (universities, public and private laboratories, consultancies, etc.) and translate into measures[3] and direct or indirect financial incentives for the establishment of strategic alliances between firms, or university spin-offs.

According to Etzkowitz *et al.* (2000), in many countries, under pressure from major scientific and technical policies, an entrepreneurial university model has emerged, despite resistance from the critics of this model and the institutional problems that it brings. One such emanates from the governments of the principal OECD countries (OECD 1999). Several governments have tried to foster an entrepreneurial culture amongst universities, which, when successful, has led to the emergence of start-ups and an increasing number of interdisciplinary research centres aimed at promoting and exploiting public research while also responding to the interests of companies and science parks in *interdisciplinary* research (Conceiçao and Heitor 1999). Universities must be prepared to contribute directly to the

[3] For example, the Bayh-Dole act in the US, the Foundations in Sweden, etc.

creation of wealth and to forge new alliances with the political and social players (Rip 2002, p. 126).

Other pressures on universities are internal to the institutions themselves. As a result of greatly increasing their applied research activities, and becoming more receptive to the idea that it is possible to make money from research (Feller 1990; Faulkner and Senker 1995), universities have been forced to build more research centers and increase their interdisciplinarity. They have been obliged to establish partnerships with companies and, to an extent, to adopt a firm model. To facilitate this universities have implemented complementary strategies (Hamdouch 2001, p. 146): entering into more and more agreements involving public and private sector researchers, adopting active policies concerning patents agreements, and promoting the rapid development of academic swarms.

These changes provoke a number of questions. For instance, what role will universities play in countries determined to improve their economic performance in a global and highly technological environment (Rosenberg 2002)? Will universities be able to undergo a further metamorphosis? After having, in the 19th century, taken on the mission of research in addition to that of education, will universities now be able to orient their activities to satisfying the demands of mass education and a free-market (Caraça 2002)? Will universities be able to cope with such seemingly opposing objectives? Although there may be a global evolution in science and technology policies, the diversity among universities, and, thus, their varying abilities to meet these challenges, will remain.

12.3 The Need for Strategic Management in Universities

The broadening of universities' missions is one way of identifying new sources of finance for university research activities. At the university level, in entering into contractual research, care must be taken to preserve scientific reputation and demonstrate financial efficiency. Strategic management is a problem in this regard: the regulation and fiscal frameworks adopted by public administrators and the direction given to research management result in a management style that is closer to the industrial model, with its quest for efficiency and its strategic plans, as opposed to the more democratic model, with its concentration on collegiality and consensus among departments (Milot 2003. p. 73.)

The aim of this section is to underline the necessity for strategic management, but also to broaden its definition. Strategic management in universities means more than just taking account of the economic imperatives and responding as firms might. In the universities, research promotion is not simply juxtaposed with the missions of research and teaching: while these various missions generally tend to mutually enrich one another, they can also be a source of conflict.

Also, universities will not all be able to adapt to their environmental constraints in the same way. The specific environment of each university, its organization, its rules and objectives all vary. The adoption of a strategic plan allows universities to act and to put in place rules and conditions that will further the development of this third mission of research promotion.

12.3.1 The Management of Emerging Conflicts

The evolution of the role of the university, which has resulted from technology policy, generates debate concerning the contradictions that emerge between different missions, challenging the efficiency of the entire mechanism. The problem is that the involvement of universities in research promotion might detract from the performance of academic research in the creation of knowledge.

Conflicts of commitment

The activities related to the promotion of research results can conflict with other commitments. For instance, a researcher may make a commitment to do some research which results in his not being able to fulfill his responsibilities towards the university and/or the company. Etzkowitz (1996) illustrates such conflicts through the experience of four researchers from the University of Colorado who established a company, but also retained their academic positions. The researchers in question recognized the effects on their academic responsibilities: the commercial promotion of research led to restrictions with regard to publication of results. One of the researchers had abandoned his teaching responsibilities in order to devote all his time to research. Another had made the decision to reduce his involvement in the company because it was affecting the time he had available for the students he was mentoring, and also meant that he was becoming less involved in their research. So, promotion of research may harm the scientific missions and must therefore be undertaken with care.

Conflicts of interest

In the traditional university system, the main measure of performance in relation to research is publications and diffusion of results to peers. The presence of private interests is not always compatible with the disclosure of the results of research. Conflicts of interest appear for two main reasons:

the necessity to keep results confidential in order to benefit economically from them, and the necessity to hide results which could have a negative impact on the commercial value of the existing products of a firm that is providing substantial financial support to the university. The growing importance of these conflicts of interest requires that universities define a specific internal policy.

Institutional conflict
There are a number of public authorities (at local, national, and international levels) that have an influence on the activities of universities. Each has different objectives, which affect their perception of the role of the university. These different perceptions can lead to conflicts within the universities themselves. Institutional conflict can arise when a university from one country collaborates with a firm or firms from another country, in relation to their mission to promote the results of their research, which is envisaged as benefiting national economic development. According to David *et al.* (1994), universities face a very sensitive situation: if it is considered to be in the national interest to exclude foreign firms from participation in industry projects, then it is questionable whether it is in the national interest if researchers completing postdoctoral studies participate in projects financed by foreign companies.

Research promotion is a continuing source of conflict for universities. It is important that its development is not detrimental to everyday research activities and does not harm the global image of the university. Each university possesses assets developed through past activities, which it can promote or dispose of as it sees fit. How each university chooses to exploit its assets is dependent on the activities that led to their accumulation in particular scientific disciplines, and on the promotional activities that have attracted resources, which may be linked to company and market knowledge, and to the formation of networks. Thus, the private funding of research must be carefully managed by each university. The strategies put in place within the university must be in line with the university's competencies, abilities and goals.

12.3.2 The Limits of a Global Model

The strategies adopted by one university to attract private support for research, cannot simply be based on what has been successful in other universities, or even within individual departments within universities; relationships between industry and universities are characterized by a significant amount of heterogeneity. Strategies adopted must be based on the particular environment of a university, taking account of its particular constraints and opportunities, its distinctive competencies, its value system and its declared objectives.

Numerous actors and various dimensions (economic, scientific, technological, political, and social) influence the activities, the organization, and the adaptability of a university. Identifying these factors will help to clarify the opportunities and the constraints that universities have to take into consideration in developing their strategies.

Historical institutional specificities

Although there appears to be a convergence of science policies in various OECD countries, national policies have also marked the evolution of their universities. The national situation, the national innovation systems, and the extent of decentralization of their management of science vary. Branscomb *et al.* (1999) demonstrate that universities cannot and do not respond in the same way to similar pressures because of institutional differences that are associated with differing national characteristics.

For example, in France, two specific features of the national science system are seen as barriers to the development of good relations with industry (Chesnay 1993): the independence of the Centre National de la Recherche Scientifique (CNRS) from the universities, and the duality of the higher education system, with universities and the major schools coexisting. In Germany there is a clear separation between applied and basic research, with the latter coming under the auspices of the Max Planck Institute (MPI), which has little connection with industry. The Fraunhofer Institutes are more focused on industry needs, but find it difficult to develop competencies in high technology fields (Etzkowitz *et al.* 2000). Efforts are being made in Germany to eliminate the separation between applied and basic research by creating transversal structures aimed at better exploiting the complementarities between existing research teams. Therefore, the position of universities in the national innovation system of a country is dependent on the country context and this naturally affects the degree of their relations with industry, both in terms of research and education.

The degree of centralization within the national science system also has an effect on how universities can be managed. Henreckson and Rosenberg (2001) demonstrate, through a comparison of Swedish and American university systems, that centralization contributes to rigidity within the system, which makes adaptation to economic changes more difficult for universities. In the Swedish system, the rules of admission, budgetary matters, and salaries are established centrally by government. The lack of latitude with regard to remuneration handicaps universities in attracting academics from other universities in other countries, and the lack of freedom to vire funds between various budgets does not allow universities in Sweden to respond quickly to external demands.

The significant decentralization within the American university system is accompanied by strong competition between universities for research financing, both for students and for academic staff. This competitive situation is one of the main factors explaining the impact of the Bayh-Dole Act in the United States and is the reason why similar measures would not have had the same effect in other countries (Mowery 2002).

According to Rosenberg (2002), the fact that American universities have become economic institutions cannot be explained only by their ability to give birth to new high technology firms, to develop prototypes or to hold patents: it is also a reflection of their ability to create new disciplines and to rapidly diffuse new knowledge through the establishment and fast development of education programmes (similar to what occurs in the field of basic scientific research). "I have been suggesting that a central feature of the American University has been its high degree of responsiveness to changing economic opportunities. I believe that this combination of responsiveness, along with the high quality of teaching and research, is attributable to a certain organizational feature of that system and the incentives that have flowed from them" (Rosenberg 2002). Rosenberg concludes that because of these organizational differences, European universities are significantly less prepared than American universities for an evolution of their missions.

Universities as a set of cumulative and localized resources

Their histories (Grossetti and Losego 2003), their scientific potential, their level of industrial network development, and the existence of significant competition in terms of education, are all elements that produce in universities their individual characteristics. The competencies-based approach (Teece 1988; Teece and Pisano 1994; Mowery *et al.* 1995) provides a framework within which to analyze the type and cumulative character of the resources[4] developed by universities. Within this framework, the university, like any other organization, has to be considered as a group of specific, cumulative, and localized resources, which are difficult to copy, and which explain the differences in the performances of different universities.

[4] Firm specific resources can be physical (production techniques, for example) or intangible (routines ensuring the organization's coordination, for example). Technological resources are often cited as demonstrating the firm's specific capabilities, which also include knowledge of markets and user needs, organizational routines, such as decision-making techniques or management systems, knowledge, and complex product marketing and distribution networks (Mowery *et al.* 1995).

The cumulative and localized nature of universities' resources leads to disparities with regard to technological opportunities. The type of company that would be a suitable partner for activities within the university will be specific to each university, because the motivation for entering into this kind of relationship lies in the different competencies and knowledge bases of the potential partner firm, that might be exploited in order to generate new competencies (Cohen and Levinthal 1990; Teece 1986; Johnson and Lundvall 1992; Ingham 1994; Roberts and Berry 1985; Stankiewicz 1989). It is therefore important that requirements can be clearly identified in order that appropriate partnerships are continued and new collaborators identified.

Unequal resistance to change

Each activity within an organization reflects some norms of performance and some acknowledged hypotheses about the behaviour that will lead to the achievement of objectives. The shared criteria of performance and acknowledged behaviour are factors that influence and shape organizational features. For many years, universities were completely autonomous – independent of politics or economic considerations. They developed organizational modes and operating principles that were sympathetic to the pursuit of their dual role of social educators and researchers.

The status associated with being a scientist brings an important degree of autonomy and freedom of choice concerning the research undertaken. The scientific community is open and receptive to new ideas concerning research, but not so in relation to the practical administration or teaching programmes that may accompany it (David *et al.* 1994). Their openness to new ideas and the competencies of university researchers are what provokes firms to turn to universities in search of solutions to problems of innovation. However, this openness also can create dysfunction in their collaboration with companies (Hagström 1983, p. 192).

The ongoing development of relations between universities and companies has led to changes in the behaviour of both parties. Such changes are evolutionary and rely on a learning process involving the members of the university, which leads them to revise their perception of their professional environment and of their role in this context[5]. The questioning of behaviours that have for long been inherent in an organization and have been seen to be of value in the past is difficult (Cyert and March 1963).

[5] Lee (1996) relies on a survey to examine the role that American academics are believed to play in economic development, what they see as their specific role, and how they might collaborate with industry. The majority of respondents believed that their university should participate in local and regional development, facilitate the commercialization of academic research, and encourage consul-

Thus, it can be seen that diversification of activities by universities is not straightforward since it involves a process of revision of knowledge and existing norms that is not easy, and becomes even more difficult if it involves major changes to the organization. Of course, the degree of change will be particular to each university and dependent on its characteristics and assets and also its existing relationships with industry.

Due to the many unique characteristics of universities, objectives for all universities cannot be imposed through public policy. The objectives that are realistic for a university are dependent on its specific assets. For example, certain types of relationships with industry may be out of the question because of the internal resources that they would demand, or because of the reluctance of researchers to meet different administrative demands. In order to develop private support for research, the university must identify a strategic orientation on the basis of its existing scientific and human potential. It must also be clear about the management requirements any choice would involve.

In this section of the chapter, we have highlighted the multiple characteristics of universities, which demonstrate that they would not be able to adhere to a single model. In the first part of the chapter we focused on the evolution of science policy within the OECD countries and how it is converging. The aim of the next section is to analyze and propose a solution to the mismatch between the focus in these countries on a single model of policy and the diversity among universities in these countries and also within each individual country.

12.4 The Challenge: Exploitation of the Diversity in the Science System

The emerging policies described in Section 12.1 were not shaped in order to build on the diversity among universities. These policies were inspired by the American model and based on theoretical models, such as Mode II (Gibbons and Limoges 1994) or the Triple Helix (Etzkowitz and Leydersdorf 1997, 2000), which integrate the interactive modes of production of knowledge in public research and industry. These models are often quoted

---

tancy with private firms. A majority of respondents rejected the idea that their university got involved with private companies through start-ups. The survey highlighted two preoccupations: universities feel that the vitality of their research is threatened by the reduction in public financing, but also that too close cooperation with industry could, in the long run, pose a threat to academic liberty to pursue disinterested basic research.

to justify the policies aimed at the development of the entrepreneurial university and the increased private funding for research. The limitations of these policies in relation to science are that they do not exploit the diversity of the system, but rather increase the gap between those universities that fit with the emerging model of entrepreneurial university and the others.

In this section we propose an alternative political approach to influencing the role of the university in society, which sees diversity as an asset. Firstly, we demonstrate that the current political framework is not shaped to exploit the diversity of universities; second, we underline why it is important to exploit this diversity; and thirdly we propose a political framework, which takes account of the diversity in the system.

12.4.1 Policies for Science and Denial of the Diversity Between Universities

As shown in Section 12.2 of this chapter, the formulation of science policies globally is taking place in a context of increasing costs of research and increasing budgetary constraints. This situation has led to the adoption of policies with two main aims: the fostering of cooperation between the public and private sectors and increasing efficiency in the public support for science. This has produced two central threads in policies towards science: firstly, measures that encourage patenting and the creation of firms to exploit results; secondly, the orientation of public support for science on the basis of socio-economic priority.

The emergent model of the university completely ignores disciplines that have no direct application to society, especially those within the humanities. Those universities that have a reputation based around these disciplines will be unable to survive in this new climate because of lack of government support and public funding. Their reputation will suffer as a result of poor student numbers and reduced research because of lack of funding.

In a context of increasing costs of research and budgetary constraints, the search for efficiency has led to public funding becoming polarized towards recognized centres of excellence. This can be justified on the basis that reputation is a great attractor, both for funding and also for highly qualified researchers and academics. At the European level the emergence of centres of excellence is a prerequisite for becoming a world leader in knowledge creation.

Those universities more oriented towards teaching, and with no huge reputation for research, suffer a major handicap. Firstly, they will not be able to replace public support for research with private funding, and, second, the emergence of centres of excellence will increase the distance in terms of research reputation between the large and the small universities. So, policy based on the American model will, in the case of Europe, benefit those universities undertaking high-level scientific activities and will increase the gap between what are perceived as the best and the other universities.

This diversity and this gap do in fact exist in the USA, and the success of the American model is exemplified only in the best institutions. Also, the more successful American universities are highly multi-disciplinary, combining both science and arts disciplines, the former of which contribute to the reputation of the university and attract private support and high student fees.

The policies emerging in Europe do not exploit the diversity among universities. Rather, they lead to the development of a Manichaean-type system, making a distinction between "good" and “bad” universities, which is not in line with the positive exploitation of diversity for several reasons.

12.4.2 The Diversity of Universities: an Asset in a Learning Economy

In a knowledge-based society, the interactions between actors from different backgrounds are an important source of knowledge creation (Lundvall 1988). A key factor in the creation of knowledge is the sharing of knowledge. The rules governing the scientific community have been shaped by the interactions within the community. Given the diversity of the links, and mutual contribution of science and technology, universities play an important role in this interactivity. To stimulate creativity, the aim of public authorities should be to favour the development of an environment that encourages the sharing of knowledge. Sharing of knowledge between universities and industry does not occur only within research collaborations: it is also achieved within forums organized by scientists for industrialists for example, or in lifelong training, which leads to the transmission of knowledge from university to industry, and vice versa, through the placement of students, and through presentations to industry from university researchers.

Nevertheless, those universities that do not fit with the emerging entrepreneurial university model also have important roles to play. The university should be a scientific and economic actor, and also a social and spatial one. For example, universities are important in influencing urban development (Grossetti and Losego 2003) in other ways than merely through scientific activities. The presence of a university has an impact on the size of the population within a town or region, and it affects local taxes, and, thus, the financial resources of the town or region. The search for urban development may foster the establishment of universities specialized in teaching, without strong research departments. At the local level, this type of university has an influence on economic growth through demographic effects and the creation of networks. It will contribute to the construction of local networks important for the sharing of knowledge between teachers, students, and industrialists.

The diversity between universities is important in terms of their contributions to the sharing of knowledge. For Conceiçao and Heitor (2001), "there is a need to promote a diversity of organizational arrangements, even at the higher education level", because this could be important in ensuring the institutional integrity of the university, as a "special type of organization specialized in producing and diffusing knowledge in unique ways" (p.85). Universities are involved in many other activities than teaching and research. A single university cannot respond to every demand and, according to Conceiçao and Heitor (2001), a diversified higher education system, encompassing institutions with different vocations, would produce a flexible system and reconcile the maintenance of centres of excellence with the irreversible expansion in demand for university education.

The challenge for policy makers is to contribute to reinforcing the role of universities in the knowledge based society, but this means exploiting their diversity. Universities come under the scrutiny of several political levels, each of which has an incomplete vision of the role of the university in the knowledge based economy. Each level sees the university as a tool to achieve economic and social objectives and considers only the facets of the university that fit with its particular objectives. This incomplete vision results in the emergence of a university model, which does not recognize, and therefore cannot benefit from, the diversity within the system.

12.4.3 A "bottom-up" Approach in the Global Framework

Since universities have lost their monopoly over knowledge production and entered the competitive sphere, they have become economic players, but, for the moment, their scope is limited. Universities are subject to a lot of political pressure and a great diversity of expectations.

This has created a paradox: the model of the entrepreneurial university, which is led by the evolution of scientific and technological policies seems to be determined more by a series of external pressures or constraints rather than by universities controlling their own development. The strategic perspective discussed above has demonstrated that it is impossible for all universities to adapt in the same way to these pressures and that this could lead to a downgrading of certain university missions and, in some cases, might even lead to the demise of some institutions. The idea of an entrepreneurial university must ultimately lead to the university taking those initiatives that will make them successful in their new competitive environment. We must expect an evolution in university management which will involve a change of direction in strategic planning and the ability to participate in the political debate about the role of universities.

At this time of institutional change, it is important that universities are able to recognize and respond to political, social and economic changes in the environment. University management must, through strategic planning, be able to preserve some coherence between commercialization, and research and training. The challenge will be to fulfil their traditional missions ever more successfully while strengthening their links with the rest of society and becoming more active socio-economic players.

Lundvall (2002) expresses this in strong terms in saying that the university must make "room for slow and in depth learning (…) [and be] a place where one can keep a long-term perspective, and reflect critically both on theory and reality" (p.6). In other words, universities may become more active on the socio economic front, but they must remain distinct from businesses. For Lundvall, universities need to adopt strategies of diversification and differentiation of knowledge production, both within the universities themselves and in collaboration with other partners in the activities of knowledge production.

This suggests a kind of "bottom-up" approach in the creation and elaboration of a new institutional framework for universities. Universities must not only respond to external pressures, but must also develop their own strategies such that they have an impact on political thinking at all levels. Rather than a deterministic and homogeneous approach, government policies should take account of institutional differentiation. The Danish approach is interesting in this regard.

The Danish government took the initiative of asking each university to construct a development plan and design instruments to enable specific goals to be achieved. These development plans were to be the result of a strategic analysis conducted by the universities to identify their specific resources, competencies, values, and objectives within the local context, while at the same time acknowledging the new competitive environment.

For example, the basic functions of each university should be positioned in relation to other national and foreign universities; its key functions should be defined, with peripheral activities being outsourced; an analysis should be made of internal routines and micro-organizational tasks to identify how lecturers and researchers could be freed from trivial tasks in order to devote more time to teaching and scientific research; a structure of incentives and evaluation principles should be proposed to ensure a balance between teaching, inhouse research, and interaction with the outside world, etc. (Lundvall 2002, pp. 17-18.).

Adoption of this type of initiative could provide an incentive for universities to develop particular strategies, in keeping with their specific resources and competencies. It should enable universities to manage the contradictions that might emerge between their different objectives. Such an initiative should make it easier for governments to formulate policies that recognize diversity amongst universities, and thus enlarge their thinking about universities as being more than just economic actors.

12.5 Conclusion

In the regime of strategic science, universities face increasing pressures to promote the results of their research in addition to carrying out their traditional tasks of research and education. Under this regime, the management model that has largely been adopted is more strategic than collegial. However, it is important that universities do not adopt the management principles of firms and make efforts to preserve some of the rules that traditionally have governed the scientific community. Commercialization creates conflicts and incoherencies in the mode of operation of universities and these conflicts need to be managed.

David *et al.* (1994) strongly criticize attempts to force universities to play a role in knowledge creation, and knowledge transfer to the local environment. They question the tendency to see universities as instruments of national R&D policies, or as players in competitive strategies, and they attempt to evaluate the consequences of these new objectives for universities' traditional missions. For these authors, universities not only possess different institutional characteristics, but their reward system (open science) is also incompatible with that of companies (protection of intellectual property, appropriation, etc.). However, universities do not all have the same organizational characteristics. Some of them, American universities, for example, seem to adapt well to the new demands being made upon them, but others do not.

This new context raises several issues that must be addressed by universities and governments. Should we go toward a global university model, or towards a differentiation of universities, based on specialization? Are the gaps growing between universities that agree to expand their traditional activities and those that do not? If so, are these gaps be related to differences in the environment and management of different universities?

In our view, it is extremely important that the diversity between universities is acknowledged and maintained. Although universities have no choice but to bow to current pressures, given the similarity in the technology policies in OECD countries, how they do so will vary, and cannot follow a single university model. University administrators will have the responsibility of maintaining institutional integrity and emphasizing their particular university's specific competencies, vocation, and social, economic and political contributions.

It is widely acknowledged that the challenge for policy makers is to reinforce the role of universities in the knowledge-based society. We have seen that the exploitation of the diversity of universities and the sharing of knowledge are important sources of knowledge creation. Policies that aim at developing cooperation between the public and private sectors in order to replace public support with private funding, are favoring only one model, namely that of the entrepreneurial university. The current political framework is not shaped to exploit universities' diversity.

Strategic analysis gives us insights into the way the universities' new mission of knowledge promotion can be accomplished. Politicians should be cognizant of the diversity of local solutions adopted by universities. It is important that these different local solutions are extensively investigated through case study research.

12.6 References

Branscomb LM; Kodama F; Florida R (eds.) (1999) Industrializing knowledge: University-industry linkages in Japan and the United States. The MIT Press, Cambridge, MA.

Callon M, Larédo P and Mustar P (1995) Gestion stratégique de la recherche et de la technologie. L'évaluation des programmes. Economica, Paris.

Caraça J (2002) Introductory note: Should universities be concerned with teaching or with learning? In: Conceicao P , Gibson DV, Heitor MV, Sirilli G, and Veloso F (eds.) Knowledge for inclusive development, Quorum Books, New York, pp. 31-34.

Chesnay F (1993) The French national system of innovation. In:Nelson RR (ed.) National innovation systems: a comparative analysis. Oxford University Press, Oxford, pp. 192-229.

Cohen WM, Levinthal DA (1990) Absorbtive capacity: a new perspective on learning and innovation. Administrative Science Quarterly 35(1): 128.

Conceiçao P and Heitor M (1999) "On the role of university in the knowledge economy" in Science and Public Policy, 26(1):37-51.

Conceiçao P and Heitor M (2001) Universities in the learning economy: balancing institutional integrity with organizational diversity. In: Archibugi D and Lundvall B-A (eds.) The globalizing learning economy. Oxford University Press, Oxford, pp. 83-96.

David PA, Mowery DC and Steinmueller WE (1994) University-industry research collaborations: managing missions in conflict. Unpublished paper prepared for presentation to the Conference on University Goals, Institutional Mechanisms, and the Industrial Transferability of Research, Center for Economic Policy Research, Stanford University, Stanford, 18-20 March.

Cozzen SE, Healey P, Rip A and Ziman J (1990) The research system in transition. Kluwer Academic Publishers, Amsterdam.

Etzkowitz H (1996) Conflicts of interest and commitment in academic science in the United-states. Minerva 34: 259-277.

Etzkowitz H, Webster A, Gebhardt C and Terra BRC (2000) The future of the university and the university of the future: evolution of ivory tower to entrepreneurial paradigm. Research Policy 29(2): 313-330.

Etzkowitz H, Webster A and Healey P (1998) Capitalizing knowledge. New intersections of industry and academia. State University of New York Press, New York.

Etzkowitz H and Leydesdorff L (2000) The dynamics of Innovation: from national systems and "Mode 22" to a Triple Helix of university-industry-government relations. Research Policy 29(2): 109-124.

Etzkowitz H and Leydesdorff L (eds.) (1997) University in the global knowledge economy: A triple helix of academic-industry-government relations. Cassell, London.

Faulkner W and Senker J (1995) Knowledge frontiers Oxford University Press, Oxford.

Feller I (1990) University as engines of economic development: They think they can Research Policy 19: 335-348.

Gibbons M, Limoges C, NowotnyH, Schwartzman S, Scott P and Trow M (1994) The new production of knowledge. The dynamics of science and research in contemporary societies. Sage Publications, London.

Grossetti M and Losego P (2003) La territorialisation de l'enseignement supérieur et de la recherche, France, Espagne et Portugal. L'Harmattan, Paris.

Gulbrandsen M and Etzkowitz H (1999) Convergence between Europe and America: The transition from industrial to innovation policy. Journal of Technology Transfer 24: 223-233.

Guston D and Keniston K (1994) The fragile contract: university science and federal government. MIT Press, Cambridge MA and London.

Hagstrom SB (1983) Cooperation and conflicts in industry-university relations. Stanford University, Research Papers.

Hamdouch A and Depret MH (2001) La nouvelle économie industrielle de la pharmacie. BioCampus, Editions Scientifiques et Médical, Elsevier.

Ingham M (1994) L'apprentissage organisationnel dans les coopérations. Revue Française de Gestion janvier-février : 105-121.
Johnson B and Lundvall B-A (1992) Closing the institutional gap. Revue d'Economie Industrielle Numéro Spécial, 59 : 111-131.
Lee Yong (1996) Technology transfer and the research university: a search for the boundaries of university-industry collaboration. Research Policy 25(6): 843-863.
Lundvall BA (1988) Innovation as an interactive process: from user-producer interaction to the national systems of innovation. In: Dosi G, Freeman C, Nelson R, Silverberg G and Soete L (eds.) Technical change and economic theory. Pinters Publishers, London and New York, pp. 349-369.
Lundvall B-A (2002) The university in the learning economy. Presentation on the Future role of Universities, Strasbourg, 26 April.
Milot P (2003) La reconfiguration des universités selon l'OCDE. Économie du savoir et politique de l'innovation. Actes de la recherche en sciences sociales 148(, juin): 70.
Mowery DC, Oxley JE and Silverman BS (1995) Firm capabilities, technological complementarity and interfirm cooperation. Conference on "Technology and the Theory of the Firm", University of Reading, May 14-16.
Rip A (2002) Regional innovation systems and the advent of strategic science. Journal of Technology Transfer 27: 123-131.
Roberts EB and, Berry CA (1985) Entering new businesses: selecting strategies for success. Sloan Management Review 26(3): 3-18.
Rosenberg N (2002) Knowledge and innovation for economic development: Should universities be economic institutions? In: Conceiçao P, Gibson DV, Heitor MV, , Sirilli G and Veloso F (eds.) Knowledge for inclusive development. Quorum Books, New York, pp. 35-47.
Salomon J-J (1977) Science policy studies and the development of science policy. In: Spiegel-Rösing I and De Solla Price D (eds.), Science, technology and society: a cross-disciplinary perspective, Sage, London, p. 43-70.
Stankiewicz R (1986) Academics and entrepreneurs. Developing university-industry relations. Frances Pinter, London.
STI Review (1998) Special Issue on "Public/Private Partnerships in Science and Technology" 23(2)
Teece DJ (1988) Capturing value from technological innovation: integration, strategic partnering, and licensing decisions interfaces. Linthicum 18 (3): 46-62.
Teece DJ (1986) Profiting from technological innovation: implications for integration, collaboration, licensing and public policy Research Policy 15(6): 285-306
Vavakova B (1998) The new social contract between governments, universities and society: Has the old one failed? Minerva 36: 209-228.
Wouters P, Elzinga A and Nelis A (2002) Contentious science - discussing the politics of science. EASST Review 21(3/4): pp. 3-5.

Contributing Authors

Arman Avadikyan
Bureau d'Economie Théorique
et Appliquée (BETA-CNRS)
61, avenue de la Forêt Noire
67085 Strasbourg Cedex
France
avady@cournot.u-strasbg.fr

Laurent Bach
Bureau d'Economie Théorique
et Appliquée (BETA-CNRS)
61, avenue de la Forêt Noire
67085 Strasbourg Cedex
France
bach@cournot.u-strasbg.fr

Patrick Cohendet
Bureau d'Economie Théorique
et Appliquée (BETA-CNRS)
61, avenue de la Forêt Noire
67085 Strasbourg Cedex
France
cohendet@cournot.u-strasbg.fr
And
Professeur Visiteur
Service de l'Enseignement
des Affaires Internationales
3000, chemin de la Côte-Sainte-Catherine
Montréal (Québec)
Canada H3T2A7
Patrick.Cohendet@hec.ca

Olivier Dupouët
Bureau d'Economie Théorique
et Appliquée (BETA-CNRS)
61, avenue de la Forêt Noire
67085 Strasbourg Cedex
France
dupouet@cournot.u-strasbg.fr

Jakob Edler
Fraunhofer Institute for Systems
and Innovation Research ISI
Breslauer Strasse 48
76139 Karlsruhe
Germany
J.Edler@isi.fraunhofer.de

Jean-Alain Héraud
Bureau d'Economie Théorique
et Appliquée (BETA-CNRS)
61, avenue de la Forêt Noire
67085 Strasbourg Cedex
France
heraud@cournot.u-strasbg.fr

Rachel Lévy
Bureau d'Economie Théorique
et Appliquée (BETA-CNRS)
61, avenue de la Forêt Noire
67085 Strasbourg Cedex
France
levy@cournot.u-strasbg.fr

Stéphane Lhuillery
Bureau d'Economie Théorique
et Appliquée (BETA-CNRS)
61, avenue de la Forêt Noire
67085 Strasbourg Cedex
France
lhuillery@cournot.u-strasbg.fr

Patrick Llerena
Bureau d'Economie Théorique
et Appliquée (BETA-CNRS)
61, avenue de la Forêt Noire
67085 Strasbourg Cedex
France
pllerena@cournot.u-strasbg.fr

Chantale Mailhot
Service de l'enseignement du management
3000, chemin de la Côte-Sainte-Catherine
Montréal (Québec)
Canada H3T2A7
chantal.mailhot@hec.ca

Mireille Matt
Bureau d'Economie Théorique
et Appliquée (BETA-CNRS)
61, avenue de la Forêt Noire
67085 Strasbourg Cedex
France
matt@cournot.u-strasbg.fr

J. Stanley Metcalfe
ESRC Centre for Research on
Innovation and Competition (CRIC),
The University of Manchester,
Harold Hankins Building,
Booth Street West,
Manchester M13 9QH,
England, UK.
Stan.Metcalfe@manchester.ac.uk

Frieder Meyer-Krahmer
Fraunhofer Institute for Systems
and Innovation Research ISI
Breslauer Strasse 48
76139 Karlsruhe - Germany
and
Bureau d'Economie Théorique
et Appliquée (BETA-CNRS)
61, avenue de la Forêt Noire
67085 Strasbourg Cedex - France
Frieder.Meyer-Krahmer@isi.fhg.de

Véronique Schaeffer
Bureau d'Economie Théorique
et Appliquée (BETA-CNRS)
61, avenue de la Forêt Noire
67085 Strasbourg Cedex
France
schaef@cournot.u-strasbg.fr

Eric Schenk
Laboratoire d'ingénierie de conception,
cognition, intelligence artificielle
(LICIA- INSA)
24, Bld de la Victoire
67084 Strasbourg Cedex
France
eric.schenk@ensais.u-strasbg.fr

Stefania Trenti
Servizio Studi e Ricerche –
Banca Intesa
Via Boito 7
20121 Milano
stefania.trenti@bancaintesa.it

Sandrine Wolff
Bureau d'Economie Théorique
et Appliquée (BETA-CNRS)
61, avenue de la Forêt Noire
67085 Strasbourg Cedex
France
wolff@cournot.u-strasbg.fr

Printing and Binding: Strauss GmbH, Mörlenbach